DIGITAL ELECTRONICS

DIGITAL ELECTRONICS

Mark Forbes

Qualogy, Inc.
San Jose, California

Barry B. Brey

DeVry Institute of Technology
Columbus, Ohio

Bobbs-Merrill Educational Publishing
A publishing subsidiary of **ITT**
Indianapolis

Copyright © 1985 by The Bobbs-Merrill Co., Inc.
A Publishing Subsidiary of ITT

Printed in the United States of America

Cover photo courtesy of Hewlett Packard.
Design and production by Publication Services.

The Bobbs-Merrill Company, Inc.
4300 West 62nd Street
Indianapolis, Indiana 46268

First Edition
First Printing 1984

Library of Congress Cataloging in Publication Data

Forbes, Mark
     Digital electronics.

     Bibliography: p. 467
     Includes index.
     1.   Digital electronics.   I.   Brey, Barry B.   II.   Title.
TK7868.D5F67   1985   621.3815   84-6457
ISBN 0-672-98490-3

CONTENTS

PREFACE

This book was written as a first text for a course in digital electronics. Prerequisite knowledge to this course is elementary circuit theory, including Ohm's law and Kirchoff's laws. A brief treatment of semiconductor theory is given, but previous exposure to transistors and diodes will be helpful. The book begins with a short history of digital electronics, and then builds a foundation with the study of number systems, especially the binary and hexadecimal hex number systems. The use of binary and hex numbers not only simplifies the explanation of most digital circuits, but it is an area the student must understand.

The main elements of digital electronics, the gate circuits, are introduced in chapter 2, along with Boolean algebra and other combinational logic fundamentals. Even as the electronics industry moves toward LSI and VLSI, as well as custom and semicustom designs in increasing number, the volume of SSI logic gates keeps growing each year. Therefore, these "old" fundamentals are as important now as ever.

Chapter 3 delves into the more practical areas, including Boolean applications, and an introduction to the Karnaugh map as a tool for reducing digital combinational circuits to the lowest form. After this foundation, several practical design examples are presented.

Following the introduction of the basic digital logic elements for combinational circuits, an in-depth look at various logic families (TTL, CMOS, MOS, ECL High Speed CMOS and Low Power Shottky) is presented in chapters 4 and 5. Actual data sheets are studied in keeping with the theme of practical applications. Each of the families is studied, and is compared and contrasted with other families.

Characteristics as well as design trade-offs are presented. Each family is studied thoroughly, including the new high-speed CMOS (74HC family introduced by National Semiconductor and Motorola in 1982). Also dealing with the logic families, a chapter on interfacing between various families is included.

Sequential logic is covered next, in chapters 6 and 7. The study includes flip-flops, counters, shift registers, multiplexers, and demultiplexers. Methods of design and analysis are covered, again focusing on the practical side of their usage.

Interfacing to the real-world environment is covered in chapters 8 and 9. Chapter 8 studies methods of interfacing to devices such as displays, switches, relays, and motors. Chapter 9 deals with analog-to-digital and digital-to-analog conversions. The most popular conversion schemes, as well as monolithic integrated circuits, are studied.

Completing the text are two chapters on memories and an introduction to microprocessors. Although the chapter on microprocessors is only an introduction, the reader should be able to gain literacy with the microprocessor after a study of this chapter. The trend is to design the microprocessor into almost any possible device, from the space shuttle to a garbage compactor. The microprocessor has achieved the status of a digital design element, much as the logic gate did twenty years ago.

At the end of each chapter is a selection of problems and questions. These should give the reader a greater familiarity with the topics covered in the chapter. The problems, in keeping with the scope of the book, are all practical in nature and many deal with actual problems encountered by engineers and technicians.

In the appendix is a comprehensive glossary of most of the digital terms that might be encountered by the reader in this and other texts. Also in the appendix is a bibliographical list of references, a selection of data sheets, and answers to odd-numbered problems.

Modern circuitry makes greater use of digital electronics each year. The first uses of digital electronics were in replacing less reliable mechanical devices. Computing devices were also among the first to see solid state digital circuits. Recently, digital circuits have been finding their way into devices which were once exclusively analog in construction. The modern philosophy is, "If it can be done digitally, do it." We truly live in a digital age.

Thanks are due to Jeff Cochran, Pete L. Culwell, and Alexander W. Artgis, who reviewed this manuscript. We appreciate their helpful comments.

1

INTRODUCTION TO DIGITAL ELECTRONICS

Digital electronics has revolutionized the way that electronic circuits are designed, packaged, serviced, marketed, and used. With the advent of the integrated digital circuit, the price of electronics applications, which cost hundreds of dollars only two decades ago, has plummeted to only tens of dollars. An example is the digital calculator. This chapter provides some historical insight into digital electronics as well as an introduction to *number systems*, both of which are fundamental to a complete understanding of digital electronics.

DIGITAL ELECTRONICS: AN HISTORICAL PERSPECTIVE

Digital electronic circuits operate in a finite number of states (usually two—on and off), as opposed to *analog* circuits, which operate on a continous range of

1

signals (as in a stereo amplifier). Digital circuits are often considered rather new in concept—a space-age innovation. However, digital circuits of sorts have been around as long as electrical circuits.

No widespread practical use of digital circuits occurred until the 1930s, when some small-scale digital computers were built using electromechanical relays as logic switches. In the 1940s vacuum tubes were applied to digital computing devices, but the problems of heat and reliability made them largely impractical. The first vacuum tube computer was the Electronic Numerical Integrator and Computer—known as ENIAC. This computer was the brainchild of John Mauchly and used more than 18,000 vacuum tubes! The ENIAC was dedicated in 1946; it occupied a 30 × 50 foot room and weighed nearly 30 tons. Several other tube-type computers were built in the next several years. They were prohibitively large, heavy, and slow, with a tendency to overheat and a deserved reputation for unreliability.

In 1947 three physicists from Bell Telephone Laboratories—William Shockley, Walter Brattain, and John Bardeen—invented the transistor. Without this breakthrough, modern electronics would never have developed. Commercial use of the transistor caught on, and Texas Instruments assembled the first integrated circuit—a flip-flop—in 1959.

Other firms followed Texas Instruments, and soon the RTL (Resistor-Transistor Logic) family was born. One of the major logic families of today, Transistor-Transistor Logic (TTL), was invented in 1961 by a small company in California called Pacific Semiconductor. Large-scale use of TTL integrated circuits didn't take place until the late 1960s, which was too late for Pacific Semiconductor.

RCA pioneered MOS (Metal Oxide Semiconductor) technology in 1963 and subsequently introduced CMOS (Complementary MOS) in 1968. This new family was heralded as the logic family of the future because of its low power consumption. However, problems with speed and die size kept the CMOS family from being widely accepted until the last few years. A new type of CMOS, called High Speed CMOS, was developed by National Semiconductor and Motorola in late 1981. This evolution in CMOS technology will make it the most important family in the future.

Three former employees of Fairchild Semiconductor founded a semiconductor company in Santa Clara in 1968 called Intel. The founders of Intel—Andrew Grove, Robert Noyce, and Gordon Moore—began making semiconductor memories and custom *integrated circuit* (IC) parts. Intel assigned Marcian Hoff to design a calculator IC for a Japanese calculator manufacturer. What he developed was a general-purpose programmable Central Processing Unit (CPU) on a single silicon chip. Intel was the first to produce commercially a computer to be used as a design component—a microprocessor—although Texas Instruments had been using a microprocessor-like device in their calculators. In 1971, Intel marketed this device, a 4-bit processor known as the 4004, with tremendous success. It then followed up with the 8-bit 8008 in 1972 and the now famous

8080 in 1973. As much as any other development—including the invention of the transistor—the advent of the microprocessor has revolutionized digital electronics and electronics in general.

DECIMAL NUMBERS REVISITED

Like football, sports cars, photography, or any other hobby, electronics has certain "buzzwords" and techniques that help those in the industry communicate as efficiently as possible. An efficient way of describing digital circuits consists, as we shall see, of a special way of numbering called the *binary number system*. In order for us to understand the binary system, it will be useful first to review the familiar *decimal number system*.

THE DECIMAL NUMBER SYSTEM

The number system we use in everyday figuring is the decimal, or base ten, number system. Fig. 1-1a shows how a decimal number is constructed. A single digit (0 through 9) is placed in each column until the required value is represented.

$$
\begin{array}{ccccc}
\text{ten-thousands} & \text{thousands} & \text{hundreds} & \text{tens} & \text{ones} \\
4 & 7 & 9 & 3 & 2
\end{array}
$$

a. Column values.

$$
\begin{array}{ccccccc}
10^4 & 10^3 & 10^2 & 10^1 & 10^0 & . & 10^{-1} & 10^{-2} \\
4 & 7 & 9 & 3 & 2 & . & 0 & 0
\end{array}
$$

$$
\begin{aligned}
4 \times 10^4 &= 40000 \\
+\ 7 \times 10^3 &= 7000 \\
+\ 9 \times 10^2 &= 900 \\
+\ 3 \times 10^1 &= 30 \\
+\ 2 \times 10^0 &= 2 \\
\hline
& 47932
\end{aligned}
$$

b. Analysis of decimal number 47932.

FIGURE 1-1. Decimal numbers.

The number in Fig. 1-1 is, obviously, forty-seven thousand, nine hundred and thirty-two. Specifically, the digits 47932 indicate that there are 4 ten-thousands, plus 7 thousands, plus 9 hundreds, plus 3 tens, plus 2 ones.

Fig. 1-1b shows another way of noting the values of the columns in the decimal number system—by using powers of the base (ten in this case). Any number to the zero power is always one, so the first column left of the decimal point is the ones column. 10^1 is 10, so the next column is the tens column. 10^2 is 100, so the third column is the hundreds, and so on. To the right of the decimal point, the powers are expressed as negative numbers. These negative exponents denote that ten raised to that power is inverted: in other words, 10^{-1} equals $1/10$, 10^{-2} equals $1/100$, and so forth. This form of labeling the columns will aid us in understanding numbers in base systems other than the familiar base ten.

THE BINARY NUMBER SYSTEM

At the beginning of the chapter, digital circuits were defined as circuits that have only two states: on and off. (Other names for "on" and "off" that appear in texts are "high" and "low," "true" and "false," and "yes" and "no.") Since all digital circuits have exactly two states, it is quite convenient to use only two numbers to represent them. The binary, or base two, system is therefore perfect. The two digits or bits used in the binary system are 0 and 1. (The term *bit* is a contraction for *bi*nary digi*t*.) When the binary system is applied to digital circuits, 1 represents on and 0 represents off. This notation has even begun to appear on appliances and light switches.

Fig. 1-1b showed how powers of ten can be used to define the values of the columns in the decimal system. Fig. 1-2 shows how the powers of two can be used to define the values of the columns in the binary system. The binary number, 10010101, that appears in Fig. 1-2 is called an 8-bit number because it contains eight binary digits. Because any number to the zero power is one, the rightmost column in binary notation, as in decimal notation, represents ones. The binary system, however, does not provide from 0 to 9 ones, but only 0 or 1. When the ones column exceeds one, it overflows into the twos column, just as the ones column in the decimal system overflows into the tens column when it exceeds nine.

The next column to the left is the 2^1 or twos column. To the left are the 2^2 or fours column, the 2^3 or eights column, and so on, with the value (or weight) of each succeeding column doubling the previous one. By adding up the decimal values of the columns, as in Fig. 1-2b, we find that the binary number 10010101 represents the same value as the decimal number 149. Often subscripts denote the base in which the number is written, as in the following equation:

$$10010101_2 = 149_{10}.$$

one hundred twenty-eights	sixty-fours	thirty-twos	sixteens	eights	fours	twos	ones
2^7	2^6	2^5	2^4	2^3	2^2	2^1	2^0
1	0	0	1	0	1	0	1

a. Column values.

$$
\begin{aligned}
1 \times 2^7 &= 128 \\
+ \ 0 \times 2^6 &= 0 \\
+ \ 0 \times 2^5 &= 0 \\
+ \ 1 \times 2^4 &= 16 \\
+ \ 0 \times 2^3 &= 0 \\
+ \ 1 \times 2^2 &= 4 \\
+ \ 0 \times 2^1 &= 0 \\
+ \ 1 \times 2^0 &= 1 \\
\hline
& \ \ 149
\end{aligned}
$$

b. Analysis of binary number 10010101.

FIGURE 1-2. **Binary numbers.**

Fig. 1-3 shows the relationship between counting in decimal and counting in binary. Note how the binary system overflows from column to column, as does the decimal system, when it runs out of symbols. While the binary system is obviously convenient for describing digital circuits, an extremely large number of bits are needed to represent a fairly large number. This problem will be dealt with later in the chapter.

Binary Fractions

As in the decimal system, binary numbers can also have digits to the right of the *binary point*. The rules for reading these fractions are the same as in the decimal system. Fig. 1-4 illustrates how to read the number 110.01101. Notice that just as in the decimal system, the numbers to the right represent the base (two) raised to a negative power. The first column is $1/2$, the next is $1/4$, then $1/8$, and so on. We can see that 110.01101 is equal to the decimal number 6-13/32.

Decimal	Binary
0	0
1	1
2	10
3	11
4	100
5	101
6	110
7	111
8	1000
9	1001
10	1010
11	1011
12	1100
13	1101
14	1110
15	1111

FIGURE 1-3. *Counting in decimal and in binary.*

fours	twos	ones		halves	quarters	eighths	sixteenths	thirty-secondths
1	1	0	.	0	1	1	0	1

$$1 \times 2^2 = 4$$
$$+ 1 \times 2^1 = 2$$
$$+ 0 \times 2^0 = 0$$
$$+ 0 \times 2^{-1} = 0$$
$$+ 1 \times 2^{-2} = 1/4$$
$$+ 1 \times 2^{-3} = 1/8$$
$$+ 0 \times 2^{-4} = 0$$
$$+ 1 \times 2^{-5} = 1/32$$

$$6\text{-}13/32$$

FIGURE 1-4. *Analysis of a binary number with a fractional part.*

CONVERTING FROM BINARY TO DECIMAL AND VICE VERSA

Binary to Decimal

Fig. 1-2 was one example of the conversion of a binary number to its decimal equivalent. The following examples should help clarify the principle involved.

Example 1-1. Convert the binary number 1100 to its decimal equivalent.

Solution.

$$
\begin{aligned}
1 \times 2^3 &= 8 \\
+ 1 \times 2^2 &= 4 \\
+ 0 \times 2^1 &= 0 \\
+ 0 \times 2^0 &= 0 \\
\hline
&= 12
\end{aligned}
$$

$$1100_2 = 12_{10}$$

Example 1-2. Convert the binary number 111011110 to its decimal equivalent.

Solution.

$$
\begin{aligned}
1 \times 2^8 &= 256 \\
+ 1 \times 2^7 &= 128 \\
+ 1 \times 2^6 &= 64 \\
+ 0 \times 2^5 &= 0 \\
+ 1 \times 2^4 &= 16 \\
+ 1 \times 2^3 &= 8 \\
+ 1 \times 2^2 &= 4 \\
+ 1 \times 2^1 &= 2 \\
+ 0 \times 2^0 &= 0 \\
\hline
&= 478
\end{aligned}
$$

$$111011110_2 = 478_{10}$$

With a little practice, most people can learn to read binary numbers almost as quickly as they read decimal numbers.

Converting from decimal to binary is not quite as simple as converting from binary to decimal, however. The process itself is not that difficult, but it does take a little more time and effort.

Decimal to Binary

One way to convert from decimal to binary is the trial-and-error method. For example, to convert the number 11_{10} to binary, we can break 11_{10} down into 1 eight, 0 fours, 1 two, and 1 one. Therefore, $11_{10} = 1011_2$. Any decimal number could be so converted, but it would be very tedious. A much better way is to use a method called *successive division by two*. To use this technique, divide the decimal number by two. Then use the remainder from this division to fill in the rightmost column. If the decimal number is even, the remainder will be zero, and if it is odd, the remainder will be one. Continue division by two on each resulting value of the previous division until either a one or zero is left. This procedure will yield the binary value of any decimal number. The following examples demonstrate this technique.

Example 1-3. Find the binary equivalent to the decimal number 22.

Solution.

$$22 \div 2 = 11 \text{ rem.} = 0 \text{ (ones column)}$$
$$11 \div 2 = 5 \text{ rem.} = 1 \text{ (twos column)}$$
$$5 \div 2 = 2 \text{ rem.} = 1 \text{ (fours column)}$$
$$2 \div 2 = 1 \text{ rem.} = 0 \text{ (eights column)}$$
$$1 \div 2 = 0 \text{ rem.} = 1 \text{ (sixteens column)}$$
$$0 \div 2$$
$$22_{10} = 10110_2$$

Example 1-4. Convert the decimal number 32 to its binary equivalent.

Solution.

$$32 \div 2 = 16 \text{ rem.} = 0 \text{ (ones column)}$$
$$16 \div 2 = 8 \text{ rem.} = 0 \text{ (twos column)}$$
$$8 \div 2 = 4 \text{ rem.} = 0 \text{ (fours column)}$$
$$4 \div 2 = 2 \text{ rem.} = 0 \text{ (eights column)}$$
$$2 \div 2 = 1 \text{ rem.} = 0 \text{ (sixteens column)}$$

$$1 \div 2 = 0 \quad \text{rem.} = 1 \text{ (thirty-twos column)}$$

$$0 \div 2$$

$$32_{10} = 100000_2$$

You may have recognized this from our previous discussion, without even going through the conversion.

Example 1-5. Convert the decimal number 117 to its binary equivalent.

Solution.

$$117 \div 2 = 58 \quad \text{rem.} = 1 \text{ (ones column)}$$

$$58 \div 2 = 29 \quad \text{rem.} = 0 \text{ (twos column)}$$

$$29 \div 2 = 14 \quad \text{rem.} = 1 \text{ (fours column)}$$

$$14 \div 2 = 7 \quad \text{rem.} = 0 \text{ (eights column)}$$

$$7 \div 2 = 3 \quad \text{rem.} = 1 \text{ (sixteens column)}$$

$$3 \div 2 = 1 \quad \text{rem.} = 1 \text{ (thirty-twos column)}$$

$$1 \div 2 = 0 \quad \text{rem.} = 1 \text{ (sixty-fours column)}$$

$$0 \div 2$$

$$117_{10} = 1110101_2$$

THE HEXADECIMAL NUMBER SYSTEM

Hexadecimal numbers (or *hex* for short) are used in electronics as a shortcut to working with binary numbers. Because it takes such a long string of ones and zeros to represent a large value in the binary system, it is convenient to group together binary numbers to form some other value with the same meaning. Since computers often operate with 4-, 8-, or 16-bit numbers, groups of four are quite handy. And since $2^4 = 16$, base sixteen or hexadecimal is a logical choice.

CONVERSION TO AND FROM HEXADECIMAL

Binary Conversion

To convert a binary number to its hex equivalent, the bits are grouped by fours, from right to left. Then we "add up" the value of this 4-bit number to get the hex equivalent. When we get past nine, we use the letters A through F to represent the decimal values of 10 through 15. The following examples illustrate this simple technique.

Example 1-6. Following is a list of the hex numbers 0 through F and their binary equivalents.

Hex	Binary	Hex	Binary	Hex	Binary
0	0000	6	0110	B	1011
1	0001	7	0111	C	1100
2	0010	8	1000	D	1101
3	0011	9	1001	E	1110
4	0100	A	1010	F	1111
5	0101				

Example 1-7. Convert 11000101_2 to its hexadecimal equivalent.

Solution.

First, group by fours: 1100 0101

Then convert to hex: C 5 = C5

Example 1-8. Convert $1B4_{16}$ to its binary equivalent.

Solution. $1_{16} = 0001_2$ $B_{16} = 1011_2$ $4_{16} = 0100_2$
Therefore, the answer is 000110110100_2 or 110110100_2.

These three examples show how simple conversion is from hex to binary and from binary to hex. They also show the value of being able to express long strings of binary digits in a compact form.

Decimal Conversion

It is frequently necessary to determine the decimal value of a hex number or to find the hex value of a decimal number. Unfortunately, these conversions are not nearly as simple as the conversions to and from binary. One way to do this, of course, would be to convert to binary as an intermediate step, but this would be time-consuming and conducive to errors. Therefore, a direct approach is the optimum way to proceed.

Referring to our earlier discussion on number systems, it should come as no surprise that in base sixteen, the columns are powers of sixteen: ones (16^0), sixteens (16^1), two-hundred fifty-sixes (16^2), four-thousand ninety-sixes (16^3), and so on. So to convert from hex to decimal, we simply multiply the value of the column by the digit in that column and then sum up all the results in decimal. Here is an example.

Example 1-9. Determine the decimal value of $3D2_{16}$.

Solution.

$$3 \times 256 = 768_{10}$$
$$13 \times 16 = 208_{10}$$
$$\underline{2 \times 1 = 2_{10}}$$
$$978_{10}$$

To convert from decimal to hexadecimal, use the same basic procedure as for converting decimal into binary, except that for hex divide by 16 instead of 2 and then use the remainders for the answer.

Example 1-10. Find the hex equivalent of the decimal number 1023.

Solution.

$$1023 \div 16 = 63 \text{ rem.} = 15 \ (F_{16})$$
$$63 \div 16 = 3 \text{ rem.} = 15 \ (F_{16})$$
$$3 \div 16 = 0 \text{ rem.} = 3$$
$$1023_{10} = 3FF_{16}$$

NUMERIC CODES

Many devices must be designed so that the user can work with decimal numbers directly. A system that makes this possible is *Binary Coded Decimal* (BCD). BCD uses the standard binary values in groups of four bits, except that the values 10 through 15 are not allowed. Fig. 1-5 illustrates the BCD scheme. Since one group of four bits is used as a code for each decimal digit, the system is called binary coded decimal. When a circuit designed to operate on BCD numbers, either in hardware (electronics) or software (computer program), reaches a value larger than nine in any given operation, the circuit generates a CARRY into the next group of four bits and adjusts the remainder. BCD circuits are especially useful in arithmetic calculation when the results must be displayed to the user, as on an LED or LCD display.

Two codes that were once used frequently but are seldom used today are the *Gray Code* and the *Excess-3 Code*. In the Gray Code (presented in Table 1-1) only *one* binary digit changes from one number to the next in counting sequence. In older electromechanical systems this was sometimes a necessary

Decimal	Binary Coded Decimal
0	0000
1	0001
2	0010
3	0011
4	0100
5	0101
6	0110
7	0111
8	1000
9	1001
10	0001 0000
11	0001 0001
12	0001 0010
20	0010 0000
50	0101 0000
123	0001 0010 0011

FIGURE 1-5. Decimal versus BCD representation of numbers.

constraint due to the mechanical design. This form of counting is still used, however, for Karnaugh mapping, which we will study in Chapter 3.

The Excess-3 Code, which was designed to facilitate decimal subtraction in binary form, is very infrequently used due to the widespread use of microprocessors. To find the Excess-3 Code value of a decimal number, simply add 3 to the number, then represent the new decimal number in binary. These values are shown in Table 1-2.

TABLE 1-1 THE GRAY CODE FOR DECIMAL VALUES 0 THROUGH 9.

Decimal	Binary	Gray Code
0	0000	0000
1	0001	0001
2	0010	0011
3	0011	0010
4	0100	0110
5	0101	0111
6	0110	0101
7	0111	0100
8	1000	1100
9	1001	1101

TABLE 1-2 EXCESS-3 CODE FOR DECIMAL NUMBERS 0–9.

Decimal	Binary	Excess-3
0	0000	0011
1	0001	0100
2	0010	0101
3	0011	0110
4	0100	0111
5	0101	1000
6	0110	1001
7	0111	1010
8	1000	1011
9	1001	1100

There are two properties associated with the Excess-3 Coded number that facilitate subtraction. First, the ones complement of the binary number can be easily found by simply inverting the bits. This inverted number then becomes the nines complement of the original decimal number. Nines complement addition then yields the correct subtraction result. Complements will be covered in detail later in this chapter.

Other numeric codes are possible, and in fact most codes are invented for a specific design problem and may be used only in that application. You can probably think of several ways to encode binary numbers to represent any of a number of variables.

BINARY AND HEXADECIMAL ARITHMETIC

BINARY ARITHMETIC

Binary arithmetic is exactly the same as arithmetic in base ten—it is just a matter of when to carry or borrow.

Addition

The addition of binary numbers is the same as the addition of decimal numbers, but the symbols are different. For example, one plus one in binary is equal to two, but the symbol for two, of course, is 10. Thus, if we remember to carry at the appropriate time, binary addition is very simple. Here is an example of the binary addition table.

Example 1-11. $0+0=0$ $1+0=1$ $1+1=10$ $1+1+1=11$ $1+1+1+1=100$

You can see that the table is straightforward.

Now, let's do a few examples using larger numbers.

Example 1-12. Add the binary numbers 1010 and 11.

Solution.

$$\text{CARRY 1}$$

$$1010 \ (10_{10})$$

$$+ \ 11 \ (3_{10})$$

$$\overline{1101 \ (13_{10})}$$

Note the CARRY in the 2^2 column, which is the result of adding $1+1$ in the 2^1 column.

Example 1-13. Add the binary numbers 110011 and 1110.

Solution.

$$\text{CARRY 1111}$$

$$110011 \ (51_{10})$$

$$+ \ 1110 \ (14_{10})$$

$$\overline{1000001 \ (65_{10})}$$

Example 1-14. Add the binary numbers 111111 and 111111.

Solution.

$$\text{CARRY 111111}$$

$$111111 \ (63_{10})$$

$$+ \ 111111 \ (63_{10})$$

$$\overline{1111110 \ (126_{10})}$$

Note how the CARRY propagates as $1+1+1=11$, or 1 plus a CARRY.

Subtraction

Binary subtraction can be carried out in one of two ways. The first works the same way as decimal subtraction. In this method, we subtract the bottom number from the top number (as written) and borrow if the top digit is less than the bottom digit. Borrowing is more frequent in binary arithmetic than in decimal, as is carrying.

Example 1-15. Subtract the binary number 10 from 1110.

Solution.

$$
\begin{array}{r}
1110\ (14_{10}) \\
-\ 10\ (2_{10}) \\
\hline
1100\ (12_{10})
\end{array}
$$

Note that this procedure didn't involve any BORROW conditions; the answer was simply "one from one is zero."

Example 1-16. Subtract the binary number 110 from 11000.

Solution.

$$
\begin{array}{l}
\text{BORROW}\quad 0\ 1\ 10 \\
\qquad\ \ 1\ 1\ 0\ 0\ 0\ (24_{10}) \\
\qquad -\quad\ \ 1\ 1\ 0\ (6_{10}) \\
\hline
\qquad\ \ 1\ 0\ 0\ 1\ 0\ (18_{10})
\end{array}
$$

Note that here, as in decimal subtraction, when a borrow is made from a zero, the borrow is propagated to the next column to the left—in the case above, the 2^3 column.

HEXADECIMAL ARITHMETIC

Arithmetic in hex is the same as in any other base; the only tricky part is remembering when to carry or borrow. A couple of quick examples of addition and subtraction should suffice in demonstrating hexadecimal arithmetic. If necessary, refer back to examples 1-12 through 1-16 to help determine when to carry or borrow.

Example 1-17. Perform the following hexadecimal additions:

175 + 274, 3D + 27, 3CD + 199, and F9F + F2

Solution.

CARRY		1	11	111
	175	3D	3CD	F9F
	+ 274	+ 27	+199	+ F2
	3E9	64	566	1091

Example 1-18. Perform the following hexadecimal subtractions:

F3 – 7, 2D – 11, 55 – 19, and 134 – EF.

Solution.

BORROW	E 13		4 15	0 12 14
	F3	2D	55	134
	– 7	– 11	– 19	– EF
	EC	1C	3C	45

RADIX AND RADIX MINUS ONE COMPLEMENTS

The word *radix* refers to the base of a number system. Binary, of course, is base two; hence, the radix is two. The radix minus one value for binary is one.

ONES AND TWOS COMPLEMENTS

As mentioned earlier, there is another way to subtract binary numbers. This method, called *twos complement*, is the method that computers use to subtract. (Since computers are digital, they of course do their arithmetic in binary.) The word *complement* is used frequently in digital electronics. It means opposite or inverse. Therefore, the complement of a 0 is a 1 and the complement of a 1 is a 0. Simply complementing a binary number in this way results in the *ones complement*. For example, the ones complement of the binary number 1101 is 0010.

To find the twos complement, we take the ones complement of the binary number and then add one to the result. For example, the twos complement of the number 1100 is the ones complement (0011) plus 1, or 0100.

To subtract, we add the twos complement to the number from which we wish to subtract. The examples below will clarify this simple process.

Example 1-19. Subtract the binary number 1010 from 1110 ($14_{10} - 10_{10}$).

Solution. First, find the twos complement of 1010:

ones complement:	0101
add one:	+ 1
twos complement:	0110

Now, add the values to get the answer:

$$1110$$
$$+\ 0110$$
$$\overline{1\ 0100}$$

Note that the CARRY beyond the number of digits in the top number, which always occurs when the top number is larger than the bottom, is disregarded or "thrown away." Thus the answer is 0100 or, as we hoped, 4_{10}.

Example 1-20. Subtract the binary number 1000 from 11100 ($28_{10} - 8_{10}$).

Solution.

$$\text{twos complement of 01000:}\quad 10111$$
$$+\ 1$$
$$\overline{11000}$$

Note that both numbers in the subtraction must have the same number of digits before we can perform the twos complement; hence, we used 01000.

Now, add the numbers:

$$11100$$
$$+\ 11000$$
$$\overline{1\ 10100\ (20_{10})}$$

THE SIGN BIT

In a digital system, some way must be devised to "tell the system" when it is working with a negative binary number. Most systems employ the *sign bit* for this purpose. The sign bit is always the highest-order bit (for example, the left-hand bit in an 8-bit number). When the sign bit is a 0, the binary number is positive; when the sign bit is a 1, the binary number is negative. A number that includes a sign bit is called a *signed binary number*.

Thus, subtraction by means of the twos complement is actually the addition of a negative number to a positive number: the negative number is the twos complement, in signed binary, of the number to be subtracted. Example 1-20 showed the twos complement of 1000 to be 11000. This result is consistent with the signed binary system, in which bit 5 is the sign bit: 11000 is thus equivalent to negative 1000.

Although we "ignore" the CARRY in twos complement subtraction, it does have importance: it tells us that the result is a positive number. If there is no final CARRY, the result is negative and is handled as shown in example 1-21.

When the larger number is subtracted from the smaller, the answer must be twos complemented, and the negative of the result becomes the answer.

Example 1-21. Subtract 1110 from 1000 ($8_{10} - 14_{10}$).

Solution. First, find the twos complement of 1110:

$$0001$$
$$+ \ 1$$
$$\overline{0010}$$

Now add:

$$1000$$
$$+0010$$
$$\overline{1010}$$

Twos complement: $0101 + 1 = 0110$

Because there was no CARRY, the answer is negative, and we must insert a sign bit. Thus the answer is 10110 (-6_{10}).

The reason that computers use twos complement arithmetic is that this system allows inversion circuits and adders to do both addition and subtraction. In this way, only one type of circuit is necessary. This subject will be examined in greater detail later in the text.

COMPLEMENTS OF OTHER BASES

Every base has a radix and a radix minus one complement. To find the radix minus one complement, we subtract the digit to be complemented from the radix minus one. Hence, the radix minus one of decimal is nine, and the nines complement of 6 is: $9 - 6$ or 3. To get the radix complement, we again add one to the radix minus one complement as we did for the twos complement. The complements of bases other than two are seldom used.

SUMMARY

1 Every number system is constructed by raising the base or radix to a power, that power being the column number. To the left of the radix

point the column numbers are positive, and to the right of the radix point the numbers are negative.

2 The binary number system has only two symbols: 1 and 0. The binary system is utilized because the 1 and 0 can readily represent the two states found in digital systems.

3 The word *bit* is a contraction of *binary digit*, and represents a single binary 1 or 0.

4 To convert from binary to decimal, simply add up the values of the columns that have a 1.

5 To convert from decimal to binary, use successive division by two: the remainders become the digits in the respective columns.

6 The hexadecimal numbering system uses sixteen as the radix. Hexadecimal is useful in representing large binary numbers.

7 To convert from binary to hexadecimal, simply group the bits by fours and represent them in the hexadecimal system.

8 Converting decimal to hexadecimal requires successive division by 16.

9 Converting from hexadecimal to decimal is accomplished by adding together the values of each column.

10 Binary arithmetic is identical to decimal arithmetic; just remember to carry and borrow at the appropriate times.

11 Hexadecimal arithmetic is also identical to decimal arithmetic.

12 Binary Coded Decimal is a system that uses a 4-bit binary number to represent only 10 values (0–9). The remaining values are not allowed.

13 The ones and twos complements can be used to simplify subtraction.

14 In signed binary, the leftmost bit denotes the sign of the number only and has no value.

KEY TERMS

analog	CMOS	hexadecimal
BCD	complement	IC
binary	CPU	integrated circuit
bit	digital	number system

QUESTIONS AND PROBLEMS

1 What are some common names for the two states that generally define a digital electronic circuit?

2 Define the acronyms TTL, MOS, CMOS, IC, and CPU.

3 In the decimal number 7364813, to what power of ten is the 7 raised to, and what is the value of this column?

4 Why are binary numbers so important in digital electronics? Why are hex numbers important?

5 Convert the following binary numbers to decimal:

a. 11001	e. 1111
b. 101010	f. 1100
c. 10010	g. 1000
d. 11000	h. 10

6 Convert the following decimal numbers to binary:

a. 45	e. 256
b. 16	f. 17
c. 255	g. 3
d. 128	h. 1

7 Add the following binary numbers. Express your answer in both binary and decimal:

a. 1001 + 111	c. 11 + 1
b. 1110 + 10	d. 1000 + 1111

8 Subtract the following binary numbers using the BORROW method:

a. 1000 − 1	c. 1010 − 101
b. 1110 − 111	d. 10 − 1

9 Find the twos complement of each of the following binary numbers:

a. 1000	c. 1010
b. 11	d. 0111

10 Repeat problem 8 using the "twos complement addition" method of subtraction.

11 Using the twos complement method, perform the following subtractions:

a. 10 − 100	c. 100 − 101
b. 11 − 110	

12 Express the following decimal numbers in BCD:

a. 48	c. 9
b. 374	d. 1000

13 Give the decimal equivalents (if they exist) for the following BCD expressions:

    a. 0010   d. 0110

    b. 1000   e. 1000 0110

    c. 1100

14 Convert the following binary numbers to hexadecimal:

    a. 101011   e. 11111000

    b. 11001110   f. 11111

    c. 101   g. 000001

    d. 01111110   h. 10101

15 Give the binary equivalents of the following hex numbers:

    a. B   f. 30

    b. 10   g. A6

    c. 12   h. 1F

    d. AA   i. 123

    e. 41   j. A3A

16 Give the hex values for the following decimal numbers:

    a. 43   d. 1024

    b. 74   e. 255

    c. 93

17 Find the decimal equivalent of each of the following hex numbers:

    a. 3F   d. FED

    b. 7FF   e. 999

    c. 100

18 Perform the following additions in hexadecimal:

    a. B + 5   e. A + A

    b. F + F   f. 74 + 47

    c. 10 + 10   g. A + 6 + 2

    d. 100 + CD   h. 64 + 64

19 Do the following subtractions in hexadecimal:

    a. B – 3   d. 3FC – FD

    b. AA – B   e. 10 – A

    c. 10 – 3

2

COMBINATIONAL LOGIC AND ARITHMETIC CIRCUITRY

The first "real" digital circuits, which will be introduced in this chapter, are quite simple, but they form the foundation of all digital circuitry. These basic gate circuits, as these bits of "glue" are called, remain as necessary in modern Large Scale Integration (LSI) as they were when they were the only way to design and construct digital circuits.

INTRODUCTION TO DIGITAL SWITCHING

Chapter 1 briefly introduced the two states that generally define digital electronics: on (1) and off (0). It did not, however, reveal in any detail how these two states even allow us to build useful circuits to do simple problems, let alone provide the basis for a computer system. This chapter will provide a first taste of real digital electronics. First we will define some additional terms and then begin to examine the digital circuit building blocks.

To most people, the terms "on" and "off" first bring to mind a light switch. Indeed, a simple light switch is an example of a digital circuit. The light that the switch controls has two states: on and off. There are no "in-betweens." In fact, many light switches are now being labeled 1 and 0 to denote on and off. These are the same symbols we will use in our description of digital logic.

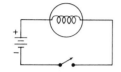

Fig. 2-1 shows a schematic diagram of the circuit containing the light bulb and the switch. When the switch is open, the light is off, and when the switch is closed, the light is on. Table 2-1 is called a *truth table.* A truth table shows every possible combination of the switch or switches in a circuit. In this particular circuit, the only combinations possible are the two just described. In truth tables, we use the binary numbers 1 and 0 to represent on and off (closed and open, as is the case for the switch).

FIGURE 2-1. Simple circuit using a switch to turn on a light.

Digital circuits are referred to as *logic circuits* because they are constructed with the building blocks, called gates, which correspond to the logic functions AND, OR, and NOT. Fig. 2-2 illustrates the AND function with switches S_1 and S_2: Obviously the light will illuminate only when *both* S_1 AND S_2 are closed. Table 2-2 describes this circuit in binary. In words, then, the AND function states that the output (light) will be 1 (on) only when both inputs (switches) are 1 (closed). Observe also that there could be any number of switches in series and this definition would still hold true.

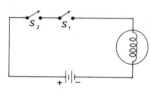

FIGURE 2-2. AND function.

The OR function is demonstrated by the circuit in Fig. 2-3. In this case the light will illuminate if either S_1 OR S_2 OR both are closed. Table 2-3 describes

TABLE 2-1 TRUTH TABLE FOR THE CIRCUIT IN FIG. 2-1.

Switch	Light
0	0
1	1

TABLE 2-2 TRUTH TABLE FOR THE CIRCUIT IN FIG. 2-2.

S_1	S_2	Light
0	0	0
0	1	0
1	0	0
1	1	1

TABLE 2-3 TRUTH TABLE FOR THE CIRCUIT IN FIG. 2-3.

S_1	S_2	Light
0	0	0
0	1	1
1	0	1
1	1	1

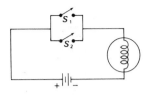

FIGURE 2-3. OR function.

TABLE 2-4 TRUTH TABLE FOR THE CIRCUIT IN FIG. 2-4.

Switch	Light
0	1
1	0

the OR circuit in binary form. In words, the OR function states that the output will be 1 if either or both of the inputs are 1. Again note that any number of switches, in parallel this time, could be used.

The third of the three building blocks is represented in Fig. 2-4. This is called the NOT (or INVERTER) function because the light will illuminate if the switch is NOT closed, or the light will NOT illuminate if the switch is closed. The truth table is presented in Table 2-4. In the case of the INVERTER function, only one switch (a single input) can be used.

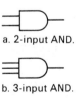

FIGURE 2-4. IN-VERTER function.

THE AND, OR, AND INVERTER GATES

Drawings that represent electronic circuits are called *schematic diagrams,* or *schematics* for short. A standard set of symbols is used to represent the various logic functions on these drawings. In addition to the three basic gates there are several other logic functions as well. These other functions and their symbols will be studied shortly.

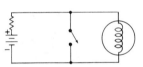

a. 2-input AND.

AND GATES

Because the AND gate can have more than the two inputs depicted in Fig. 2-2, there are commercially available gates with as many as eight or more inputs. The logic symbols for 2-, 3-, and 4-input AND gates are shown in Fig. 2-5.

If a situation calls for more inputs than the standard gates proved, the outputs of two or more gates can function as inputs to an additional AND gate to form a new AND gate, as in Fig. 2-6.

b. 3-input AND.

c. 4-input AND.

FIGURE 2-5. AND gates.

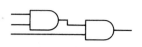

FIGURE 2-6. A 3-input AND formed from 2-input ANDS.

The AND function is *logical multiplication*. Therefore, another way of describing the AND gate is by forming an equation, much like an algebraic equation, in which the symbol denoting the AND function is the multiplication dot (·). The equation for the 2-input AND function is:

$$A \cdot B = X$$

The AND function can be represented in equation form in two other ways, just as in algebra there are three ways to write a multiplication equation. The other two forms of the 2-input AND equation are:

$$(A)(B) = X$$

and

$$AB = X$$

a. 2-input OR.

In this text, the form $AB = X$ will be used, unless a compound equation is written, such as $(A+B)C = X$.

b. 3-input OR.

c. 4-input OR.

FIGURE 2-7. OR gates.

OR GATES

Because OR gates, like AND gates, can have more than two inputs, they are also commercially available with eight or more inputs. Logic symbols for 2-, 3-, and 4-input OR gates are shown in Fig. 2-7.

Like the outputs of AND gates, the gate outputs of OR gates can be connected to the inputs of an additional OR gate to increase the number of inputs and form a new OR gate function, as shown in Fig. 2-8.

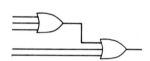

FIGURE 2-8. A 5-input OR formed from two 3-input ORs.

The OR function is *logical addition*. Therefore, the OR function is represented in equation form by the plus (+) symbol. Please note, however, that the symbol for OR should not be confused with addition in arithmetic, nor should the symbol for AND be confused with multiplication. The 3-input OR gate above can thus be represented by the equation:

$$A+B+C = X$$

THE INVERTER OR NOT GATE

FIGURE 2-9. The INVERTER.

The INVERTER complements or reverses the state of the input. The truth table in Table 2-4 completely defines all INVERTERs. Because the INVERTER is only realized with one input and one output, the only symbol to remember for the INVERTER is that depicted in Fig. 2-9.

The equation that describes the INVERTER function is:

$$A = \overline{X}$$

Note that this is unlike any algebraic expression. This equation is read as "*A* equals *X* bar" or "*A* equals NOT *X*." These two expressions are used interchangeably. Note that the equation can also be written as $\overline{A} = X$, with exactly the same meaning. This equation would be read as "X equals A bar" or "X equals NOT A."

INTERCONNECTION OF GATES

In order to make useful circuits, various gates must be connected to produce the proper output for a given set of inputs. All the gates we have studied can be put together in any combination to form new functions. We saw that the AND and OR gates could be connected to form a new AND or OR. It is also possible—in fact necessary most of the time—to mix AND, OR, and NOT functions to perform new functions. A few examples of gate combinations will emphasize this point.

Example 2-1. Make a truth table and write an equation for the circuit shown schematically in Fig. 2-10.

$W = A + B$

$X = (A + B) C$ **FIGURE 2-10.** *Circuit for example 2-1.*

Solution. First, we make a truth table for each of the gates. These are shown in Tables 2-5 and 2-6. The truth table for the entire function (Table

TABLE 2-5 TRUTH TABLE FOR EXAMPLE 2-1: THE OR GATE.

A	B	W
0	0	0
0	1	1
1	0	1
1	1	1

TABLE 2-6 TRUTH TABLE FOR EXAMPLE 2-1: THE AND GATE.

W	C	X
0	0	0
0	1	0
1	0	0
1	1	1

TABLE 2-7 TRUTH TABLE FOR THE ENTIRE FUNCTION IN EXAMPLE 2-1.

A	B	C	X
0	0	0	0
0	0	1	0
0	1	0	0
0	1	1	1
1	0	0	0
1	0	1	1
1	1	0	0
1	1	1	1

2-7) is the combination of Tables 2-5 and 2-6. Observe that the entire function is a 3-input circuit (as it was schematically). (Note that the truth tables are constructed with the inputs written in ascending binary numbers.) The output of the AND gate will be 1 only if both inputs are 1. Input W to the AND gates is the output of an OR gate, so it will be 1 if either A or B or both are 1. In words, the truth table states that "X will be 1 when either A or B or both are 1, and C is 1." The equation to represent the circuit is constructed by combining the equations for the individual gates:

$$OR: A + B = W$$

$$AND: WC = X$$

$$Function: (A + B)C = X$$

Since the term $A+B$ is a single input to the AND gate, we put parentheses around it to aid in reading the equation.

Example 2-2. Make a truth table and write an equation for the circuit shown schematically in Fig. 2-11.

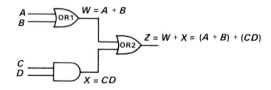

FIGURE 2-11. Circuit for example 2-2.

Solution.   Again, we first make a truth table for the individual gates. This has been done in Tables 2-8, 2-9, and 2-10. In this example, combining the truth tables yields a 4-input function as the result, as indicated in Table 2-11. Since the two OR gates are *cascaded,* the output Z will be 1 whenever

TABLE 2-8 TRUTH TABLE FOR EXAMPLE 2-2: OR1.

A	B	W
0	0	0
0	1	1
1	0	1
1	1	1

TABLE 2-9 TRUTH TABLE FOR EXAMPLE 2-2: OR2.

W	X	Z
0	0	0
0	1	1
1	0	1
1	1	1

TABLE 2-10 TRUTH TABLE FOR EXAMPLE 2-2: AND.

C	D	X
0	0	0
0	1	0
1	0	0
1	1	1

either A or B or both are 1. (Two similar gates are cascaded when the output of one serves as an input of the other.) This accounts for the large number of 1 outputs. The only other time that Z will be 1 is when both C and D are 1 and the X input is 1. In words, we can say that "Z will be 1 when A or B or both are 1 or when C and D are 1." Our equation for this function is:

$$\text{OR1}: A + B = W$$

$$\text{AND}: CD = X$$

$$\text{OR2}: W + X = Z$$

$$\text{Function}: (A+B) + (CD) = Z$$

Example 2-3. Make a truth table and write an equation for the circuit shown schematically in Fig. 2-12.

TABLE 2-11 TRUTH TABLE FOR THE ENTIRE FUNCTION IN EXAMPLE 2-2.

A	B	C	D	Z
0	0	0	0	0
0	0	0	1	0
0	0	1	0	0
0	0	1	1	1
0	1	0	0	1
0	1	0	1	1
0	1	1	0	1
0	1	1	1	1
1	0	0	0	1
1	0	0	1	1
1	0	1	0	1
1	0	1	1	1
1	1	0	0	1
1	1	0	1	1
1	1	1	0	1
1	1	1	1	1

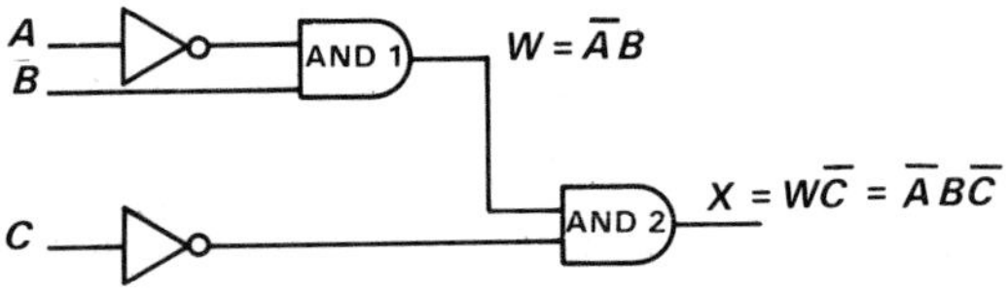

FIGURE 2-12.
Circuit for example 2-3.

Solution. This circuit is a little more thought-provoking, since two of the signals are inverted. Our truth tables are Tables 2-12 and 2-13. Now combine the truth tables into a 3-input function as shown in Table 2-14. The equations for this function are as follows:

$$\text{AND1: } \overline{A} B = W$$

$$\text{AND2: } W\overline{C} = X$$

$$\text{Function: } \overline{A} \, B \, \overline{C}$$

In this case, where there are only AND gates and an INVERTER (no mixing of ANDs and ORs), we can omit the parentheses.

As an additional exercise, prove, using a truth table, that the circuit in Fig. 2-13 is identical to the one used in example 2-3.

TABLE 2-12 TRUTH TABLE FOR EXAMPLE 2-3: AND1.

A	B	W
0	0	0
0	1	1
1	0	0
1	1	0

TABLE 2-13 TRUTH TABLE FOR EXAMPLE 2-3: AND2.

W	C	X
0	0	0
0	1	0
1	0	1
1	1	0

TABLE 2-14 TRUTH TABLE FOR THE ENTIRE FUNCTION IN EXAMPLE 2-3.

A	B	C	X
0	0	0	0
0	0	1	0
0	1	0	1
0	1	1	0
1	0	0	0
1	0	1	0
1	1	0	0
1	1	1	0

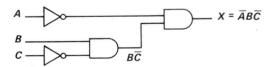

$X = \overline{A}B\overline{C}$

$B\overline{C}$

FIGURE 2-13. Circuit for additional exercise in example 2-3.

Sometimes, the equation of a function is known and the circuit must be derived that will electrically perform that function. This process is simply the reverse of that used in previous examples: we can either draw the circuit and then make a truth table or make a truth table and then draw the circuit. Several problems at the end of this chapter illustrate this process, so one example should be sufficient here.

Example 2-4. Make a truth table and draw the schematic for the circuit represented by the equation below. You need not necessarily make the truth table first.

$$(A + B)(\overline{B} + C) = X$$

Solution. This is an AND function of two ORs, so both of the OR elements must be 1 for X to be 1. For $A + B$ to be 1, either A or B or both must be 1. For $\overline{B} + C$ to be 1, B must be 0, C must be 1, or both. Since B and $\overline{B}$ are complementary, A and C cannot both be 0 for the output to be 1. Table 2-15 shows the result of our analysis. Fig. 2-14 is the circuit diagram for the function. Notice that B must pass through an INVERTER before it becomes an input in the OR gate that C is connected to. Fig. 2-14 satisfies the constraints of the truth table.

TABLE 2-15 TRUTH TABLE FOR EXAMPLE 2-4.

A	B	C	X
0	0	0	0
0	0	1	0
0	1	0	0
0	1	1	1
1	0	0	1
1	0	1	1
1	1	0	0
1	1	1	1

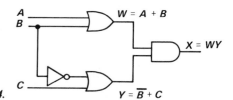

FIGURE 2-14. Function in example 2-4.

THE NOR AND NAND GATES

In addition to the three basic gates—AND, OR, and INVERTER—other important gates are also used extensively in digital electronics. In this section, we will consider two: the NAND and the NOR gates.

THE NOR GATE

If an OR gate is followed by an INVERTER, the equation for the resulting function is:

$$\overline{A + B} = X$$

The entire OR function is complemented because the output of the gate rather than the inputs are inverted. This combination of an OR and an INVERTER forms a new gate, which is called the NOR (NOT OR) gate. As Chapter 4 will show, the NOR gate is easier to construct than the OR. And, as we will see later, the NOR along with the NAND are *universal gates*; that is, any function can be constructed using only NORs or NANDs. The symbol for the NOR gate is shown in Fig. 2-15.

FIGURE 2-15. The NOR gate.

Like the OR gate, the NOR can have any number of inputs. The truth table for a 2-input NOR appears in Table 2-16. As might have been expected, the output for a NOR is exactly opposite to that of an OR: the output is a 1 only when *all* inputs are 0. Note the importance of the word *all*. No matter how many inputs a NOR gate has, the output will be 1 only when *all* inputs are 0.

THE NAND GATE

Because the NAND gate, like the NOR, is a universal gate, you have probably already deduced that a NAND gate is an AND gate followed by an INVERTER (NOT AND). Also, like the NOR, the NAND gate is easier to construct than the AND gate. The equation for the NAND is:

$$\overline{A \cdot B} = X$$

The symbol for the NAND (see Fig. 2-16) follows the convention of the NOR, and the truth table (Table 2-17) shows that the output is exactly opposite to that of the AND gate. For the NAND gate, irrespective of the number of inputs, the output will be 1 at all times except when *all* inputs are 1. Again, note the importance of the word *all*. By keeping this principle in mind, we can instantly determine, without knowledge of any other input, that if one of the inputs of a NAND gate is 0 the output is 1.

FIGURE 2-16. The NAND gate.

TABLE 2-16 TRUTH TABLE FOR THE NOR FUNCTION.

A	B	X
0	0	1
0	1	0
1	0	0
1	1	0

TABLE 2-17 TRUTH TABLE FOR THE NAND FUNCTION.

A	B	X
0	0	1
0	1	1
1	0	1
1	1	0

THE EXCLUSIVE OR AND EXCLUSIVE NOR GATES

Two other gates that are of particular interest are the Exclusive OR and Exclusive NOR gates. These gates are especially useful in arithmetic circuits, but they have other uses as well.

THE EXCLUSIVE OR

One special type of OR gate is the Exclusive OR. Although the standard OR gate is sometimes called the Inclusive OR for purposes of differentiation, here we will always refer to the two gates as simply OR and Exclusive OR. One important use of the Exclusive OR is in binary addition. Table 2-18 shows that the 1/1 input yields a 0 output. In binary addition, 1 + 1 = 10, or, a zero and a CARRY. The Exclusive OR function gives the zero necessary in addition. The equation for the Exclusive OR utilizes a symbol that is unfamiliar to most people—a plus sign (+) with a circle around it, as the following equation shows:

$$A \oplus B = X$$

Table 2-18 and Fig. 2-17 are the truth table and the schematic symbol for the Exclusive OR gate.

FIGURE 2-17. The Exclusive OR gate.

TABLE 2-18 TRUTH TABLE FOR THE EXCLUSIVE OR FUNCTION.

A	B	X
0	0	0
0	1	1
1	0	1
1	1	0

TABLE 2-19 TRUTH TABLE FOR THE EXCLUSIVE NOR FUNCTION.

A	B	X
0	0	1
0	1	0
1	0	0
1	1	1

In words, the Exclusive OR gate has the following function: "When either A OR B, *but not both* are 1, the output is 1." Besides binary addition, Exclusive OR gates are used when it is desired to have the output be 0 when both inputs are 1. The Exclusive OR is available with only two inputs.

THE EXCLUSIVE NOR

As you might expect, an Exclusive NOR gate is also available; however, it has rather limited application. The symbol and truth table for the Exclusive NOR are shown in Fig. 2-18 and Table 2-19. The equation for the Exclusive NOR is:

$$\overline{A \oplus B} = X$$

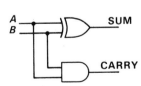

FIGURE 2-18. The Exclusive NOR gate.

HALF AND FULL ADDERS

The half and full adders are slightly more complex than the three basic gates, but they can be extremely useful. The two *adder* circuits are constructed from basic gates; however, they are available as a single building block on an IC. Although we will study their makeup at the gate level, they are used as complete blocks in most designs.

THE HALF ADDER

Binary addition, as seen in Chapter 1, differs from decimal addition in only one respect: when to carry. A review of the truth table for the Exclusive OR gate will reveal that it is identical to binary addition except for the CARRY, which results when both inputs are 1. The circuit that carries out the addition, plus CARRY, is called a *half adder*. The truth table for a half adder is shown in Table 2-20.

The CARRY function, as shown in Table 2-20, is simply the AND of the two inputs. The logic diagram for a half adder is shown in Fig. 2-19. The inputs are A and B, and the outputs are the SUM, which is the Exclusive OR of A and B, and the CARRY, which is the AND of A and B.

FIGURE 2-19. Logic function for the half adder.

TABLE 2-20 TRUTH TABLE FOR THE HALF ADDER.

A	B	SUM	CARRY
0	0	0	0
0	1	1	0
1	0	1	0
1	1	0	1

Although a half adder circuit can accept only two bits as inputs, any number of half adders can be connected in parallel. To find the sum of three bits, for example, two half adders are used: one input to the second adder is the CARRY-OUT of the first, and the other input is the third bit. Each additional bit requires another half adder. Use of the full adder eliminates the necessity of requiring a half adder for each bit.

THE FULL ADDER

The *full adder* accepts three inputs: the two input bits and a CARRY input bit. It produces a SUM and a CARRY output. The CARRY input bit is called the CARRY-IN, and the CARRY output is the CARRY-OUT. This circuit is somewhat more complex than the half adder but results in a more compact solution when more than a few bits must be added.

The truth table for a full adder is depicted in Table 2-21. Chapter 3 will show how to construct a circuit from a truth table, but for now notice that the SUM output is the Exclusive OR of the CARRY-IN with the sum of A and B (the sum of A and B is, of course, their Exclusive OR). Since only two input Exclusive OR gates are possible, two of these gates must be cascaded to give us the SUM of the three inputs.

TABLE 2-21 TRUTH TABLE FOR THE FULL ADDER.

A	B	CARRY-IN	SUM	CARRY-OUT
0	0	0	0	0
0	0	1	1	0
0	1	0	1	0
0	1	1	0	1
1	0	0	1	0
1	0	1	0	1
1	1	0	0	1
1	1	1	1	1

The CARRY-OUT is somewhat more difficult to obtain. The equation for the CARRY-OUT is:

$$AB + (A \oplus B)\text{CARRY-IN}$$

You can verify this from the truth table. Putting all of these gates together gives us the circuit in Fig. 2-20. An alternate implementation of the full adder is to construct it with half adders and an OR gate, as in Fig. 2-21.

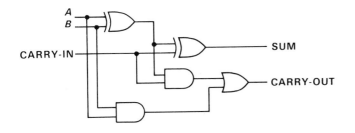

FIGURE 2-20.   Full adder circuit.

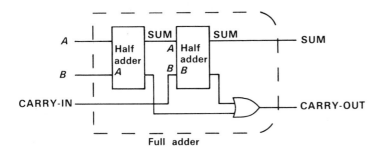

FIGURE 2-21.   Equivalent circuit to that of Fig. 2-20.

HALF AND FULL SUBTRACTORS

HALF SUBTRACTOR

Like the half adder, the inputs to a *half subtractor* are *A* (called the *minuend*) and *B* (the *subtrahend*), and the outputs are the DIFFERENCE (*A-B*) and the BORROW-OUT. Referring back to the binary arithmetic section in Chapter 1 will aid in understanding the half subtractor truth table, constructed in Table 2-22. Coincidentally, the DIFFERENCE is the same function as the SUM in the half adder, that is, the Exclusive OR function. The BORROW function is 1 only when the subtrahend is larger than the minuend, which is only true when $A = 0$ and $B = 1$. Thus, the BORROW is $\overline{A} B$. The half subtractor logic circuit is pictured in Fig. 2-22, as is the functional block logic symbol.

TABLE 2-22 TRUTH TABLE FOR THE HALF SUBTRACTOR.

A	B	DIFFERENCE	BORROW-OUT
0	0	0	0
0	1	1	1
1	0	1	0
1	1	0	0

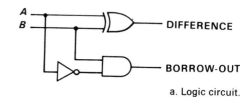

a. Logic circuit.

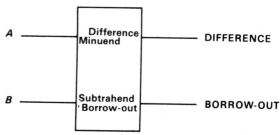

FIGURE 2-22. Half subtractor circuit.

b. Logic block.

THE FULL SUBTRACTOR

The *full subtractor*, as with the full adder, allows a BORROW from a previous subtractor to be input along with the subtrahend and minuend. The outputs, of course, are the DIFFERENCE and the BORROW-OUT. The truth table for the full subtractor is detailed in Table 2-23. You should take a minute to study this function as it is somewhat more complicated than any of the other adder or subtractor functions. The function for the DIFFERENCE, as with the half subtractor, is the same as the full adder SUM function. The expression is as follows:

$$(A \oplus B) \oplus (\text{BORROW-IN}) + AB(\text{BORROW-IN})$$

The function for the BORROW-OUT is:

$$B(\text{BORROW-IN}) + \overline{A}(\text{BORROW-IN}) + \overline{A}B$$

The logic circuit for the full subtractor is shown in Fig. 2-23a, and the functional block logic symbol is shown in Fig. 2-23b. We can, as you might suspect, construct a full subtractor from two half subtractors, as depicted in Fig. 2-24.

TABLE 2-23 TRUTH TABLE FOR THE FULL SUBTRACTOR.

A	B	BORROW-IN	DIFFERENCE	BORROW-OUT
0	0	0	0	0
0	0	1	1	1
0	1	0	1	1
0	1	1	0	1
1	0	0	1	0
1	0	1	0	0
1	1	0	0	0
1	1	1	1	1

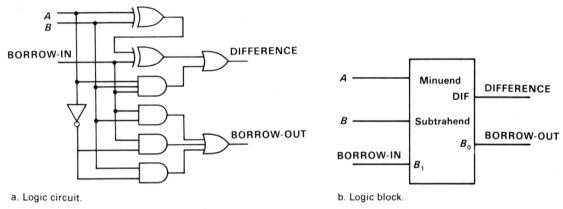

a. Logic circuit.

b. Logic block.

FIGURE 2-23.   Full subtractor circuit.

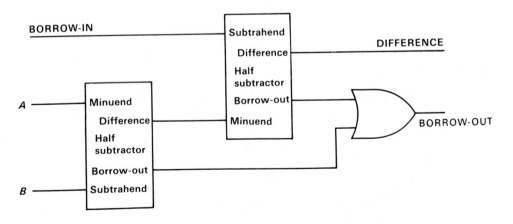

FIGURE 2-24.   Full subtractor constructed from two half subtractors.

As mentioned previously, the adder and subtractor circuits are treated as functional blocks, just like an OR or AND gate. The common usage of microprocessors has diminished the importance of adder and subtractor circuits, but they still have many uses. Although called "adders," these circuits can be used in any function specified by the truth table.

THE SCHMITT TRIGGER

Although digital circuits generally use only two states, some signals used in digital circuits aren't quite digital. For example, if we were to transform the AC line voltage to 5 volts (5 V), we would have a 60 hertz (Hz) sine wave of 5 V, but it would be shaped as in Fig. 2-25a. Since our digital circuits operate at 5 V and ground (generally), the times when the sine wave is *between* 5 V and ground causes the gates to be confused. A device called a *Schmitt trigger* can solve this problem.

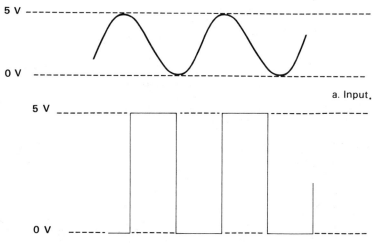

a. Input.

b. Output.

FIGURE 2-25. Schmitt trigger.

The Schmitt trigger has an upper threshold and a lower threshold (also called the *upper* and *lower trip points*). When the input voltage exceeds the upper threshold, the output "snaps" to a 1 state. When the input voltage falls below the lower threshold, the output "snaps" to the 0 state. The difference between the two thresholds is generally small, and the resulting shape of the 60 Hz waveform is illustrated in Fig. 2-25b.

Almost all the gates we have studied are available in the Schmitt trigger configuration. To denote use of the Schmitt trigger, a squared-off *S* is inserted inside the gate symbol, as Fig. 2-26 shows.

THE BASIC PRINCIPLES OF BOOLEAN ALGEBRA

FIGURE 2-26. *The symbols used for the Schmitt trigger.*

In the mid-nineteenth century, George Boole tackled the problem of expressing logic in mathematical terms by developing the branch of mathematics called *Boolean algebra*. Although based on standard algebraic principles, Boolean algebra is much easier to understand and use because it has only two possible values for every variable—1 and 0. The notation used in Boolean algebra and the Boolean identities are fundamental to the further study of digital electronics. In fact, we have been working with Boolean algebra in the preceding sections of this book: the equations used to design and understand basic circuitry are *Boolean* equations. The *properties* of Boolean algebra, which are presented later in this section, are useful in understanding some digital functions, but it is not necessary to study them intensively nor to memorize them.

BOOLEAN IDENTITIES

Table 2-24 presents some fundamental Boolean identities. Proofs in Boolean algebra are relatively simple because there are only two variables. We will use the truth tables of the previous sections to prove the various identities.

The first identity deals with the NOT function and is sometimes called the *double inversion property*. Because only two variables are possible, 1 and 0, the double inversion property is quite simple. A 0 inverted once becomes a 1. A second inversion returns the value to 0, thus proving this identity. Fig. 2-27 illustrates this identity with gates. Chapter 3 will explain the importance of the double inversion identity.

$$A \quad \overline{A} \quad \overline{\overline{A}}$$
$$\begin{pmatrix} 0 \\ 1 \end{pmatrix} \quad \begin{pmatrix} 1 \\ 0 \end{pmatrix} \quad \begin{pmatrix} 0 \\ 1 \end{pmatrix}$$

$$A \qquad\qquad A$$
$$\begin{pmatrix} 0 \\ 1 \end{pmatrix} \qquad\qquad \begin{pmatrix} 0 \\ 1 \end{pmatrix}$$

FIGURE 2-27. *Proof of* $\overline{\overline{A}} = A$ *with gates.*

TABLE 2-24 FUNDAMENTAL BOOLEAN IDENTITIES.

NOT function	AND function	OR function
$\overline{\overline{A}} = A$	$A \cdot 0 = 0$	$A + 0 = A$
	$A \cdot 1 = A$	$A + 1 = 1$
	$A \cdot A = A$	$A + A = A$
	$A \cdot \overline{A} = 0$	$A + \overline{A} = 1$

TABLE 2-25 THE TRUTH TABLE FOR THE AND IDENTITIES.

a. Zero AND A is zero.

0	A	X
0	0	0
0	1	0

b. One AND A is A.

1	A	X
1	0	0
1	1	1

c. A AND A is A.

A	A	X
0	0	0
1	1	1

d. A AND $\overline{A}$ is zero.

A	$\overline{A}$	X
0	1	0
1	0	0

The AND identities are equally easy to understand when only two variables are used. Any variable ANDed with 0 will yield a 0. Likewise, a variable ANDed with a 1 will be a 1 if the variable is a 1 and a 0 if the variable is a 0: this is the AND identity. The last two AND properties are easy to understand in the 1–0 context. Fig. 2-28 shows the gate configurations, and Table 2-25 contains the truth tables for the four AND identities.

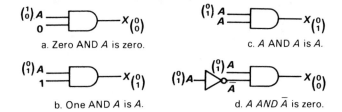

FIGURE 2-28. *Diagrams proving the four AND identities.*

a. Zero AND A is zero.

b. One AND A is A.

c. A AND A is A.

d. A AND $\overline{A}$ is zero.

TABLE 2-26 THE TRUTH TABLE FOR THE OR IDENTITIES.

a. Zero OR A is A.

0	A	X
0	0	0
0	1	1

b. One OR A is one.

1	A	X
1	0	1
1	1	1

c. A OR A is A.

A	A	X
0	0	0
1	1	1

d. A OR $\overline{A}$ is one.

A	$\overline{A}$	X
0	1	1
1	0	1

The OR identities follow from the above discussion. The four OR identities are illustrated in Fig. 2-29. Table 2-26 shows the truth tables for the OR identities.

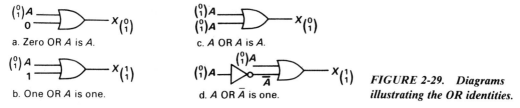

a. Zero OR A is A.

b. One OR A is one.

c. A OR A is A.

d. A OR $\overline{A}$ is one.

FIGURE 2-29. Diagrams illustrating the OR identities.

OTHER BOOLEAN PROPERTIES

The following properties are presented to aid in the understanding of Boolean algebra and digital circuits. Methods presented in Chapter 3 accomplish the same results as those accomplished with these properties, but in a more expedient manner. Therefore, as mentioned earlier, comprehensive study and memorization of these properties are not necessary.

TABLE 2-27 ADDITIONAL BOOLEAN PROPERTIES.

Name	AND		OR	
Distributive Theorem	$A(B+C)=AB+AC$		$(A+B)(A+C)=A+BC$	
Commutative Theorem	$AB=BA$		$A+B=B+A$	
Associative Theorem	$(AB)C=A(BC)$		$(A+B)+C=A+(B+C)$	
Absorption Theorem	$A+AB=A$	$A+\overline{A}B=A+B$	$\overline{A}+AB=\overline{A}+B$	

The Boolean properties presented in Table 2-27 are very similar to their algebraic counterparts in most cases.

The *Commutative Theorem* states that, for either the AND or the OR function, the inputs can be operated on in any order. This property applies for any number of inputs. The Commutative Theorem is illustrated with gates in Fig. 2-30a (for the AND function) and in Fig. 2-30b (for the OR function). The proofs of these two functions appear in truth table form in Tables 2-28 and 2-29, respectively. Note that adding more variables to either truth table has no effect on the output.

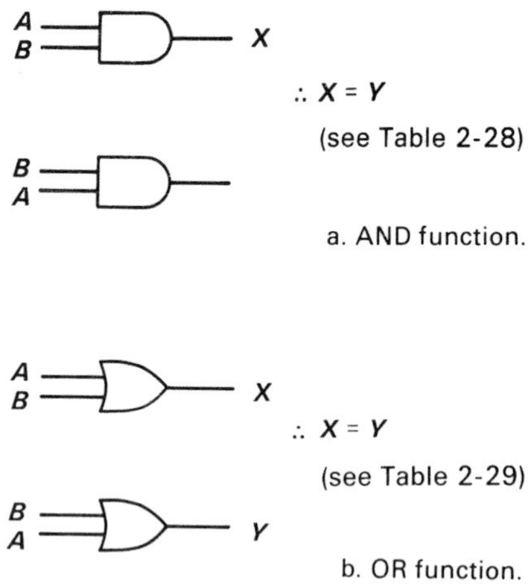

$\therefore X = Y$

(see Table 2-28)

a. AND function.

$\therefore X = Y$

(see Table 2-29)

FIGURE 2-30. Commutative property.

b. OR function.

Association, as in algebra, implies that the manner in which variables are grouped has no effect on the result. In Boolean algebra, this means that the inputs to two or more cascaded gates can be connected in any order, and the output will remain the same. Two examples of circuits that make use of association appear in Fig. 2-31. Proof of association is shown in Table 2-30.

Distribution differs from standard algebra in that it works for both AND and OR. For the OR function, this is a somewhat unfamiliar operation. The proof of the two distribution properties is shown with both equations and truth

TABLE 2-28 AND COMMUTATIVE THEOREM PROOF.

	$AB = X$			$BA = Y$	
A	B	X	B	A	Y
0	0	0	0	0	0
0	1	0	0	1	0
1	0	0	1	0	0
1	1	1	1	1	1

$$\therefore X = Y$$

TABLE 2-29 OR COMMUTATIVE THEOREM PROOF.

	$A + B = X$			$B + A = Y$	
A	B	X	B	A	Y
0	0	0	0	0	0
0	1	1	0	1	1
1	0	1	1	0	1
1	1	1	1	1	1

$$\therefore X = Y$$

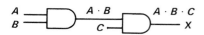

$\therefore X = Y$
(see Table 2-30a)

a. AND function.

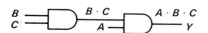

$\therefore X = Y$
(see Table 2-30b)

b. OR function.

FIGURE 2-31.   Associative property.

TABLE 2-30 ASSOCIATIVE PROPERTY PROOFS.

a. AND Function.

	(AB) C = X				A (BC) = Y		
A	B	C	X	B	C	A	Y
0	0	0	0	0	0	0	0
0	0	1	0	0	0	1	0
0	1	0	0	0	1	0	0
0	1	1	0	0	1	1	0
1	0	0	0	1	0	0	0
1	0	1	0	1	0	1	0
1	1	0	0	1	1	0	0
1	1	1	1	1	1	1	1

$$\therefore X = Y$$

b. OR Function

	(A + B) + C = X				A + (B + C) = Y		
A	B	C	X	B	C	A	Y
0	0	0	0	0	0	0	0
0	0	1	1	0	0	1	1
0	1	0	1	0	1	0	1
0	1	1	1	0	1	1	1
1	0	0	1	1	0	0	1
1	0	1	1	1	0	1	1
1	1	0	1	1	1	0	1
1	1	1	1	1	1	1	1

$$\therefore X = Y$$

tables in Tables 2-31. Circuit illustrations, as shown in Figs. 2-32a and 2-32b, should help to explain the distribution property.

The *Absorption Theorem* will prove to be very important in Chapter 3. The three equations listed as examples of the Absorption Theorem are not quite as obvious as those used to illustrate the other properties. The absorption property has no real parallel in algebra, either. The name, of course, comes from the fact that one of the variables can be absorbed to form a simpler equation. Tables 2-32 proves each of the three absorption properties with truth tables and equations. Fig. 2-33 depicts examples with gates to show the saving of circuit components.

TABLE 2-31 DISTRIBUTIVE PROPERTY PROOFS.

a. AND Function.

	$X = A\,(B + C)$				$Y = AB + AC$		
A	B	C	X	A	B	C	Y
0	0	0	0	0	0	0	0
0	0	1	0	0	0	1	0
0	1	0	0	0	1	0	0
0	1	1	0	0	1	1	0
1	0	0	0	1	0	0	0
1	0	1	1	1	0	1	1
1	1	0	1	1	1	0	1
1	1	1	1	1	1	1	1

$$\therefore X = Y$$

b. OR Function.

	$(A + B)\,(A + C) = X$				$A + BC = Y$		
A	B	C	X	A	B	C	Y
0	0	0	0	0	0	0	0
0	0	1	0	0	0	1	0
0	1	0	0	0	1	0	0
0	1	1	1	0	1	1	1
1	0	0	1	1	0	0	1
1	0	1	1	1	0	1	1
1	1	0	1	1	1	0	1
1	1	1	1	1	1	1	1

$$\therefore X = Y$$

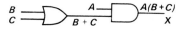

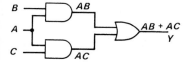

a. AND function.

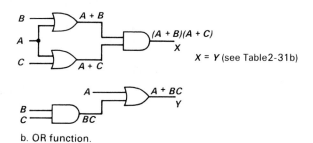

b. OR function.

FIGURE 2-32. Distributive property.

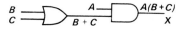

X = Y(see Table 2-31a)

X = Y (see Table2-31b)

TABLE 2-32 THE ABSORPTION PROPERTY.

a.

A	B	$(A+AB)$
0	0	0
0	1	0
1	0	1
1	1	1

$$\therefore A + AB = A$$

b.

A	$\overline{A}$	B	$\overline{A} + AB$	$\overline{A} + B$
0	1	0	1	1
0	1	1	1	1
1	0	0	0	0
1	0	1	1	1

$$\therefore \overline{A} + AB = \overline{A} + B$$

c.

A	$\overline{A}$	B	$A + \overline{A}B$	$A+ B$
0	1	0	0	0
0	1	1	1	1
1	0	0	1	1
1	0	1	1	1

$$\therefore A + \overline{A}B = A + B$$

DE MORGAN'S THEOREM

Boolean theorems can sometimes help to simplify circuit equations, resulting in a simpler circuit with fewer gates. This process is explored in detail in Chapter 3. One of the most important of the Boolean theorems is De Morgan's Theorem (sometimes called De Morgan's Law), which the following example defines.

Example 2-5. Construct truth tables for the two equations:

$$\overline{A \cdot B} = X$$

$$\overline{A} + \overline{B} = X$$

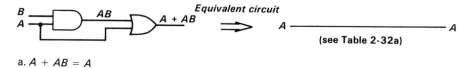

a. $A + AB = A$

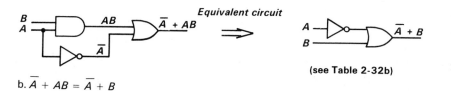

b. $\overline{A} + AB = \overline{A} + B$

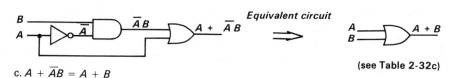

c. $A + \overline{A}B = A + B$

(see Table 2-32c)

FIGURE 2-33. Absorption theorem.

TABLE 2-33 TRUTH TABLE FOR $\overline{A \cdot B} = X$.

A	B	X
0	0	1
0	1	1
1	0	1
1	1	0

TABLE 2-34 TRUTH TABLE FOR $\overline{A} + \overline{B} = X$.

A	B	X
0	0	1
0	1	1
1	0	1
1	1	0

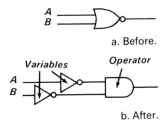

a. Before.

Variables   Operator

b. After.

FIGURE 2-34. To perform De Morgan's theorem, complement the variables and change the operator.

Solution. Inspection of Tables 2-33 and 2-34 reveals that they are identical—somewhat of a surprise. Since they are identical, we have exhaustively proven that $\overline{A \cdot B} = \overline{A} + \overline{B}$, which is De Morgan's Theorem. In words, De Morgan's Theorem states, "To remove the bar from an entire equation, complement (invert) the individual variables and change the operator (from an AND to an OR, or vice versa)." The operation is shown symbolically in Fig. 2-34.

The reverse of the operation in example 2-5 is also true:

$$\overline{A + B} = \overline{A} \cdot \overline{B}$$

Thus we can construct an OR gate from a NAND and possibly some INVERTERS. The AND gate and the Exclusive OR, as well as the INVERTER, can also be constructed using the NAND gate (refer to the problems at the end of this chapter). The NOR gate can also be used in the same manner. It is for this reason that the NAND and the NOR are called *universal gates*.

For the NOR gate, the application of De Morgan's Theorem produces the following results:

$$\overline{A + B} = \overline{A} \cdot \overline{B}$$

This equation, and the one above for the NAND gate, are called *duals* because one is the OR function and one is the AND function. Several examples of De Morgan's Theorem are found in Table 2-35. Two special examples are also given using the double inversion identity from Table 2-24.

SUMMARY

1   In digital switching, circuits generally have only two states: *on* (1) and *off* (0).

2   Truth tables are useful for defining and analyzing digital circuits. Several examples appear in the chapter.

3   The AND gate performs the function: "The output is only a 1 when all inputs are 1" (see Table 2-36).

4   The OR gate performs the function: "The output is 1 when any or all inputs are 1" (see Table 2-36).

5   The INVERTER performs the complement function: "The output is the opposite of the input" (see Table 2-36).

6   The NAND gate is an AND followed by an INVERTER (see Table 2-36).

7   The NOR gate is an OR followed by an INVERTER (see Table 2-36).

TABLE 2-35 EXAMPLES OF THE APPLICATION OF DE MORGAN'S THEOREM.

Initial equation	First Step (Complement) Variables & expression	Second Step (Change the Operator)	Result	Circuits
$\overline{A+B+C}=Y$	$\overline{A}+\overline{B}+\overline{C}$	$\overline{A}\cdot\overline{B}\cdot\overline{C}$	$\overline{A}\cdot\overline{B}\cdot\overline{C}=Y$	
$\overline{A\cdot B\cdot C}=Y$	$\overline{A}\cdot\overline{B}\cdot\overline{C}$	$\overline{A}+\overline{B}+\overline{C}$	$\overline{A}+B+\overline{C}=Y$	
$\overline{\overline{A}+\overline{B}}=Y$	$\overline{\overline{A}+\overline{B}}$	$\overline{\overline{A}\cdot\overline{B}}$	$\overline{\overline{A}\cdot\overline{B}}=Y$	
$(A+B=Y)$				
$\overline{\overline{A\cdot B\cdot C}}=Y$	$\overline{\overline{A}\cdot\overline{B}\cdot C}$	$\overline{\overline{A}+\overline{B}+C}$	$\overline{\overline{A}+\overline{B}+C}$	
$(A\cdot B\cdot\overline{C}=Y)$				

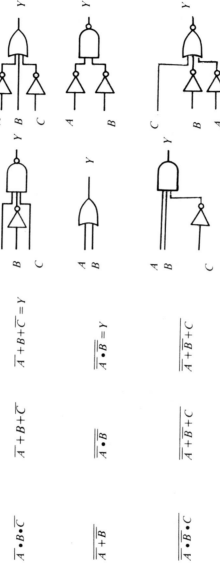

TABLE 2-36 LOGIC FUNCTION SUMMARY.

Function	Notation	Logic symbol
Inverse or NOT	$\overline{}\ (\overline{A}=B)$	$A \longrightarrow\!\!\!\rhd\!\!\circ\!\longrightarrow B$
AND	$\bullet\ (A \bullet B=C)$ $(A\,B=C)$	$\begin{array}{c}A\\B\end{array}\!\!=\!\!\!\sqsupset\!\!-C$
OR	$+\ (A+B=C)$	$\begin{array}{c}A\\B\end{array}\!\!=\!\!\!\rhd\!\!-C$
Exclusive OR	$\oplus\,(A\oplus B=C)$	$\begin{array}{c}A\\B\end{array}\!\!=\!\!)\!\!\rhd\!\!-C$
Exclusive NOR	$\overline{\oplus}\ \ (\overline{A\oplus B}=C)$	$\begin{array}{c}A\\B\end{array}\!\!=\!\!)\!\!\rhd\!\!\circ\!\!-C$
NOR	$\overline{+}\ \ (\overline{A+B}=C)$	$\begin{array}{c}A\\B\end{array}\!\!=\!\!\!\rhd\!\!\circ\!\!-C$
NAND	$\overline{\bullet}\ \ \overline{A\bullet B}=C)$ $(\overline{AB}=C)$	$\begin{array}{c}A\\B\end{array}\!\!=\!\!\!\sqsupset\!\!\circ\!\!-C$

8 The Exclusive OR performs the function: "The output is 1 only when either input is 1 (see Table 2-36).

9 The Schmitt trigger is a special type of digital gate that has "snap action" switching. It is used with slowly changing signals.

10 The basic principles of Boolean algebra are presented in Tables 2-24 and 2-27. De Morgan's Theorem is an extremely important part of digital electronics (see Table 2-35).

KEY TERMS

AND	Exclusive OR	NOT
Boolean algebra	gates	OR
combinational logic	INVERTER	Schmitt trigger
De Morgan's Theorem	NAND	truth table
Exclusive NOR	NOR	

QUESTIONS AND PROBLEMS

1 The symbols 1 and 0 are used to represent the two states in digital logic. What are some other names for the two states?

2 Name the function represented by the logic symbol in Fig. 2-35 and make a truth table of its function.

FIGURE 2-35. Function for problem 2.

3 Draw the logic symbols for the following functions and construct a 3-input truth table for each:

  a. NOR

  b. AND

  c. OR

  d. NAND

4 Draw a circuit diagram that will represent the equation: $Y = \overline{A} + (\overline{BCA})$. Verify the circuit with a truth table.

5 Using truth tables, prove that the equations $(A + B)C = X$ and $A + (BC) = Y$ are not equivalent.

6 Using only NAND gate(s), construct an INVERTER function. Repeat with the NOR gate.

7 Using only NAND gate(s), construct the AND function. Repeat with the NOR gate.

8 Using only NAND gate(s), construct the OR function. Repeat with the NOR gate.

9 Now, use the NAND to construct the Exclusive OR function. Repeat with the NOR gate.

10 Draw circuit diagrams for the following equations (a truth table is not necessary):

  a. $A + B + CD = Z$

  b. $(A + B)(B + C)(A + C) = Y$

  c. $(\overline{\overline{A + B}}) + (C + D) = X$

11 Repeat problem 10b using only NAND gates.

12 Repeat problem 10b using only NOR gates.

13 Complete the following equations:

  a. $A \cdot 0 =$ d. $\overline{\overline{A}} =$

  b. $\overline{A} + A =$ e. $A + 1 =$

  c. $\overline{A} \cdot A =$ f. $A \cdot 1 =$

14 Make a truth table for a 4-input NAND.

15 Make a truth table for a 5-input NOR.

16 Write the equations for the logic circuits shown in Fig. 2-36.

17 Find the equivalent logic circuits for the circuits depicted in Fig. 2-37. Use De Morgan's Theorem.

18 Make truth tables for the following equations:

    a. $C(D + E) = W$ b. $CD + CE = X$

What property does this illustrate?

19 Repeat problem 18 for the following equations:

    a. $(A + B) + C = Z$ b. $A + (B + C) = Y$

20 Using only properties presented in the chapter, prove each of the three absorption properties. Check the results with truth tables.

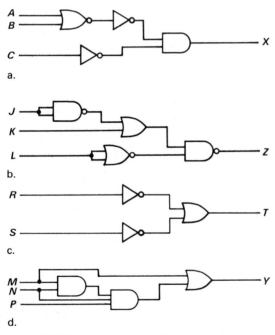

a.

b.

c.

d.

FIGURE 2-36. Circuits for problem 16.

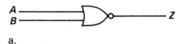

a.

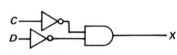

b.

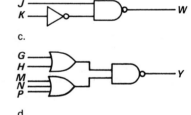

c.

d.

FIGURE 2-37. Circuits for problem 17.

3

DEVELOPING COMBINATIONAL LOGIC CIRCUITRY

In Chapter 2 several examples and problems demonstrated that more than one circuit can often satisfy the same truth table. In designing circuits for production, it is most desirable to find the simplest circuit, the one that uses the fewest gates, since it usually costs the least. This chapter will show how to simplify circuits and how to apply combinational circuits (circuits utilizing a combination of functions) to implement a given truth table.

SIMPLIFICATION USING BOOLEAN ALGEBRA

One of the chief uses of Boolean algebra is in the simplification of logic circuits to less expensive forms—less expensive circuits being those with fewer gates. Although methods presented later in this chapter are often easier to use than Boolean algebra, anyone who is well acquainted with Boolean algebra should be able to simplify logic circuits by inspecting the circuit itself or its representative equation.

Simplification via Boolean algebra entails finding an equation for the digital circuit of interest and then applying the Boolean identities (Table 2-24) and the Boolean properties (Table 2-27), as well as De Morgan's Theorem (Table 2-35). A few examples will illustrate this procedure. However, it takes practice to be able to apply the principles through inspection.

Example 3-1. Simplify the logic circuit shown in Fig. 3-1. The equation of the circuit is:

$$A(A + B) = C \qquad (3\text{-}1)$$

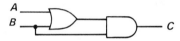

FIGURE 3-1. Logic circuit for example 3-1.

Solution. First let's expand this equation to make it easier to apply Boolean properties. Expansion yields:

$$AA + AB = C \qquad (3\text{-}2)$$

In Table 2-24, one of the AND identities states that $AA = A$, so Eq. 3-2 can now be written as:

$$A + AB = C \qquad (3\text{-}3)$$

Turning to Table 2-27, we see that the absorption property $A + AB = A$ becomes applicable to the equation. Now we can rewrite Eq. 3-3 as:

$$A = C \qquad (3\text{-}4)$$

Thus, by applying the principles of Boolean algebra, we find that we can achieve the same results as those of the logic circuit in Fig. 3-1, by simply using a "wire" to connect input A to output C. The latter circuit is also certainly cheaper than the two gates of Fig. 3-1!

Example 3-2. Fig. 3-2 shows a circuit with two AND gates, an OR, and a NAND. Apply the Boolean theorems to reduce this circuit.

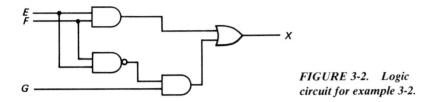

FIGURE 3-2. Logic circuit for example 3-2.

Solution. First, the equation for the entire circuit is:

$$X = EF + \overline{EF}G \qquad (3\text{-}5)$$

Intermediate equations in the circuit are:

$$W = EF \qquad (3\text{-}6)$$

$$Z = \overline{EF} \qquad (3\text{-}7)$$

Note also that:

$$W = \overline{Z} \qquad (3\text{-}8)$$

Now, if we rewrite Eq. 3-5 using our intermediate equations, one of the properties from Table 2-27 becomes easily applicable:

$$X = W + \overline{W}G \qquad (3\text{-}9)$$

Eq. 3-9 now fits the form of the third Absorption Theorem in Table 2-32, $A + \overline{A}B = A + B$. Finally, by applying the absorption property, the circuit equation is reduced to:

$$X = EF + G \qquad (3\text{-}10)$$

This reduced circuit is shown in Fig. 3-3. Notice how we have reduced the gate count from four to two.

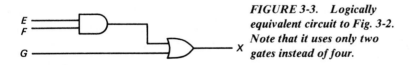

FIGURE 3-3. Logically equivalent circuit to Fig. 3-2. Note that it uses only two gates instead of four.

Example 3-3. Fig. 3-4 shows a relatively complex logic circuit. Apply reduction techniques to see if it can be simplified.

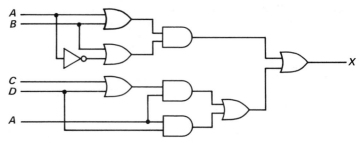

FIGURE 3-4. *Logic circuit for reduction in example 3-3.*

Solution. The circuit equation is:

$$(B + A)(B + \overline{A}) + A(C + D) + DA = X \qquad (3\text{-}11)$$

First, apply the Distributive Theorem to the leftmost pair of terms:

$$(B + A)(B + \overline{A}) = B + A\overline{A} \qquad (3\text{-}12)$$

From the identity table (Table 2-24), note that $A\overline{A} = 0$ and also that $B + 0 = B$. Thus, Eq. 3-12 reduces to:

$$B + A\overline{A} = B \qquad (3\text{-}13)$$

Now let's work on the rightmost terms of Eq. 3-11. Again using Table 2-27, we may apply the distributive and commutative properties for AND:

$$A(C + D) + DA = AC + AD + AD \qquad (3\text{-}14)$$

From Table 2-24 we find that $AD + AD = AD$. So the equation further reduces, after redistributing, to:

$$AC + AD + AD = A(C + D) \qquad (3\text{-}15)$$

The combination of all our reduction efforts yields:

$$B + A(C + D) = X \qquad (3\text{-}16)$$

The reduced circuit is drawn in Fig. 3-5. In this example, the number of gates has been reduced from eight, plus one INVERTER, to three. These three examples show the importance of reduction in the design of logic circuits.

FIGURE 3-5. *Logically equivalent circuit of Fig. 3-4. Note
that it saves six gates.*

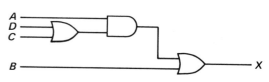

KARNAUGH MAPPING

Fortunately, there is an easier way to reduce a logic circuit than by manipulating equations. It is called *Karnaugh mapping*. This section introduces the Karnaugh map (K map), and the next section details circuit simplication using K maps.

KARNAUGH MAP STRUCTURE

The Karnaugh map can be used for simplification of Boolean expressions of any number of variables, but is virtually never used for less than three variables and becomes difficult to use as the number of variables increases past four. The Karnaugh map has several advantages over Boolean principles in the simplification of logic circuits:

1. Karnaugh mapping is systematic and easy to learn; therefore, theorems and their application need not be memorized.
2. When Karnaugh mapping is systematically applied, the solution is *always* the simplest form of the logic function.
3. Karnaugh maps are easier to use than Boolean algebra and therefore lead to fewer errors.

Figs. 3-6a, 3-6b, and 3-6c show K maps of three, four, and five variables, respectively. Notice that in Fig. 3-6c there are two maps in three dimensions; in other words, one "floats" on top of the other. This three-dimensionality is what makes maps with more than four variables difficult to use. Notice also that the variables are not labeled in binary progression, as is the case in truth tables. In a Karnaugh map, *one and only one* variable can change in progression from right to left or from top to bottom (gray code).

C\AB	00	01	11	10
00	0	1	1	1
01	0	0	1	0
11	1	1	0	0
10	0	1	1	1

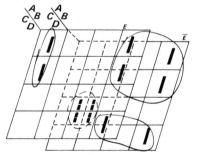

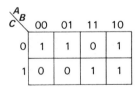

C\AB	00	01	11	10
0	1	1	0	1
1	0	0	1	1

a. A 3-variable Karnaugh map.

b. A 4-variable Karnaugh map.

c. A 5-variable Karnaugh map. Three dimensions are necessary in constructing maps with more than four variables. The top map represents variable E, and the bottom map represents variable $\bar{E}$.

FIGURE 3-6. Karnaugh maps.

	$\overline{AB}$	$A\overline{B}$	AB	$\overline{A}B$
$\overline{CD}$	1	1	0	0
$\overline{C}D$	0	0	0	0
CD	1	1	1	0
$C\overline{D}$	1	0	0	0

FIGURE 3-7. An alternate way of labeling a Karnaugh map using the variable names instead of binary numerical values.

Fig. 3-7 shows an alternate method of labeling a Karnaugh map, using the variable names. The use of variable names can be effectively applied to any map, not just the 4-variable map illustrated.

STANDARD FORM FOR LOGIC EXPRESSIONS

To facilitate the use of Karnaugh maps, logic expressions need to be written in one of two standard forms. The standard forms are called *sum of products* and *product of sums*. Generally, the sum of products form is more often used.

The *sum of products* (SOP) form is an "ORing of ANDed expressions." As an example, let's put the expression $A + BC + A(C + BD) = X$ into standard SOP form. This is easily accomplished by ANDing the A through the parenthetical term, which yields: $A + BC + AC + ABD = X$. As you can see, the expression is now an OR function of several AND expressions, which is the standard SOP form.

The *product of sums* (POS) form is, as you might have guessed, an ANDing of OR expressions. Here, we will put the following expression into standard form: $(A + B)C + (A + B)CD$. When we factor out the $(A + B)$ term, the expression becomes the POS standard form, $(A + B)(C + CD)$.

To use an SOP expression with a Karnaugh map, a 1 is inserted into the square whose value would make a corresponding term in the SOP expression true. This 1 is called a *minterm*. (If 0s were used to form a POS expression, each 0 would be called a *maxterm*.) For example, 101 would make the term $A\overline{B}C$ true; hence, the square in a Karnaugh map labeled 101 would receive a 1.

Example 3-4. Enter the expression $AC + A\overline{D} + \overline{A}BC + C = X$ into a Karnaugh map.

Solution. A blank 4-variable map has been drawn in Fig. 3-8a. To make the first term (AC) true, or 1, A and C must be 1. However, the first term tells

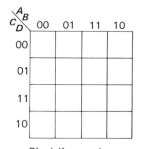

a. Blank Karnaugh map.

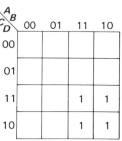

b. Karnaugh map where the squares representing AC are filled in with ones.

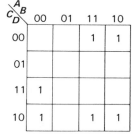

c. The squares in this map are filled in for the terms $A\overline{D}$ and $\overline{A}BC$.

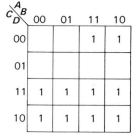

d. Completed Karnaugh map for the expression $AC + A\overline{D} + \overline{A}BC + C = X$.

FIGURE 3-8. Karnaugh maps for example 3-4.

nothing about B and D. Therefore, the possible values that make AC true are: 1010, 1011, 1110, and 1111. Fig. 3-8b shows the map with these four squares filled in. The second and third terms ($A\overline{D}$ and $\overline{A}BC$) are handled in a similar manner, as shown in Fig. 3-8c. Notice that the last term is simply C. This means that *both rows* where C appears as 1 are filled in. Fig. 3-8d shows the completed map filled in correctly.

The POS expression is transferred to the Karnaugh map in the same manner, except that 0s are inserted for value that would make each OR term false.

A truth table is used more often than an equation as the data source for the map. The next example illustrates this.

Example 3-5. Construct a Karnaugh map that corresponds to the truth table in Table 3-1.

TABLE 3-1 TRUTH TABLE FOR EXAMPLE 3-5.

A	B	C	D	X
0	0	0	0	0
0	0	0	1	1
0	0	1	0	1
0	0	1	1	0
0	1	0	0	0
0	1	0	1	0
0	1	1	0	1
0	1	1	1	0
1	0	0	0	1
1	0	0	1	1
1	0	1	0	1
1	0	1	1	1
1	1	0	0	0
1	1	0	1	1
1	1	1	0	1
1	1	1	1	1

$C\!D$＼$^{A}\!B$	00	01	11	10
00	0	0	0	1
01	1	0	1	1
11	0	0	1	1
10	1	1	1	1

FIGURE 3-9. Karnaugh map constructed from the truth table in Table 3-1.

Solution. The Karnaugh map is shown in Fig. 3-9. Notice how the lines of the truth table completely define the squares in which a 1 (or a 0) must be placed. For example, the truth table line 1001 corresponds to the box in the Karnaugh map where $A = 1$, $B = 0$, $C = 0$, and $D = 1$.

SIMPLIFICATION USING KARNAUGH MAPS

Once a Karnaugh map has been constructed, simplification is done by grouping 1s or 0s. If 1s are grouped, an SOP equation is the simplified solution; if 0s are grouped, a POS equation is the solution. The rules for grouping are simple:

1. Terms to be grouped must be either horizontally or vertically adjacent, or both; diagonals cannot be grouped.

2. The number of terms in a group must be a power of two: 1, 2, 4, 8, 16, etc.

3. A term may be in more than one group.

4. The map "wraps around" on itself on both the sides and the top and bottom. That is, two right-column 1s could combine with two left-column 1s to form a group of four.

5. The simplified form of the equation is written from the resulting groups. If a variable remains constant within a group, it remains in the equation; if the variable changes within the group, it drops out. If 1s are grouped, the equation is written in SOP form; if 0s are grouped, the equation is written in POS form.

Example 3-6. Using the truth table and Karnaugh map from example 3-5, simplify the expression and write the simplified equation.

Solution. First, let's write the equation in SOP form directly from the truth table:

$$\overline{A}BCD + \overline{A}\,\overline{B}C\overline{D} + \overline{A}\,BC\overline{D} + ABCD + A\overline{B}\overline{C}\,D +$$
$$A\overline{B}\,C\overline{D} + A\overline{B}\,CD + ABC\,D + ABC\overline{D} + ABCD = X \qquad (3\text{-}17)$$

As you can see, this expression could definitely use simplifying! A circuit for this function would require 11 gates (assuming that a 10-input OR gate was available).

Fig. 3-10 is the Karnaugh map with all possible groupings made. Clearly, three groups of four are possible, as well as one group of two, which uses the "wrap around" principle of the map. Notice that the terms 1111 and 1110 were not grouped. When all terms are grouped in the largest possible groups, the table is said to be *completely covered*. These groups are called the *essential prime implicants*. Any further groups would be *redundant*. Note, however, that the term 1010 is included in more than one group. In order to make the largest possible groups that completely cover the map, it is included in two 4-term groups. This, then, is *not* redundant, since a completely covered map would not result if it were not included in two groups.

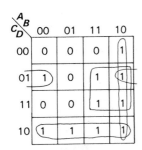

FIGURE 3-10. Karnaugh map from Fig. 3-9 with all groupings made.

To write the equation, follow rule 5 above. Since we have grouped 1s in the map, the equation will be in SOP form. In the bottom row, we find that within the group of four, A and B change but C and D remain constant. Since C is constant at 1, it is written as is, but since D is constant at 0, it is written as the complement ($\overline{D}$), resulting in the term $C\overline{D}$. Next, notice that the opposite occurs in the rightmost column: A and B are constant, but C and D change. Thus the product term for the rightmost group of four is $A\overline{B}$. In the final group of four, A remains constant but B changes, and D is constant while C changes. Thus the product term for this group of four is AD. Finally, looking at the group of two, we see that $\overline{B}$, $\overline{C}$, and D remain constant, and only A changes. The product term for this group is $\overline{B}\,\overline{C}D$.

If the four individual terms are combined through ORing, the simplified SOP equation results:

$$C\overline{D} + A\overline{B} + AD + \overline{B}\,\overline{C}D = X \qquad (3\text{-}18)$$

This is certainly simpler than Eq. 3-17!

You may have noticed that the groups of four reduced to 2-variable terms, and the group of two reduced to a 3-variable term. In general, the number of variables eliminated in a term will be the power to which two is raised to equal the size of the group: A group of two eliminates one variable ($2^1 = 2$), a group of four eliminates two variables ($2^2 = 4$), a group of eight eliminates three variables ($2^3 = 8$), and so on.

Example 3-7. Construct a truth table from the equation below. Then, using Karnaugh mapping techniques, simplify the circuit, write the simplified equation, and draw a schematic of the circuit.

$$Z = A\overline{B}C + A(\overline{B} + D) + \overline{A}BD + ABD + A\overline{B}CD + \overline{B}D + BC\overline{D} \qquad (3\text{-}18)$$

Solution. To construct the truth table we insert 1s to satisfy each term of the equation, expanding the second term to $A\overline{B} + AD$. Table 3-2 is the completed truth table.

Next, we can construct the Karnaugh map by inserting 1s in the cells corresponding to the truth table. After circling all combinations as shown in Fig. 3-11, we can construct the terms for the equation (indicated by arrows) from the encircled groups. If a variable changes within the encircled group, it is not included in the term. Thus the group in the far right-hand column yields $A\overline{B}$, since C and D both change within the grouping (and B is complemented). The group at the top yields $B\overline{C}$, since both A and D change within the grouping. Finally, the group of eight in the center results simply in D, since all other

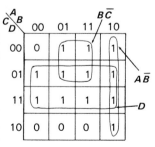

FIGURE 3-11. The completed map for example 3-7. The arrows show the equation for each circled term.

TABLE 3-2 TRUTH TABLE FOR EXAMPLE 3-7.

A	B	C	D	Z
0	0	0	0	0
0	0	0	1	1
0	0	1	0	0
0	0	1	1	1
0	1	0	0	1
0	1	0	1	1
0	1	1	0	0
0	1	1	1	1
1	0	0	0	1
1	0	0	1	1
1	0	1	0	1
1	0	1	1	1
1	1	0	0	1
1	1	0	1	1
1	1	1	0	0
1	1	1	1	1

variables change. Combining all the terms into an SOP expression (because the 1s were used to form groups) yields:

$$A\overline{B} + B\overline{C} + D \qquad (3\text{-}20)$$

DON'T CARES

In a number of instances, the value of the output (1 or 0) of a certain input state is irrelevant to the design of a circuit. This generally results because that particular input state will never occur (this is called a *disallowed state*) or because the output resulting from that given input doesn't matter. The example below illustrates the disallowed state.

Example 3-8. Table 3-3 is a truth table for a hypothetical function. In Fig. 3-12, the Karnaugh map has been constructed and the essential prime implicants circled. The equation resulting from this map is:

$$A\overline{C}D + A\overline{B}D + \overline{A}C + BC\overline{D} = X$$

Now, assume that the inputs to the function come from a circuit which outputs only binary coded decimal (BCD) values. In other words, the terms 1100, 1101, 1110, and 1111 will never occur—they are *don't cares:* We don't care what the output will be, since that input can never occur. The truth table for the function

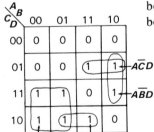

FIGURE 3-12. The map for Table 3-3 (not using don't cares).

TABLE 3-3 TRUTH TABLE FOR EXAMPLE 3-8.

A	B	C	D	X
0	0	0	0	0
0	0	0	1	0
0	0	1	0	1
0	0	1	1	1
0	1	0	0	0
0	1	0	1	0
0	1	1	0	1
0	1	1	1	1
1	0	0	0	0
1	0	0	1	1
1	0	1	0	0
1	0	1	1	1
1	1	0	0	0
1	1	0	1	1
1	1	1	0	1
1	1	1	1	0

now becomes that depicted in Table 3-4. Notice that "x" is used wherever a *don't care* appears. Transferring the truth table to a Karnaugh map results in Fig. 3-13; again, "x" has been used for *don't cares*. The importance of *don't cares* is that they can be used for *either* a 1 or a 0—whichever is most convenient. In Fig. 3-13, all the *don't cares* are used as 1s. The equation for this map is:

$$AB + AD + C = X$$

This shows the importance of examining the problem carefully and utilizing *don't cares* where possible.

COMBINATIONAL LOGIC CIRCUIT DESIGN

A clear understanding of the basic equations and the ability to make truth tables and simplify expressions by using Karnaugh maps are all that is necessary to proceed to the next step: the actual design of a combinational logic circuit.

Let's begin by designing a simple burglar alarm for the house shown in Fig. 3-14. The house has two windows (W_1 and W_2), a door (D), and a control switch (CS). The alarm has the following constraints:

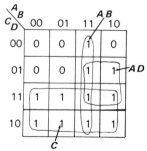

FIGURE 3-13. The map from Table 3-4, including don't cares. Using all of the don't cares as 1s gives us the largest groups for this map.

TABLE 3-4 TRUTH TABLE FOR EXAMPLE 3-8 (IF THE INPUTS ARE BCD, WITH THE RESULTING *DON'T CARE* VALUES).

A	B	C	D	X
0	0	0	0	0
0	0	0	1	0
0	0	1	0	1
0	0	1	1	1
0	1	0	0	0
0	1	0	1	0
0	1	1	0	1
0	1	1	1	1
1	0	0	0	0
1	0	0	1	1
1	0	1	0	x
1	0	1	1	x
1	1	0	0	x
1	1	0	1	x
1	1	1	0	x
1	1	1	1	x

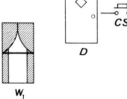

FIGURE 3-14. Parts of house for which burglar alarm must be designed. The switch allows the door to be opened without triggering the alarm.

1. If the switch is closed (1), the door may be opened (1) and entry made without setting off the alarm.
2. If both windows are open (1) at the same time (for ventilation), the alarm will not sound.
3. Otherwise, if the door or windows are opened (1) the alarm (ALARM) is sounded (1).

Even with a simple digital system like this alarm, the best method of design to follow is the systematic approach that we have learned:

1. Make a truth table.
2. Simplify with a Karnaugh map.

3. Write the equation.
4. Draw the circuit schematically.

Table 3-5 is the truth table for the digital burglar alarm. Note that the inputs and outputs are labeled W_1, W_2, D, CS, and ALARM. This allows us to see more readily what the inputs and outputs actually represent.

Fig. 3-15 is the Karnaugh map with the essential prime implicants encircled. The equation resulting from the Karnaugh map is:

$$D\overline{CS} + W_1\overline{W}_2 + \overline{W}_1 W_2 = \text{ALARM}$$

Finally, Fig. 3-16 shows the circuit drawn schematically. While this circuit is fairly simple, it illustrates the systematic approach to combinational circuit design that will yield a minimal, functionally correct circuit.

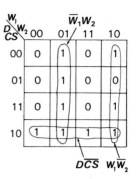

FIGURE 3-15. *Minimal equations for our burglar alarm.*

TABLE 3-5 THIS TRUTH TABLE DEFINES THE LOGIC FOR THE BURGLAR ALARM EXAMPLE.

W_1	W_2	D	CS	ALARM
0	0	0	0	0
0	0	0	1	0
0	0	1	0	1
0	0	1	1	0
0	1	0	0	1
0	1	0	1	1
0	1	1	0	1
0	1	1	1	1
1	0	0	0	1
1	0	0	1	1
1	0	1	0	1
1	0	1	1	1
1	1	0	0	0
1	1	0	1	0
1	1	1	0	1
1	1	1	1	0

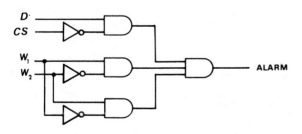

FIGURE 3-16. *Circuit design for our burglar alarm.*

MULTIPLE-OUTPUT COMBINATIONAL CIRCUITS

The examples so far have shown digital circuits with only one output. However, quite frequently a digital circuit will have two or more outputs for a given set of inputs. The example below of a digital combination padlock is an example of a circuit with more than one output.

Example 3-9. A digital padlock provides a good example of the multiple-output combinational circuit. The padlock to be designed here will have four binary numbers as its combination—in other words, it is a 4-bit padlock. Let's say that the combination will be 1011.

Solution. The design of this circuit is so obvious that we will simply write the equation and draw the circuit. It should be clear that the only input that will generate a 1 output is 1011; hence, the equation is:

$$A\overline{B}CD = \text{UNLOCK}$$

A 4-bit padlock would not provide much security, since all a person would have to do to "pick" the lock is try a maximum of 16 combinations. To add security to the lock, we can add a second output. This second output will freeze the padlock unless the 4-digit code is entered the first time. In this way, a person attempting to pick the lock would freeze it in the locked position by randomly trying combinations. A new input is introduced here, called the STROBE input. In a combinational circuit containing a STROBE input, all inputs are ignored except those coincident with the STROBE input. (Of course, there would have to be a secret "unfreeze" combination too, or the owner could never open the lock!)

Table 3-6 is the truth table for this function. Notice that any input except the valid combination of 1011 causes the FREEZE output to go high. Remember that each of the inputs in the truth table occurs only coincident with the STROBE input.

The equations are still simple to write for both the UNLOCK and the FREEZE outputs:

$$\text{UNLOCK} = A\overline{B}CD$$

$$\text{FREEZE} = \overline{A\overline{B}CD}$$

Before drawing the circuit, however, we must determine how the STROBE input works. In words, when the STROBE input is 0, it disables all inputs; when 1, it allows inputs. This is clearly an AND function. So, if we AND each of the inputs with the STROBE, we will have the desired function. Also, in order for the FREEZE output actually to freeze the padlock, it must disable the STROBE and allow no inputs. This can also be accomplished with an AND gate by first complementing the FREEZE output. The entire

TABLE 3-6 TRUTH TABLE FOR THE DIGITAL PADLOCK.

A	B	C	D	UNLOCK	FREEZE
0	0	0	0	0	1
0	0	0	1	0	1
0	0	1	0	0	1
0	0	1	1	0	1
0	1	0	0	0	1
0	1	0	1	0	1
0	1	1	0	0	1
0	1	1	1	0	1
1	0	0	0	0	1
1	0	0	1	0	1
1	0	1	0	0	1
1	0	1	1	1	0
1	1	0	0	0	1
1	1	0	1	0	1
1	1	1	0	0	1
1	1	1	1	0	1

circuit is drawn in Fig. 3-17. You can verify its operation by mentally trying values to determine the output. As an additional exercise, devise a way to "unfreeze" the lock. Also, do you see a startup problem with the lock, and if so, how would you fix it?

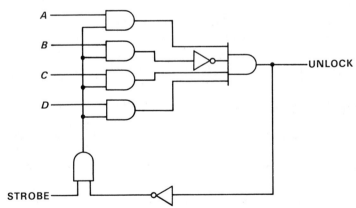

FIGURE 3-17. *Digital padlock circuit diagram.*

GATE ARRAYS

One of the newest tools in the design of combinational logic circuits is the *gate array*. Gate arrays allow the user to place many logic functions on a single IC rather than buying many "off the shelf" ICs and connecting them.

Many kinds of circuits are reproduced and sold in enormous quantities—tens of thousands, even millions of copies. One of the foremost design considerations when dealing with these quantities is cost. The elimination of a single 10-cent component could save a company thousands of dollars. Often, when cost and quantity warrant, it is desirable to use special parts in combinational circuits in place of commercially available parts to save money and reduce the parts count. The gate array is one such component.

If a company or semiconductor manufacturer designs a circuit and makes ICs that contain this special circuit, the ICs are referred to as *custom* or *full-custom* ICs. The development costs and quantities of such circuitry are usually quite high. Designers can, however, take advantage of custom circuits at lower costs by using a semicustom IC, such as a gate array. Gate arrays are available in two versions: factory-configurable and user-configurable.

FACTORY-CONFIGURABLE GATE ARRAYS

To produce integrated circuits, each individual circuit is etched onto a wafer of silicon semiconductor material in a process similar to photography. The individual gates formed in the silicon are connected through a metal mask, which is the last step in the fabrication process.

Gate arrays are formed by making one silicon gate mask for a particular gate array configuration. Then the user defines how the gates are to be interconnected for his circuit via the metal mask. In this way, manufacturers can keep the cost down by mass production techniques, up to the metal mask stage. At the same time, the user can tailor the integrated circuit to his need.

Standard factory-programmed gate arrays, sometimes called *uncommitted logic arrays* (ULAs), are available with as few as 300 gates and more than one thousand. Available gates include NORs, NANDs, INVERTERs, Exclusive ORs, and Schmitt triggers, as well as several sequential circuits (see Chapter 4).

USER-CONFIGURABLE GATE ARRAYS

The user-configurable gate array, also called Programmable Array Logic (PAL) *or* Programmable Logic Array (PLA), is similar to the ULA, except that the connection—or, more appropriately, disconnection—of the gates is performed by the user.

In the PAL, all gates are interconnected. The circuit designer then "blows" tiny fuses to disconnect the appropriate gates until his circuit is complete. In this way, one particular PAL can be quickly configured into an almost infinite number of circuits.

Both the manufacturer-configurable arrays and the user-configurable arrays have become increasingly popular in recent years. With them, designers can minimize space used on printed circuit boards, cut costs, and protect circuits from being copied by others without authorization.

PROGRAMMED LOGIC

Memories "remember" and hold digital information, usually in groups of one, four, or eight bits. A given piece of information is recalled by asking for the information located at a certain "address." These addresses enable the circuit to find particular information at any time necessary. More thorough descriptions of memories are contained in Chapters 10 and 11; however, since *Read Only Memories* (ROMs) have a special use in combinational logic circuits, they will be introduced here.

The data information in a ROM is permanently etched into the memory and cannot be changed. When truth table inputs function as addresses and truth table outputs as data for those addresses, a ROM can be used as a combinational logic circuit. Because a single ROM can be programmed to realize *any* truth table, in most cases a single ROM can replace many combinational gate packages, resulting in a more compact and often cheaper circuit.

SUMMARY

1   Digital circuits that satisfy a given truth table or equation can often be simplified to use fewer gates. One way of simplifying digital circuits is by applying the principles and theorems of Boolean algebra.

2   Karnaugh mapping involves placing the outputs of a truth table onto a standard form map.

3   Once the truth table has been transferred to the Karnaugh map, circling implicants in groups of 2^n allows simplification of the truth table and digital circuit.

4   If only essential prime implicants are grouped, the resulting circuit will be the simplest possible for a given set of constraints.

5 It is easiest to design actual combinational circuits using a systematic approach. This approach follows:

    a. Define all inputs and outputs.

    b. Construct a truth table based on inputs and outputs.

    c. Use a Karnaugh map to find the simplest equation.

    d. Write the equation for the circuit.

    e. Draw the schematic diagram for the circuit.

6 In the output of a truth table, a *don't care* will result either when the inputs necessary for that output can never occur or when the output under those input conditions does not matter. Inclusion of *don't cares* in Karnaugh maps often results in simpler circuits. A *don't care* can be used as either a 1 or a 0, whichever facilitates a simpler design.

7 Gate arrays allow a designer to place a fairly complex circuit on a single IC package. Gate arrays save space on printed circuit boards and often save money.

8 ROMs can be used like gate arrays. Any truth table can be programmed into a ROM, thus replacing the combinational logic needed to realize that truth table.

KEY TERMS

circuit simplification	Karnaugh map	product of sums
completely covered	maxterm	redundant
disallowed state	memory	ROM
don't care	minterm	STROBE
essential prime implicants	PAL	sum of products
gate array	PLA	ULA
implicant		

QUESTIONS AND PROBLEMS

1 Simplify the following expressions using the techniques of Boolean algebra:

    a. $A(A + B)(\overline{A} + C)$

    b. $(A + AB)(A + \overline{A}B)$

    c. $AB + \overline{A}C + BC$

    d. $\overline{A}(A + B)(A + \overline{B})$

2 Simplify the following expressions using the techniques of Boolean algebra:

    a. $(A + \overline{A})(A + AB)A(A + B)A(\overline{A} + B)$

    b. $(A + A(\overline{A} + B))A(A + B)$

    c. $(A\overline{A})(A + B + C + D)(AB + CD)(ACD)$

    d. $A(A + B) + (A + AB)$

3 Is a group of three implicants allowed in a Karnaugh map? What sizes are allowed?

4 What variables are eliminated in writing equations from a grouped Karnaugh map?

5 Does a Karnaugh map always result in the simplest circuit?

6 Use the truth table in Table 3-7 to construct a Karnaugh map. Write both the minimal SOP equation (by grouping 1s) and the minimal POS equation (by grouping 0s).

7 Use the truth table in Table 3-8 to construct a Karnaugh map. Write a minimal equation for the output. Draw the circuit schematically.

8 List the systematic steps in designing a digital combinational logic circuit.

9 Use Table 3-7 to construct a Karnaugh map and minimize the function. Assume that the inputs to the truth table are now BCD only (use *don't cares*).

10 Use Table 3-8 to construct a K map and minimize the function. Assume that the inputs 0010 and 0011 cannot occur in the circuit (use *don't cares*).

11 List reasons for a gate array or ROM to be used to realize a digital circuit.

12 Fig. 3-18 shows the labeling convention for a seven-segment display. Utilizing logic design procedures learned in this chapter, design a digital circuit with inputs of BCD values and outputs to the seven-segment display. Notice that there are four inputs and seven outputs.

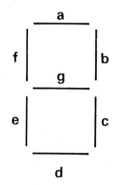

FIGURE 3-18. Labeling convention for a seven-segment display.

TABLE 3-7 TRUTH TABLE FOR PROBLEMS 8 AND 11.

A	B	C	D	X
0	0	0	0	0
0	0	0	1	1
0	0	1	0	1
0	0	1	1	1
0	1	0	0	1
0	1	0	1	0
0	1	1	0	0
0	1	1	1	0
1	0	0	0	1
1	0	0	1	0
1	0	1	0	0
1	0	1	1	1
1	1	0	0	0
1	1	0	1	0
1	1	1	0	0
1	1	1	1	0

TABLE 3-8 TRUTH TABLE FOR PROBLEMS 9 AND 12.

A	B	C	D	X
0	0	0	0	0
0	0	0	1	0
0	0	1	0	1
0	0	1	1	0
0	1	0	0	0
0	1	0	1	0
0	1	1	0	1
0	1	1	1	0
1	0	0	0	0
1	0	0	1	0
1	0	1	0	0
1	0	1	1	1
1	1	0	0	1
1	1	0	1	1
1	1	1	0	0
1	1	1	1	1

13 Design a 4-bit "odd number" detector. This circuit has four inputs and one output. The output is a 1 when the binary value of the inputs is odd. Otherwise, the output is 0.

14 Expand problem 15 to five bits.

15 Design a 4-bit "prime number" detector. This circuit also has four inputs and one output. The output is a 1 when the binary value of the inputs is a (decimal) prime number (1, 2, 3, 5, 7, 11, 13, etc.), and a 0 otherwise.

16 Why are Karnaugh maps labeled "00, 01, 11, 10" instead of using straight binary progression?

4

THE BIPOLAR LOGIC FAMILIES

Bipolar logic families are popular systems and most likely will remain so well into the future, mainly because of their speed of operation and reliability. This chapter covers the theory of operation of some of the more common bipolar logic families—TTL, ECL, and IIL—and explains how to interpret a manufacturer's data sheet, an extremely important skill.

THE DIODE AS A SWITCH

The semiconductor diode makes an excellent switch because it can be turned on (by *forward biasing*) and off (by *reverse biasing*). Because the semiconductor diode has two different states of operation, it is very useful for binary or 2-state logic circuitry.

DIODE OPERATION

The transfer curve of a typical diode is illustrated in Fig. 4-1. When the diode is reverse biased—by placing a more positive voltage on its cathode with respect to its anode—it operates in the left-hand portion of the curve. This portion of the curve is labeled "OFF" because virtually no reverse current will flow. When the diode is forward biased—by placing a more negative voltage on its cathode with respect to its anode—it operates in the right-hand portion of the curve. The right-hand portion of the curve is labeled "ON" because in this state current will readily flow through the diode. The diode operates like a voltage-controlled switch, which conducts current when turned on and acts like an open circuit when turned off.

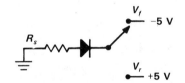

a.   Switch determines the ON and OFF conditions.

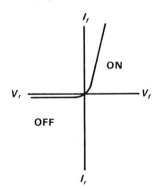

FIGURE 4-1.   Diode switch.   b.   The transfer curve.

SWITCHING SPEEDS OF DIODES

Switching diodes are different from *power supply rectifier diodes* because of the physical size of the junctions. Because it must dissipate a great deal of heat, a power supply diode must have a very large junction area, which naturally exhibits a considerable amount of capacitance. This large *junction capacitance* must be charged through the input circuit's resistance to about 0.7 V before forward biasing can occur. (Remember that forward biasing is the process by which the diode is activated.) The amount of time required to complete this process is equal to a portion of the *RC* (resistance times capacitance) time constant of this circuit. As a result, the large junction exhibits a limited switching speed. A switching diode, on the other hand, contains a very small junction area with a very small amount of capacitance, which allows it to switch at very high speeds. Many switching diodes will switch in a few nanoseconds. (A nanosecond, or 1 ns, is 10^{-9} seconds or one-billionth of a second.)

SIGNAL DEGRADATION WITH SWITCHING DIODES

Fig. 4-2 illustrates the loss of signal level when diodes are cascaded in a circuit. In this case, silicon diodes, which lose approximately 0.7 V, have been used. If this circuit were expanded by adding more diodes in series, the entire signal would be lost. This is a major problem with this type of device, but it can be overcome by amplifying the signal.

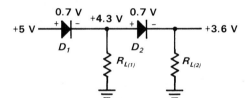

FIGURE 4-2.   Circuit illustrating signal loss when diodes are connected in cascade.

DIODE LOGIC CIRCUITRY

One early application of the semiconductor diode was the diode logic gate. It may seem peculiar to include this type of logic circuitry in a modern digital electronics textbook, but diode logic is still found on occasion. It is often less expensive to include a few diodes instead of an extra integrated circuit, and it requires less printed circuit board area.

THE DIODE OR GATE

Fig. 4-3 depicts a diode logic OR gate that contains two input connections. A truth table describing its operation is shown in Table 4-1. Whenever a logic level of 1, or positive voltage, is placed at either the *A* or the *B* input, a 1 will appear at the output. This happens because one or both of the diodes will conduct, and the current through the 470 ohm (Ω) load resistor develops a voltage at the output.

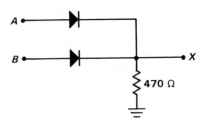

FIGURE 4-3. A diode logic OR gate.

TABLE 4-1 TRUTH TABLE FOR A 2-INPUT DIODE LOGIC OR GATE.

A	B	X
0	0	0
0	1	1
1	0	1
1	1	1

THE DIODE AND GATE

The diode logic AND gate is illustrated in Fig. 4-4. Its truth table is shown in Table 4-2. Whenever a logic level of 1, or positive voltage, is placed at both the A and B inputs, the output becomes 1. This occurs because both diodes are reverse biased and no current flows through the 470 Ω load resistor. Because no current flows through the load, no voltage is dropped across it. This causes the output to become a logic 1 level. The output will assume a logic 0 level if any of the inputs becomes a 0. This occurs because a diode will conduct, and the current through the load will drop most of the applied voltage, causing the output to become a logic 0.

TABLE 4-2 TRUTH TABLE FOR THE 2-INPUT AND GATE.

A	B	X
0	0	0
0	1	0
1	0	0
1	1	1

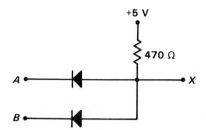

FIGURE 4-4. A 2-input diode logic AND gate.

ADDING ADDITIONAL INPUTS TO DIODE LOGIC GATES

Additional inputs can be added to diode logic gates by merely connecting extra diodes across an existing diode. This is illustrated in Fig. 4-5, which shows a 2-input OR gate converted to a 3-input OR gate.

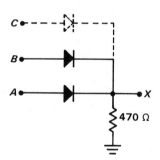

FIGURE 4-5. Circuit illustrating how an additional input can be added to a diode logic gate.

SWITCHING SPEEDS OF DIODE LOGIC GATES

The switching speed of the diode logic gate is determined by the junction capacitor and the series circuit resistance. If the input to a gate is a perfect voltage source, the switching speed is determined by the junction capacitance and the value of the load resistance. It is advisable to keep the load resistor as small as is practical, if the *rise* and *fall times* of the output are critical. (The rise time is the amount of time it takes an output to change from a logic 0 to a logic 1, and the fall time is the time it takes an output to change from a logic 1 to a logic

0.) A smaller load resistance reduces the RC time constant and increases the switching speed by reducing the rise and fall times of the output waveform.

THE TRANSISTOR AS A SWITCH

The diode switch, as mentioned earlier, is not an ideal switch, mainly because of the possibility of signal loss. The use of an amplifier can overcome this problem. The *transistor switch* is such an amplifier.

THE TRANSISTOR SWITCH

Fig. 4-6 illustrates a typical NPN *transistor switch* used as an INVERTER. (An NPN transistor is one in which the emitter and collector are of *n*-type semiconductor material, and the base is of *p*-type.) Whenever a logic 1, or +5 V, is input to the *base resistor,* base current will flow. The value chosen for the base resistor must allow enough current to flow to saturate the transistor. When the transistor is saturated, the output becomes a logic 0, or 0 V.

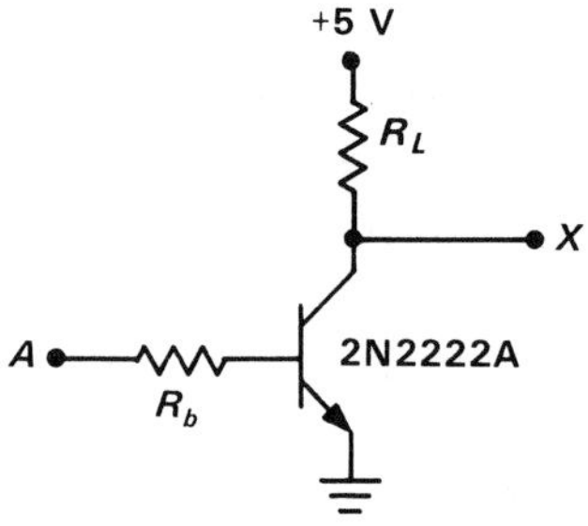

FIGURE 4-6.   *A typical NPN transistor INVERTER.*

If a logic 0 is placed at the input, no base current will flow. This causes the transistor to cut off, and no current flows through the load resistor. As a result, no voltage is dropped across the load, and output voltage rises toward +5 V.

BASE RESISTOR SELECTION

The value of the base resistor is calculated by taking the minimum value of the logic 1 voltage at the input as a source. This is often equal to about +2.1 V, as illustrated in Fig. 4-7. Since the base-to-emitter junction drops about 0.7 V, the voltage across the base resistor is 1.4 V. This drop, along with the amount of current required to saturate the transistor, is used to calculate the value of the base resistor.

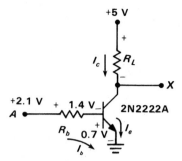

FIGURE 4-7. *NPN transistor biased into saturation. Notice that 2.1 V is applied to the A input connection. This level causes the transistor to conduct and saturate.*

The amount of base current required to saturate the device depends on two factors: the minimum current gain of the transistor and the amount of required emitter saturation current. Suppose the minimum current gain is 60 and the amount of required emitter current (I_e) is 16 mA (milliamperes, or thousandths of an ampere). In this example, the base current I_b is approximately equal to 267 μA (microamperes, or millionths of an ampere), as shown in the following equation:

$$I_b = \frac{16\,\text{mA}}{60} = 267\ \mu\text{A}$$

The value of the base resistor R_b can now be determined by the use of Ohm's Law. In this example R_b would be equal to 5.2 kΩ (1 kΩ = 1000 Ω), which is the standard resistance value nearest to 5.24 kΩ:

$$R_b = \frac{1.4\ \text{V}}{267\ \mu\text{A}} = 5.24\ \text{k}\Omega$$

THE COLLECTOR LOAD RESISTOR

The *collector load resistor* can now be determined by using the saturation current and the maximum logic 0 output voltage. Let's assume that the maximum logic 0 output voltage for this example is +0.4 V. In addition to this voltage, we also need to know the maximum output current for this circuit. Let's assume that this switch will sink 10 mA maximum. (*Sink current* is the amount of current that will flow into an output when the output becomes a logic 0.) The equation for the maximum output current is:

$$I_c = I_e - I_{\text{sink}}$$
$$I_c = 16\ \text{mA} - 10\ \text{mA}$$
$$I_c = 6\ \text{mA}$$

The Bipolar Logic Families 83

Because both the value of the current through the load resistor and the voltage across the resistor are now known, it is possible to calculate the value of the load resistor:

$$R_L = \frac{5.0\,\text{V} - 0.4\,\text{V}}{6\,\text{mA}} = 776\,\Omega$$

In this example a 750 Ω resistor is used because it has the closest standard resistance value to 766 Ω.

TTL CIRCUIT TOPOLOGY

The *TTL* or *Transistor-Transistor Logic* circuit is an important portion of many digital systems. It is the glue that interconnects the many modern *LSI (Large-Scale Integration)* components, such as microprocessors. TTL circuitry is easily distinguished from other types of logic circuitry by the *multiple-emitter* transistors used in its construction, as illustrated in Fig. 4-8. The truth table for this gate is shown in Table 4-3.

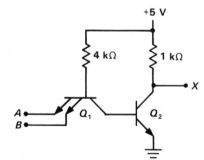

FIGURE 4-8.   Two-input TTL NAND gate.

TABLE 4-3   TRUTH TABLE FOR THE 2-INPUT TTL NAND GATE.

A	B	X
0	0	1
0	1	1
1	0	1
1	1	0

THE BASIC TTL NAND GATE

Although the circuit in Fig. 4-8 is not normally found in practice, it will help to illustrate basic TTL operation. Notice from the truth table that if both inputs are logic 1s, the output is a logic 0. Since A and B are each at a +5 V level, or a logic 1, no current will flow through either emitter-base junction of Q_1, and thus the base potential will rise toward +5 V. When the base of Q_1 reaches about 1.4 V, the base-to-collector junction is forward biased. Current from the collector of Q_1 now causes the emitter-base junction of Q_2 to be forward biased. The forward-biased output transistor Q_2 saturates and drops the output voltage X to very near 0 V, or a logic 0.

If either or both inputs are placed at a logic 0 level, or 0 V, current flows through the base resistor of Q_1 and lowers the base voltage to 0.7 V. This is not enough voltage to forward bias the base-to-collector junction of Q_1, which is in series with the base-to-emitter junction of transistor Q_2. Because no current flows through the emitter-base junction of Q_2, this transistor is cut off. The voltage at X now rises toward +5 V, or a logic 1, because no current flows through the collector load resistor.

Current flows when an input of Q_1 is grounded, but if an input is placed at a logic 1 level, almost no current flows. Because gate inputs are connected to outputs of other gates, current will flow only when the output is a logic 0. Due to this characteristic, TTL is called a *current-sinking* logic family.

Unused Input Connections

The preceding description of the operation of a NAND gate showed that a +5 V level on an input causes no emitter current on the input transistor. If the input is left open-circuited, or *floating,* it would be treated as a logic 1.

The problem with using an open circuit as a logic 1 input is that "noise" can cause problems. The open-circuited connection acts as an antenna that picks up stray noise. How, then, can an unused input be effectively used as a logic 1 level? The most common practice is to tie it *high* through a *pullup resistor,* as illustrated in Fig. 4-9. Here a 2-input NAND gate has been connected to

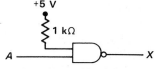

a. TTL NAND gate connected to function as an INVERTER by "pulling up" the unused input.

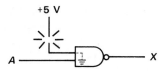

b. Effect of a shorted input if the input is tied to 5V without a pull-up resistor.

FIGURE 4-9.   Effect of a pullup resistor.

function as an INVERTER by "pulling up" one of the inputs. The value of this resistor can range anywhere from about 470 Ω to 47 kΩ. In practice, either a 1 kΩ or a 2 kΩ resistor is commonplace.

Why couldn't the input be connected directly to 5 V? This connection is fine in many cases—until the input short-circuits. Because an input short usually occurs to ground, the amount of current flow is excessive if a pullup resistor is not used (see Fig. 4-9b). Thus the pullup resistor prevents a fire hazard in most cases.

THE 7400 QUAD, 2-INPUT NAND GATE

Fig. 4-10 illustrates a more practical NAND gate: one-fourth of the 7400 Quad 2-input NAND gate. Notice that the basic circuit of Fig. 4-8 is present, along with some additional components.

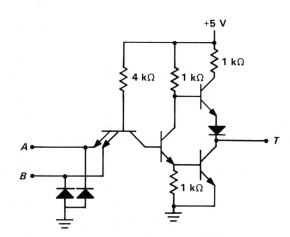

FIGURE 4-10. *Schematic of a 2-input 7400 TTL NAND gate.*

The Input Diodes

The *input diodes* act as clamps to prevent the inputs from dropping below −0.7 V. If the inputs go below this value, the device will be damaged. Under normal operation, a negative voltage appears at an input because the wire connected to the input is a transmission line. Fig. 4-11 illustrates the effect of applying a *squarewave* to a piece of wire. If the wire is long enough, the waveform at the output end will be severely distorted due to the inductance and capacitance of the line. This distortion occurs when an abrupt change in the voltage level causes the line to oscillate as indicated in Fig. 4-11.

Fig. 4-12 illustrates the effect of the diode clamp on this distortion: it has reduced the amount of *undershoot* to a minimum. The clamp is primarily used

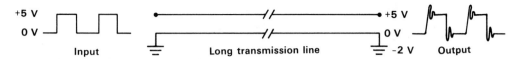

FIGURE 4-11. *Effect of a long transmission line on the waveshape of a TTL-compatible square.*

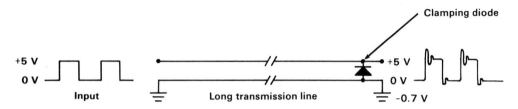

FIGURE 4-12. **Clamping diode placed on the output terminals of a long transmission line reduces the amount of undershoot.**

to prevent damage to the device from extremely large undershoots generated by long lines or erroneous inputs.

The Input Circuitry

The input circuitry of the 7400 TTL NAND gate is almost the same as that shown in Fig. 4-8, except that it contains a few more components. Refer to Fig. 4-13 for a detailed picture of the input circuit.

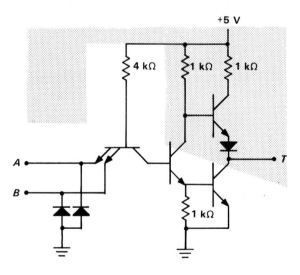

FIGURE 4-13. *Input circuitry of the 2-input 7400 TTL NAND gate.*

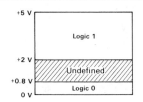

The additional components are used to set the logic 0 and logic 1 levels at the input. These levels are illustrated graphically in Fig. 4-14. The logic 0 level is defined as 0 to +0.8 V, and the logic 1 input level is defined as +2 to +5 V. Because any voltage between +0.8 V and +2 V may cause the device to function unpredictably, this region of operation is undefined and must be avoided.

The Output Circuitry

The output circuitry of the 7400 TTL NAND gate is typical of many TTL gates because it is a *totem-pole* circuit. The totem-pole output, as illustrated in Fig. 4-15, uses two transistors stacked one upon the other. This type of output circuit differs from the one illustrated in Fig. 4-8 in the shorter rise time it provides at the output.

The rise time is reduced in this circuit because of the *active pullup transistor*. Since a load normally contains some capacitance, a simple resistive pullup would form a fairly long *RC* time constant. When the pullup transistor is turned on, or saturated, however, it exerts a very low impedance from the emitter to the collector. Thus the pullup transistor allows the load capacitance to charge rapidly, which reduces the amount of time required for the output voltage to rise toward +5 V, or a logic 1.

Fig. 4-16 illustrates a graph of the output logic voltage levels. The logic 1 level is +2.4 to +5 V, and the logic 0 output voltage level is 0 to +0.4 V. Any potential between +0.4 V and +2.4 V does not normally appear at an output and therefore is undefined.

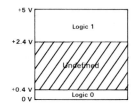

FIGURE 4-16. Logic 0
and logic 1 output levels
of any TTL logic gate.

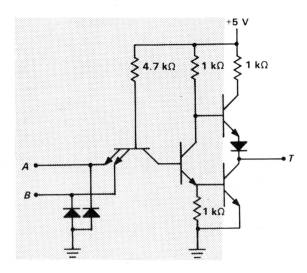

FIGURE 4-15. Totem-pole output circuitry of a 2-
input 7400 TTL NAND gate.

The Propagation Delay Time

The *propagation delay time* can be reduced by the transistor highlighted in Fig. 4-17. Let's suppose the inputs to the circuit are at a logic 1 level for quite some time. In this condition, current flows through the input transistor's collector-to-base junction and also through the two series' emitter-base junctions, as illustrated. The output is a logic 0 because the bottom transistor in the totem-pole is saturated and the top transistor is cut off.

If one of the inputs suddenly drops to 0 V, the base current will be steered through the emitter of the input transistor, causing it to conduct. Because the input transistor exhibits a very low impedance when it conducts, the charge on the base of the transistor connected to the input transistor's collector is rapidly dissipated. This, in turn, causes the bottom transistor in the totem-pole to turn off and the top transistor to turn on. The output voltage rises toward +5 V more rapidly because of the active pullup transistor, as well as the active input transistor.

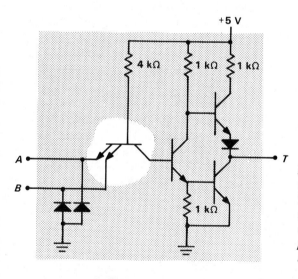

FIGURE 4-17. *Input transistor of a 2-input 7400 TTL NAND gate. This transistor is used to decrease the amount of propagation delay time through the gate.*

The Diode in the Output Circuit

The diode in the output circuit is important because it ensures that the pullup transistor is turned off when the output becomes a logic 0. Without this component, both transistors would conduct for a logic 0 output condition.

Fluctuations of I_{CC}

The power supply current, I_{CC}, can fluctuate widely in response to a change in the output logic levels of TTL components with totem-pole output circuits. Fig. 4-18 illustrates the fluctuations that would exist for a squarewave at the input of a TTL INVERTER. Notice the surges in I_{CC} that take place whenever the output changes. These surges occur because both transistors are partially on or off for a short period of time (1 or 2 ns).

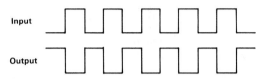

Input

Output

a. The waveforms found at the input and output of a standard TTL INVERTER.

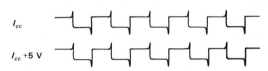

I_{cc}

I_{cc} +5 V

b. The fluctuations that appear in both the power supply voltage and current waveforms.

FIGURE 4-18.   The wave forms and fluctuations for a squarewave.

These surges (*glitches*) can cause tremendous problems in a digital system if they are not bypassed with a capacitor at the power supply connections to the integrated circuit. Because the printed circuit trace to the power supply voltage (V_{CC}) pin and the ground pin contains some resistance, current glitches produce comparable glitches in the supply voltage, and it is these voltage glitches that cause the trouble. If glitches exceed the ±5 percent tolerance level on the power supply voltage, the device will not function reliably. These glitches, which are of short duration (or high frequency), can also cause quite a bit of *electromagnetic interference* (EMI). According to the guidelines of the Federal Communications Commission (FCC), any digital equipment that operates above 30 kHz (30,000 Hertz) must be shielded in order to reduce this unwanted radiation.

Fig. 4-19 depicts a bypass capacitor connected between +5 V and ground at the circuit. A bypass capacitor is generally a 0.1 μF (microfarad) capacitor and is normally found across most TTL components in a system.

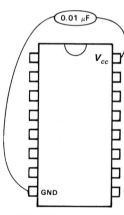

0.01 μF

V_{cc}

GND

FIGURE 4-19.   TTL logic circuit bypassed to reduce problems that might arise from power supply glitches (transients).

Fanout

The *fanout* of a logic gate consists of the number of input connections that can be driven from or connected to an output. The fanout of a TTL logic gate has been designed at ten. That is, ten TTL inputs can be safely connected to one TTL output pin. If a fanout greater than ten occurs, the device will not function reliably because the output voltage levels will be out of the tolerance range. Table 4-4 illustrates the maximum input and output currents for a standard TTL logic gate. From this table, the fanout can be calculated by dividing the input current into the output current. Because one input connection causes up to 1.6 mA of current to flow when at a logic 0 and the output can supply a maximum of 16 mA of current, the fanout is ten.

TABLE 4-4 MAXIMUM INPUT AND OUTPUT CURRENTS FOR A STANDARD 74XX LOGIC GATE.

Logic Level	Input Current	Output Current
0	−1.6 mA	16 mA
1	40 μA	−400 μA

TTL Noise Immunity

The *noise immunity* in a TTL system is the difference between the output logic voltage levels and the input logic voltage levels, as illustrated in Fig. 4-20. For example, the minimum logic 1 output voltage is +2.4 V, while the minimum logic 1 input voltage is +2.0 V. Thus a 0.4 V margin exists between an output and an input. Because this margin allows for noise in a system, it is also called the *noise margin.* The same margin exists for the logic 0 output. The maximum logic 0 output voltage is +0.4 V, and the maximum logic 0 input voltage is +0.8 V. The difference between these voltages is 0.4 V, the same noise margin that exists in the logic 1 input.

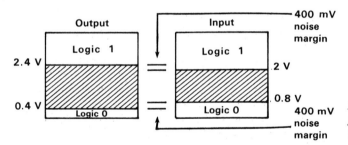

FIGURE 4-20. *Noise margin, or immunity, between a TTL output and input.*

TTL AND, OR, AND NOR GATES

Fig. 4-21 contains the internal schematic diagrams for the 7408 AND gate, the 7432 OR gate, and the 7402 NOR gate. Notice that these gates require some additional circuitry in relation to the 7400 TTL NAND gate of Fig. 4-10. The AND gate is more complicated than the NAND gate because the AND function is produced by the NAND gate followed by an internal INVERTER. As a result, the amount of propagation delay time is greater for the AND gate. The NOR and OR gates are also more complicated than the NAND gate and, in consequence, have the same problem with propagation delay time. In addition to increased propagation delay time, these gates require more power to operate because of the extra components. It is important that the amount of power consumed by a system be held to a minimum, because power supplies are expensive, generate a considerable amount of heat, and can cause maintenance problems.

schematics (each gate)

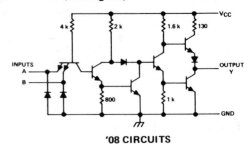

'08 CIRCUITS

a. The 7408 TTL AND gate.

schematics (each gate)

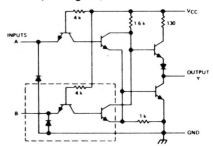

The portion of the schematic within the dashed
lines is repeated for the C input of the '27.

'02, '27 CIRCUITS

b. The 7402 TTL NOR gate.

schematics (each gate)

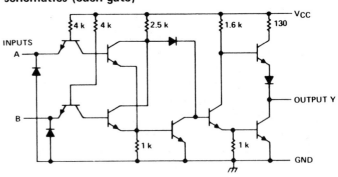

*FIGURE 4-21. Schematics. (Courtesy of
Texas Instruments Incorporated.)*

'32 CIRCUITS

c. The 7432 TTL OR gate.

INTERPRETING THE TTL DATA SHEET

Before a component can be effectively used for any purpose, the manufacturer's *data sheet* for the component must be understood. This applies to any component, at almost any level of usage, from design to maintenance. This section will explain the TTL data sheet in such a way that you will be able to extract information for any TTL device from any manufacturer's data sheet.

TTL DATA SHEET

Fig. 4-22 illustrates a data sheet for a few common TTL devices. This sheet has been copied from an actual manufacturer's data book so that it can be examined in detail. The data illustrated here cover the 7400 Quad 2-input NAND, the 7404 hex INVERTER, the 7410 Triple 3-input NAND, the 7420 Dual 4-input NAND, and the 7430 8-input NAND.

recommended operating conditions

		54 FAMILY 74 FAMILY	SERIES 54 SERIES 74		
			'00, '04, '10, '20, '30		
			MIN	NOM	MAX
Supply voltage, V_{CC}	54 Family		4.5	5	5.5
	74 Family		4.75	5	5.25
High-level output current, I_{OH}	54 Family				−400
	74 Family				−400
Low-level output current, I_{OL}	54 Family				16
	74 Family				16
Operating free-air temperature, T_A	54 Family		−55		125
	74 Family		0		70

FIGURE 4-22. Recommended operating conditions for a few 74XX logic gates. (Courtesy of Texas Instruments Incorporated.)

Recommended Operating Conditions

Fig. 4-22 lists the recommended operating conditions for both the 74XX and 54XX series of these devices. The 54XX series is the industrial duty version of the 74XX series integrated circuit. The 54XX series of devices has extended operating temperatures and a wider tolerance on the power supply voltages than the 74XX series. In addition to operating temperatures, Fig. 4-22 also indicates the maximum recommended sink and source currents for these devices. With a logic 1 at any output, −400 μA of current can be sourced, and with a logic 0 at an output, a maximum of 16 mA of current can be sunk. (The minus sign on a current indicates that the direction of current flow is out of the device.)

Electrical Characteristics

Fig. 4-23 lists the electrical characteristics of some 74XX TTL devices over the recommended free-air temperature range.

V_{IH} and V_{IL} specify the input voltage conditions for a logic 1 and a logic 0, respectively. The logic 1 level must be at least +2 V, and the logic 0 level must be less than +0.8 V.

V_{OH} and V_{OL} specify the output voltage conditions for a logic 1 and a logic 0, respectively. The logic 1 level will be no less than +2.4 V, and the logic 0 level will be no greater than +0.4 V. These levels only apply if the maximum output currents are not exceeded.

I_{IH} and I_{IL} specify the maximum input currents for the logic 1 input condition and the logic 0 condition, respectively. The logic 1 maximum current level is 40 μA, and the logic 0 maximum current level is -1.6 mA.

Specifications for the amount of supply current are illustrated in Fig. 4-24. The typical supply current has been depicted for gates with all outputs high, for gates with all outputs low, and for gates operating at a 50 percent duty cycle.

electrical characteristics over recommended operating free-air temperature range

PARAMETER		TEST FIGURE	TEST CONDITIONS†		SERIES 54 SERIES 74 '00, '04, '10, '20, '30		
					MIN	TYP	MAX
V_{IH}	High-level input voltage	1, 2			2		
V_{IL}	Low-level input voltage	1, 2		54 Family			0.8
				74 Family			0.8
V_{IK}	Input clamp voltage	3	V_{CC} = MIN, I_I = §				−1.5
V_{OH}	High-level output voltage	1	V_{CC} = MIN, V_{IL} = V_{IL} max, I_{OH} = MAX	54 Family	2.4	3.4	
				74 Family	2.4	3.4	
V_{OL}	Low-level output voltage	2	V_{CC} = MIN, V_{IH} = 2 V	I_{OL} = MAX 54 Family		0.2	0.4
				I_{OL} = MAX 74 Family		0.2	0.4
				I_{OL} = 4 mA Series 74LS			
I_I	Input current at maximum input voltage	4	V_{CC} = MAX	V_I = 5.5 V			1
				V_I = 7 V			
I_{IH}	High-level input current	4	V_{CC} = MAX	V_{IH} = 2.4 V			40
				V_{IH} = 2.7 V			
I_{IL}	Low-level input current	5	V_{CC} = MAX	V_{IL} = 0.3 V			
				V_{IL} = 0.4 V			1.6
				V_{IL} = 0.5 V			
I_{OS}	Short-circuit output current•	6	V_{CC} = MAX	54 Family	20		55
				74 Family	18		55
I_{CC}	Supply current	7	V_{CC} = MAX				

†For conditions shown as MIN or MAX, use the appropriate value specified under recommended operating conditions.

‡All typical values are at V_{CC} = 5 V, T_A = 25°C.

§I_I = −12 mA for SN54'/SN74', −8 mA for SN54H'/SN74H', and −18 mA for SN54LS'/SN74LS' and SN54S'/SN74S'.

•Not more than one output should be shorted at a time, and for SN54H'/SN74H', SN54LS'/SN74LS', and SN54S'/SN74S', duration of short-circuit should not exceed 1 second.

FIGURE 4-23. *Electrical characteristics of a few 74XX TTL logic gates (Courtesy of Texas Instruments Incorporated).*

supply current¶

TYPE	I_{CCH} (mA) Total with outputs high		I_{CCL} (mA) Total with outputs low		I_{CC} (mA) Average per gate (50% duty cycle)
	TYP	MAX	TYP	MAX	TYP
'00	4	8	12	22	2
'04	6	12	18	33	2
'10	3	6	9	16.5	2
'20	2	4	6	11	2
'30	1	2	3	6	2

FIGURE 4-24.   *Power supply consumption for a few 74XX TTL logic gates. (Courtesy of Texas Instruments Incorporated.)*

With a 50 percent duty cycle or the application of a symmetrical squarewave, the average is 2 mA per gate. In other words, a 7400 Quad 2-input NAND gate would draw an average 8 mA of current.

Switching Characteristics

The switching characteristics of some typical TTL gates are illustrated in Fig. 4-25. These times were measured with a load resistance of 400 Ω and a load capacitance of 15 pF (picofarad, one-trillionth of a farad). Two different propagation delay times have been listed for each gate: T_{PLH} and T_{PHL}.

As mentioned earlier, T_{PLH}, or rise time, is the amount of time that it takes an output to change from a logic 0 to a logic 1. T_{PHL}, or fall time, is the amount of time that it takes an output to change from a logic 1 to a logic 0. Notice that the times are not exactly the same. This disparity can cause some distortion in the output waveshape. Fig. 4-26 shows how a signal can be distorted by passing it through two 7404 TTL INVERTERs. Notice how the output waveshape of the first INVERTER differs somewhat from the original input waveshape. The final output is the same as the original input, but it is shifted in time (typical delay times are used in this example). This distortion may or may not be critical to a particular application.

switching characteristics at V_{CC} = 5 V, T_A = 25°C

TYPE	TEST CONDITIONS#	t_{PLH} (ns) Propagation delay time, low-to-high-level output			t_{PHL} (ns) Propagation delay time, high-to-low-level output		
		MIN	TYP	MAX	MIN	TYP	MAX
'00, '10			11	22		7	15
'04, '20	C_L = 15 pF, R_L = 400 Ω		12	22		8	15
'30			13	22		8	15

FIGURE 4-25.   *Switching characteristics of a few 74XX TTL logic gates. (Courtesy of Texas Instruments Incorporated.)*

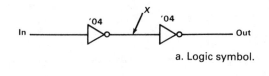

a. Logic symbol.

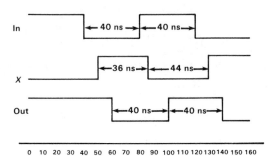

0 10 20 30 40 50 60 70 80 90 100 110 120 130 140 150 160

T—ns

b. Timing diagram.

FIGURE 4-26. Effect of rise and fall times on the relative position and shape of the output waveform.

Schematic from the Manufacturer's Data Sheet

Fig. 4-27 shows the schematic diagram of the 7400, 7404, 7410, 7420, or 7430 and lists the nominal value of the internal resistors. As you can see, this diagram is essentially the same schematic presented in earlier figures.

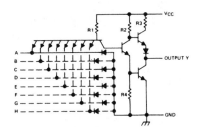

CIRCUIT	R1	R2	R3	R4
'00, '04, '10, '20, '30	4 k	1.6 k	130	1 k
'L00, 'L04, 'L10, 'L20, 'L30	40 k	20 k	500	12 k

'00, '04, '10, '20, '30
'L00, 'L04, 'L10, 'L20, 'L30, CIRCUITS
Input clamp diodes not on
SN54L'/SN74L' circuits.

FIGURE 4-27. Schematic diagrams of the 7400, 7404, 7410, 7420, and 7430 logic gates. (Courtesy of Texas Instruments Incorporated.)

OTHER TTL FAMILY MEMBERS

Up to this point, the discussion in this chapter has concerned the standard TTL family of circuits, the 74XX family. This is not the only type of TTL circuitry available, however. Newer, and perhaps better, lines of TTL logic circuitry have been introduced in the past few years. In fact, in many applications, standard TTL circuitry has been replaced by some newer series.

Table 4-5 illustrates a chart comparing the standard TTL logic circuitry with some of the newer family members. The 74LXX series and the 74HXX series are no longer being manufactured and are considered obsolete except for replacement purposes. In systems repair, the same type of TTL circuit must be used for a replacement because, in certain applications, the propagation delay time of the circuit is important and the substitution of a higher-speed version or another type that seems equivalent may cause a malfunction.

TABLE 4-5 A COMPARISON OF CURRENTLY AVAILABLE
TTL LOGIC FAMILY MEMBERS.

Type	Propagation Delay Time	Power per Gate	Clock Frequency
74XX	10 ns	10 mW[a]	DC to 35 MHz[b]
74LXX	33 ns	1 mW	DC to 3 MHz
74HXX	6 ns	22 mW	DC to 50 MHz
74LSXX	9.5 ns	2 mW	DC to 45 MHz
74SXX	3 ns	19 mW	DC to 125 MHz
74ALSXX	5 ns	1 mW	DC to 50 MHz
74ASXX	2.5 ns	20 mW	DC to 160 MHz

[a] mW = milliwatt = one-thousandth of a watt
[b] MHz = megahertz = one million Hertz

74LSXX TTL LOGIC GATES

The 74LSXX logic family uses the integrated *Schottky-barrier diode clamped transistor*, which has virtually no junction capacitance and therefore can switch at very high rates of speed. Fig. 4-28 illustrates the standard schematic symbol for this device, along with some detail on its internal construction. This device functions as a diode because the aluminum region contains an abundance of free electrons and looks for all practical purposes like *n*-type material. The aluminum region forms a junction with the semiconductor *p*-type material, creating a diode. This junction switches faster than a standard diode because it has been constructed with aluminum, which has a lower resistance that allows the junction capacitance to be more quickly charged or discharged.

Fig. 4-29 illustrates a Schottky-clamped transistor. The Schottky diode is attached from the base to the collector. This means of attachment prevents the transistor's base region from completely saturating because the Schottky diode begins to conduct at about 0.4 V. This reduces the amount of charge stored in the transistor's emitter-to-base junction, allowing it to switch rapidly from the saturated state to the cutoff state.

a. Schematic.

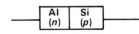

b. Internal structure showing the aluminum (Al) and silicon (Si) regions.

FIGURE 4-28.
Schottky-barrier diode.

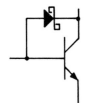

a. Placement of the Schottky-barrier diode.

b. Schematic.

FIGURE 4-29. Schottky-clamped transistor.

Because the 74LSXX series uses the Schottky diode, the amount of power required to operate it can be reduced, and a relatively high switching speed can be maintained. 74LSXX logic gates switch at about the same speed as standard 74XX TTL gates but consume only 20 percent as much power. In most new designs, the 74LSXX gates have replaced the 74XX gates.

74SXX TTL LOGIC GATES

The 74SXX TTL gate is a high-speed Schottky-clamped version of the standard TTL gate. It consumes about twice as much power as a 74XX gate and performs at about three times the speed. The 74SXX series gate finds its application in circuits that either require a very short propagation delay time or operate with very high frequency signals.

74ALSXX AND 74ASXX TTL LOGIC GATES

Two recent TTL logic family members—the 74ASXX, or advanced Schottky, and the 74ALSXX, or advanced low-power Schottky—may eventually replace all the other gate types. These new family members require less power and switch at a higher speed than the types mentioned previously.

INTERCONNECTING TTL FAMILY MEMBERS

Table 4-6 illustrates the fanout from each TTL family member to various other members. Remember that the fanout for a particular type of gate is always ten unless it is interconnected with other types. Table 4-7 indicates the input and output current loading for each family member for cases when various members must be connected. Example 4-1 illustrates how to use the table.

Example 4-1. If two 74LSXX loads and one 74SXX load were connected to the output of a standard TTL logic gate, how many additional standard loads could be connected to the same output?

Solution. The standard TTL gate can supply 16 mA of sink current and -400 μA of source current. The 74LSXX gates each require -0.4 mA of current if the input is a logic 0 and 20 μA of current if the input is a logic 1. After the connection of two 74LSXX gates 15.2 mA of sink current and -360 μA of source current remain available at the standard gate's output.

The 74SXX gate requires -2 mA of current with a logic 0 at its input and 50 μA of current with a logic 1 at its input. Following the connection of one 74SXX gate in addition to the two 74LSXX gates, 13.2 mA of sink current and -310 μA of source current remain available. Since a standard TTL gate requires -1.6 mA of logic 0 current and 40 μA of logic 1 current, seven more

TABLE 4-6 FANOUTS FOR INTERCONNECTING TTL FAMILY MEMBERS.

	74XX	74HXX	74LXX	74LSXX	74SXX	74ALSXX	74ASXX
74XX	10	8	20	20	8	20	2
74HXX	12	10	25	25	10	25	2
74LXX	5	4	10	10	4	10	4
74LSXX	5	2	10	10	2	10	1
74SXX	12	10	25	25	10	25	2
74ALSXX	5	4	10	20	4	10	2
74ASXX	12	10	25	50	10	100	10

TABLE 4-7 INPUT AND OUTPUT CURRENT LOADS FOR EACH TTL FAMILY MEMBER.

	Input		Output	
Type	0	1	0	1
74XX	−1.6 mA	40 μA	16 mA	−400 μA
74HXX	−2.0 mA	50 μA	20 mA	−500 μA
74LXX	−0.8 mA	20 μA	8 mA	−200 μA
74LSXX	−0.4 mA	20 μA	4 mA	−400 μA
74SXX	−2.0 mA	50 μA	20 mA	−1000 μA
74ALSXX	−0.2 mA	20 μA	8 mA	−400 μA
74ASXX	−2.0 mA	200 μA	20 mA	−2000 μA

74XX gates may be safely connected. This connection will leave an extra 2 mA of sink current and −30 μA of source current at the output connection. But remember, if the recommended loading is exceeded, the output voltages will begin to creep into the undefined region and, in most cases, cause unreliable operation.

OPEN-COLLECTOR AND TRI-STATE®* GATES

Two special-purpose gates used for interfacing or buffering are open-collector and Tri-state® TTL gates. *Open-collector gates,* as the name implies, have no

*Tri-state® is a trademark of the National Semiconductor Corporation.

internal collector pullup resistor. *Tri-state® gates* differ from standard gates because they have a 3-state output. The output of a Tri-state® gate can be a logic 1, a logic 0, or an open circuit. This feature is useful for interfacing.

OPEN-COLLECTOR GATES

Fig. 4-30 illustrates the typical schematic diagram of a 2-input open-collector TTL NAND gate. This circuit is similar to the 2-input NAND gate introduced in Fig. 4-8 except that it has no collector resistor. This type of gate finds application whenever the amount of source current must be greater than -400 μA or the output voltages are other than 0 and 5 V.

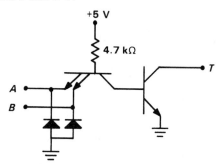

FIGURE 4-30.   Schematic of an open-collector 7401 TTL NAND gate.

The Open-Collector Gate as a Relay Driver

Fig. 4-31 illustrates the open-collector 7401 NAND gate interfaced to a 10 V low-current DC magnetic reed relay. If a standard TTL gate is connected to the relay, it will be destroyed the first time that the output goes to a logic 1 because a standard TTL gate cannot have more than 7 V applied to any point with respect to ground.

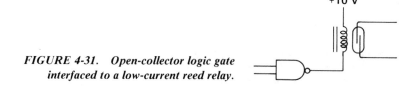

FIGURE 4-31.   Open-collector logic gate interfaced to a low-current reed relay.

The Wired AND Function

Another, somewhat uncommon, use of the open-collector gate is the *wired AND,* or *dot AND,* logic function. Since there is no internal collector resistor, the outputs of two or more open-collector gates may be connected to form a wired (dot) AND connection. (The dot AND connection is sometimes erroneously called a dot OR connection.) The wired AND function is illustrated in Fig. 4-32. It is very important not to connect standard TTL gates in this configuration because of the damage that may occur.

The dot connection functions as an AND gate because, if either of the open-collector outputs is at a logic 0 level, the voltage at the dot connection will be 0 V. This occurs even when the other output is attempting to force the *node,* or dot connection, to a logic 1; the logic 0 output shorts the node to ground. The only way that the dot connection can assume a logic 1 level is when the outputs of both open-collector gates assume a logic 1 level—the AND function.

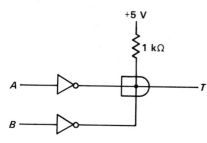

a. Circuit.

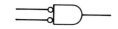

b. Equivalent logic symbol.

FIGURE 4-32. Two open-collector TTL INVERTERs connected to form a wired AND gate at their outputs.

TRI-STATE® LOGIC GATES

Tri-state® or three-state logic is a very important portion of many modern digital systems. This type of gate is used to place information onto the *bus* of a computer system without the use of a multiplexer. The term *bus* is used to denote the network of conductors connecting various sources to various destinations.

The Tri-state® logic gate has three unique output conditions: the logic 0, the logic 1, and the high-impedance state ∞. A simple Tri-state® noninverting buffer is illustrated in Fig. 4-33 along with the mechanical equivalent. Its truth table is presented in Table 4-8. This gate has a special input called OUTPUT ENABLE that allows the output of the buffer to be turned on or off. The mechanical equivalent of this process would be the opening or closing of the switch. Most Tri-state® gates use an active-low ENABLE input or an active-high

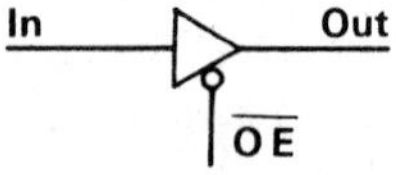

a. Logic symbol.

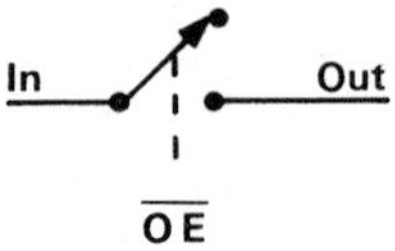

b. Mechanical realization.

FIGURE 4-33. Tri-state® noninverting buffer.

TABLE 4-8 TRUTH TABLE OF THE TRI-STATE® NONINVERTING BUFFER. NOTICE THAT WHEN $\overline{OE}$ IS A LOGIC 1, THE OUTPUT OF THE BUFFER IS AT ITS HIGH-IMPEDANCE STATE.

$\overline{OE}$	IN	OUT
0	0	0
0	1	1
1	0	∞
1	1	∞

DISABLE input (Active-high indicates that the input is activated by a high logic level, and active-low refers to an input activated by a low logic level.) The circuit pictured has its control input labeled $\overline{OE}$ for "OUTPUT ENABLE" (the overbar signifies that the input is active-low).

Schmitt Trigger Inputs

In addition to functioning as a switch in a digital system, the Tri-state® buffer is often used to drive a bus line or receive information from a bus line. Because bus lines are often fairly long, many systems require a gate or buffer with a higher than normal noise immunity. Standard TTL circuitry has a 400 mV noise immunity, while the Schmitt trigger inputs on a Tri-state® buffer have about 1.2 V of noise immunity.

The upper trip point is typically 1.6 V and the lower trip point 1.2 V for a typical hysteresis of 400 mV. *Hysteresis* is the difference between the upper and lower trip points. The trip points and a typical bus signal are shown in Fig. 4-34. Notice that the input signal has quite a bit of noise and the output signal has none. This feature is extremely important when a digital system drives a line or wire of more than 15 inches. With longer lines, Schmitt trigger circuits should be used to remove any noise that might appear.

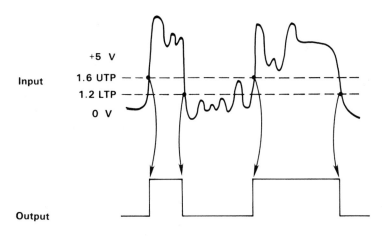

FIGURE 4-34. *Effect of a Schmitt trigger input on a noisy bus waveform. Notice how the Schmitt trigger gate clears up the input waveform.*

Low-Impedance Outputs

The outputs of a typical Tri-state® buffer can drive heavily loaded lines. Typically, the buffer's output can sink 32 mA and source or supply -5.2 mA of current. The outputs have also been designed with an output impedance low enough for the device to drive a low-impedance bus line or transmission line.

Tri-State® Buffers Used as Multiplexers

Figure 4-35 shows two Tri-state® buffers used to multiplex a single line. *Multiplex* means share, in this case, two signals share the same signal line. When a logic 1 is applied to the SELECT input (*S*), input *A* is passed to the output line. When a logic 0 is applied to the SELECT input, input *B* is passed on to the output line. This switching occurs because a logic 1 on the $\overline{OE}$ pin causes

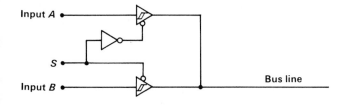

FIGURE 4-35. *Two Tri-state® logic gates connected to perform as a digital multiplexer. SELECT (S) determines which input is passed through the circuit to the output.*

the output of a buffer to "open circuit" and a logic 0 on $\overline{OE}$ causes the output of the buffer to be connected to the input.

ECL LOGIC CIRCUITRY

In certain applications, TTL circuitry is not fast enough. This problem can be overcome by using *Emitter Coupled Logic* (ECL), or, as it is sometimes termed, *Current Mode Logic* (CML). This type of circuitry is typically faster than TTL because it does not use saturated switching. When a transistor is saturated, the emitter-to-base junction will store a small charge in its junction capacitance. Since it takes time to discharge the junction capacitance, TTL circuitry can be constructed to switch in only about 2 ns or longer. Higher switching speeds require some form of nonsaturating logic circuit.

THE BASIC ECL LOGIC CIRCUIT

Fig. 4-36 depicts a typical ECL OR gate. Notice that this type of logic circuit requires a power supply voltage of -5.2 V, which is *not* TTL-compatible. The logic levels are also *not* TTL-compatible. The logic 0 level is normally -1.625 to -1.99 V, and the logic 1 level is normally -0.695 to -0.795 V.

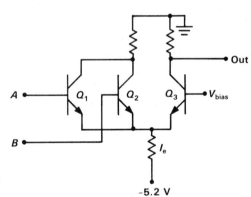

FIGURE 4-36. *Schematic of a simplified ECL OR gate.*

Operation

When both the *A* and *B* inputs are placed at their logic 0 levels, the output becomes a logic 0. This condition occurs because most of the current I_e is steered through Q_3, which conducts heavily because Q_1 and Q_2 are turned off. Because it conducts heavily, the collector voltage of Q_3 becomes more negative, which causes the output to become more negative, or a logic 0. When a logic 1 (a more positive voltage) is placed at either input A or B, or both of them, less current (I_e) will flow through Q_3, causing the output voltage to become less negative, or a logic 1.

The OR/NOR Gate Circuit

In practice, a combination of OR and NOR gates is provided in one IC package. This type of circuit is illustrated in Fig. 4-37. Notice the addition of an emitter bias network and a temperature-compensated internal reference voltage. This particular ECL circuit has a propagation delay time of less than 2 ns.

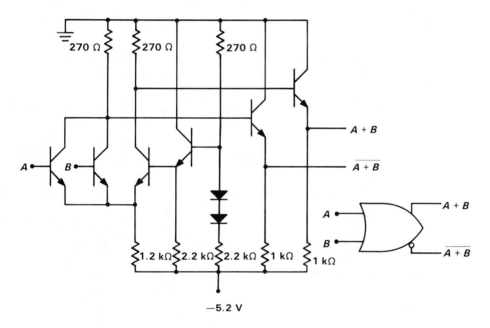

FIGURE 4-37.   ECL NOR/OR gate schematic showing the basic circuitry.

IIL LOGIC CIRCUITRY

Another type of logic family that is finding some application is *Integrated Injection Logic* or IIL, which is also known as *Merged-Transistor Logic* or MTL. This type of logic is often called a *current-hogging* or *current-steering* logic due to its pattern of operation, which will be explained shortly. It is distinguished by a constant current source included in the base circuitry of the input transistors and by multiple-collector transistors.

THE BASIC IIL LOGIC CIRCUIT

Fig. 4-38 depicts a combination NOR gate and dual INVERTER. The supply voltage can range from about 1.5 to 9 V for normal operation. The interface of IIL circuitry to TTL circuitry would require a 5 V power supply connection due to the power specifications of TTL.

The logic levels at an input or an output are either 0 V for a logic 0 condition or an open circuit for a logic 1 level. As you can imagine, the inputs and outputs are not directly TTL-compatible, but they can be made compatible quite easily by using an open-collector switch on the inputs and a transistor buffer on the outputs.

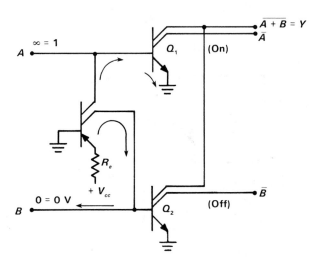

FIGURE 4-38. Schematic diagram of a combination NOR and dual INVERTER using IIL logic circuitry.

Operation

If both inputs A and B are at their logic 0 conditions, or shorted to ground, all the current from the current source is "hogged" by these inputs. Because no current is available to forward bias the base-to-emitter junction of either Q_1 or Q_2, both transistors are cut off and the common output Y becomes a logic 1, or an open circuit (0 NORed with 0 produces a logic 1 condition). In addition to the Y output becoming a logic 1, both the $\overline{A}$ and $\overline{B}$ outputs also become logic 1s, or open circuited.

If input A is placed at a logic 1 condition, or an open circuit, current is no longer hogged and flows through the base-to-emitter junction of Q_1, saturating it. This state produces a logic 0 at output Y and also at output $\overline{A}$. Output $\overline{B}$

remains a logic 1 because input B still hogs all the base current of transistor Q_2. This state, which produces the logic 1 level at output $\overline{B}$, keeps transistor Q_2 cut off.

SPEED OF OPERATION
VERSUS POWER-CONSUMPTION

One of the most important features of IIL logic is this circuit's ability to function at many different speeds and power-consumption levels. Speed changes are accomplished by varying the magnitude of the constant current source connected to the base of Q_1 and Q_2 in Fig. 4-38. If high speeds are required, the amount of current for each base circuit is increased, allowing the transistors to switch at a higher rate of speed. If speed is not important but power consumption is, the amount of base circuit current can be reduced to a very low level. This will conserve system power but reduce the switching speed. This feature is useful in some applications where a standby mode of operation is important because the value of bias current can be reduced for standby operation and increased for normal operation.

SUMMARY

1   A diode is called a *binary switching device* because it has two states: *forward bias,* which functions as a closed switch or an ON condition; and *reverse bias,* which functions as an open switch or an OFF condition.

2   Diodes can be arranged to form two of the basic digital gates: the AND gate and the OR gate.

3   The limitations of diode logic circuitry include loss in signal level for each level of logic and also a limited speed due to the diode's large junction capacitance.

4   A transistor can function as a switching device because it can be saturated to represent a closed switch (or ON condition) and cut off to represent an open switch (or OFF condition).

5   TTL circuitry, or Transistor-Transistor Logic, is distinguishable from other forms of circuitry because of its multiple-emitter transistors.

6   A logic 1 level in TTL circuitry is a voltage of between +2.4 and +5.0 V, and a logic 0 level is a voltage of between 0 and +0.4 V.

7   Unused TTL input connections must be connected to 5 V through a pullup resistor for a logic 1 level, or to ground if a logic 0 level is desired.

8   TTL source current is that which flows out of an output when it is at a logic 1 level, and TTL sink current is that which flows into an output when it is at a logic 0 level.

9   The noise immunity of TTL logic circuitry is 400 mV.

10   The fanout, or number of input pins that can be connected to one output, for any TTL logic family is always ten when other members of the same family are connected to an output.

11   Many types of TTL logic circuitry are available, and it is very important always to replace a component in an existing circuit with the same type of circuit.

12   Open-collector gates can both interface TTL logic circuitry to circuits of other than 5 V logic and implement a wired or a dot AND gate.

13   Tri-state® logic circuits have three output states: a logic 1, a logic 0, and an open circuit. In other words, the output is either a valid TTL output or an open circuit.

14   Schmitt trigger inputs are used in TTL circuits to increase the noise immunity to approximately 1.2 V instead of the normal 400 mV.

15   ECL, or Emitter Coupled Logic, is used in high-speed digital application because of its short propagation delay times.

16   IIL, or Integrated Injection Logic, uses a short circuit to ground for a logic 0 level and an open circuit for a logic 1 level. Its speed and power consumption can also be changed by altering the amount of internal bias current.

KEY TERMS

bipolar transistor	hysteresis	Schottky diode
bypass capacitor	IIL	Schottky-clamped transistor
diode logic	junction capacitance	
ECL	merged-transistor logic	sink current
electromagnetic interference	noise margin	source current
fanout	open-collector	totem-pole
FCC	propagation delay	Tri-state®
glitch	pullup resistor	TTL

QUESTIONS AND PROBLEMS

1 The semiconductor diode is a 2-state device. What are these two states?

2 What is the major difference between a power-supply rectifier diode and a switching diode?

3 If a 5 V level is used to forward bias a silicon diode, what will be the resulting maximum output voltage?

4 Draw a 3-input diode logic OR gate and develop a voltage table if the logic 1 input voltage is 3.6 V and the logic 0 input voltage is 0 V.

5 Draw a 3-input diode logic AND gate and develop a voltage table if the logic 1 input voltage is 9 V and the logic 0 input voltage is 0 V.

6 What is the main problem with diode logic circuitry?

7 Why would diode logic circuitry be used in an age when digital integrated circuits seem to be used in almost all applications?

8 A saturated transistor switch will produce an output voltage of very near 0 V. What would be a typical value of the logic 0 output voltage from a saturated transistor switch?

9 Design a saturated transistor switch that will produce a maximum logic 0 output current of 16 mA at 0.3 V maximum and a maximum logic 1 output current of 1 mA at 3 V minimum. The input logic 1 voltage is 2.4 V minimum, and the logic 0 intput voltage is 1 V maximum. Assume that the minimum current gain of the transistor is 80.

10 Design a saturated transistor switch that will produce a maximum logic 0 output current of 20 mA at 0.4 V maximum and a maximum logic 1 output current of 3 mA at 4 V minimum. The input logic 1 voltage is 3 V minimum, and the logic 0 input voltage is 0.8 V maximum. Assume that the minimum current gain of the transistor is 100.

11 What limits the switching speed of a transistor switch?

12 How is it possible to have current flow from base to collector in an NPN transistor?

13 How are additional inputs added to a TTL logic gate such as a NAND gate?

14 Explain the operation of the circuit in Fig. 4-8.

15 What is the noise immunity of a typical TTL logic circuit?

16 How long can the output of a TTL logic gate withstand a direct short circuit to ground?

17 What would be the effect of connecting 11 TTL input circuits to one TTL output?

18 How many 74LSXX TTL logic gates can be safely connected to one 74SXX TTL output?

19 How many 74SXX TTL logic gates can be safely connected to one 74LSXX TTL output?

20 What is a Schottky-clamped transistor, and how does this device allow a TTL circuit to function at a higher speed?

21 What is an open-collector TTL logic gate?

22 When open-collector outputs are connected, what new logic function is created at this connection?`

23 If you have a TTL data book at your disposal, find a 13-input NAND gate and write down the part number.

24 If you have a TTL data book at your disposal, find a 5-input NOR gate and write down the part number.

25 If you have a TTL data book at your disposal, find the OR gate with the greatest number of input connections and write down the part number.

26 List the three states available at the output of a Tri-state® logic circuit.

27 Why would a Schmitt trigger input circuit be found in a Tri-state® TTL buffer?

28 Draw a 4-input multiplexer using four Tri-state® buffers. Don't forget to include the circuitry required to address each buffer.

29 What is another name for ECL circuitry?

30 What are the typical logic voltages for ECL circuitry?

31 What is one advantage of ECL circuitry?

32 What is another common name for IIL circuitry?

33 Why is IIL called a *current-steering* or *current-hogging* logic family?

34 What are the logic 1 and the logic 0 levels in IIL circuitry?

35 Draw the schematic diagram of an IIL OR gate.

5

MOSFET LOGIC FAMILIES

Unlike the bipolar transistor, the MOSFET logic family is controlled with an electric field set up by an applied voltage. MOSFET logic families share several characteristics. First and perhaps foremost, the MOSFET families offer extremely low power consumption. Second, the noise immunity and fanout are much better for MOSFET than for the bipolar counterparts. However, MOSFET families are slower than bipolar families and can be destroyed by static electricity. This chapter will cover two MOSFET families—NMOS and CMOS—in detail.

THE MOSFET AS A SWITCH

Chapter 4 covered the bipolar logic families—the diodes and bipolar transistors used as switches. However, there is a different type of transistor that makes a more effective switch for many digital applications. This transistor is the Metal Oxide Semiconductor Field Effect Transistor (MOSFET).

While the bipolar transistor switches in response to a current flowing into its base, the MOSFET transistor is controlled by a voltage applied to the terminal called the *gate*. The other two terminals in the MOSFET transistor are the *drain* and the *source* (there is, however, a special type of MOSFET, called the dual-gate MOSFET, that has two gate inputs). Fig. 5-1 shows the schematic representations of an *n-channel* and a *p-channel enhancement-mode* MOSFET. There is also a *depletion-mode* MOSFET, but it is seldom used in digital circuitry.

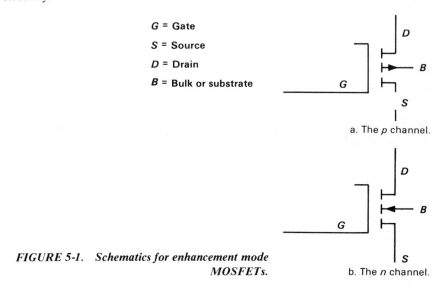

G = Gate
S = Source
D = Drain
B = Bulk or substrate

a. The *p* channel.

b. The *n* channel.

FIGURE 5-1. Schematics for enhancement mode MOSFETs.

Fig. 5-2 shows the mechanism by which the MOSFET operates. The width of the *channel* (either of *n*-type material or of *p*-type material) is controlled by the field set up by the *voltage*, applied to the *gate*, with respect to the *source*. This voltage is thus termed V_{gs}. The substrate is connected to the source. In the absence of V_{gs} ($V_{gs}=0$), no channel exists between the drain and the source, so no current can flow (see Fig. 5-2a). With the *n*-channel MOSFET, as V_{gs} is increased, a channel forms between the drain and source as electrons are attracted to the area. Current can then flow in the channel from drain to source (see Fig. 5-2b). Higher voltages enhance the channel—hence the term enhancement-mode MOSFET. The voltage at which the MOSFET conducts, or is turned on, is called the *threshold voltage*, or V_t.

Although the switching is brought about by voltage rather than current, the MOSFET can be compared to the bipolar transistor because the *n*-channel MOSFET (all MOSFETS referred to in this chapter will be enhancement-mode) operates in basically the same way as the NPN transistor.

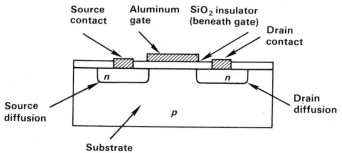

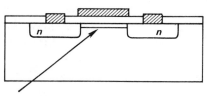

a. The threshold voltage (V_t) has not been exceeded, and no channel for conduction exists between the source and drain.

b. The gate voltage has exceeded V_t, which has caused electrons to be attracted to the gate area, forming an n-type channel for conduction—hence the name "n-channel."

FIGURE 5-2.   Mechanics of an n-channel MOSFET.

Fig. 5-3 illustrates the switching modes of the n-channel MOSFET. When *ground potential* (V_{gs} = 0) is applied to the gate, no channel exists and the transistor is switched off (see Fig. 5-3a). When a positive voltage is applied to the gate, the channel is formed and the transistor is switched on (see Fig. 5-3b). In MOSFET transistors, virtually no current flows into the gate; thus the input impedance is nearly infinite.

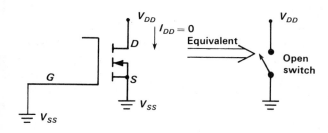

a. Gate at ground potential (V_{gs} = 0), which is less than the threshold voltage. No channel is formed and an open circuit exists between the drain and source.

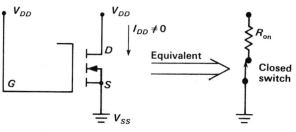

b. Gate is connected to V_{DD}, which is assumed to be greater than V_t, creating a channel. This results in a path between the drain and the source, and essentially a short circuit. There is a small amount of resistance, called the "on-resistance" (R_{on}), which is represented in the switch model.

FIGURE 5-3.   Switching modes of n-channel MOSFET.

MOSFET Logic Families 113

The p-channel MOSFET operates in the converse manner. The threshold voltage is defined here as V_{sg}. In other words, a *negative* potential from gate to source (with substrate connected to source) is necessary to set up the field for the transistor to conduct.

Unlike bipolar transistors, whose impedance is relatively constant and whose saturation voltage is fixed, in MOSFETs these parameters can be controlled in the manufacturing process. The width and length of the channel can be controlled to vary the on-resistance (R_{on}) of the MOSFET. R_{on} can vary from less than $1\ \Omega$ to several hundred kilohms. This is an important feature, as a look at the NMOS switch will reveal.

MOS CIRCUIT TOPOLOGY

MOS logic circuits are very important digital circuits. The majority of LSI and microprocessor chips manufactured today use the NMOS (n-channel MOS) topology. In addition to consuming less power than a bipolar transistor, an NMOS transistor requires only about 15 percent of the area of a bipolar transistor; hence, much more complex circuits can be fabricated on an IC chip of a given size using NMOS rather than bipolar circuitry.

NMOS AND PMOS

Most early MOS devices were made from only p-channel MOSFETs (PMOS) because the fabrication process for PMOS is simpler than for NMOS. NMOS devices, however, are about three times faster than their PMOS counterparts. As a result, much research went into developing cost-effective NMOS processes. Once these developments were made, virtually all MOS circuits were (and still are) made with n-channel devices.

THE NMOS CIRCUIT

The Active Pullup

Recall from Chapter 4 that in the basic bipolar switch, a load (or pullup) resistor is needed to pull the output to the high state when the transistor is turned off. In bipolar circuitry, due to the fairly large amount of current flowing, only a small-value resistor is needed (around $1-2$ kΩ). Because of the much smaller amounts of current flow in MOS circuits, a rather large resistance is needed (as much as 100 kΩ).

This large resistance would be of little consequence in a discrete circuit, since the value of resistance has nothing to do with the physical package size. However, in integrated circuits, the physical size of the resistor is directly proportional to the value of resistance, and the size of resistor needed in MOS circuits is so large that it would offset much of the size advantage of MOS circuitry.

Therefore, another NMOS transistor is used in place of the pullup resistor. Fig. 5-4 is the schematic of the basic NMOS switch. If the *load transistor* is fabricated with a specific channel length and width, its resistance may be known. Thus the upper transistor in Fig. 5-4 (Q_1) is usually called the *active pullup* and has an equivalent resistance in the neighborhood of 100 kΩ.

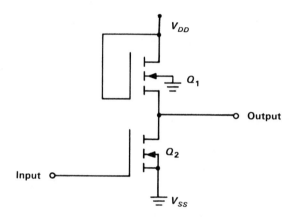

FIGURE 5-4. *Schematic of an NMOS INVERTER, the basic NMOS circuit element. Q_1 is the load transistor and Q_2 is the driver.*

In bipolar circuitry, the power supply voltage is referred to as V_{CC} since the positive connection is to the collector (for an NPN). In MOS circuitry, the supply is connected to the drain of the transistor and is thus referred to as V_{DD} (as in Fig. 5-4). Ground in the MOS circuit is usually connected to the source and so is usually called V_{SS}. This terminology may be somewhat confusing at first, but it will become natural after a time. One word of caution: If a PMOS circuit is used, V_{SS} is the supply and V_{DD} is ground!

Circuit Operation

In Fig. 5-4, notice that when the gate of Q_1 is connected to V_{DD} directly, Q_1 is always on, but, as mentioned above, it acts as a resistor. If the gate of Q_2 (Input) is connected to a "low" (ground), Q_2 is off and appears as an open switch. With Q_2 off, the output assumes a "high" since it is connected to V_{DD} through the active pullup.

If the gate of Q_2 is connected to a high, Q_2 is on. When this occurs, the output is pulled low since the on-resistance of Q_2 is quite low compared to that of Q_1. Notice that the operation of this circuit is the NOT or INVERTER function.

What happens if the input is connected to neither a high nor a low (i.e., left floating)? As mentioned in Chapter 4, the TTL gate's input will be pulled high with no input connection. In NMOS and PMOS (as well as CMOS, as we will see shortly) an input *cannot* be left floating, because it will be pulled neither high nor low. All unused inputs must be tied either to a high or to a low or unpredictable operation will result.

NMOS Gates

Fig. 5-5 is the schematic representation of a 2-input NMOS gate. Let's analyze the circuit to determine what type of gate it is.

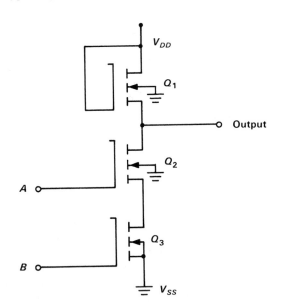

FIGURE 5-5.   Two-input NMOS gate.

Q_1 is, of course, the active pullup for this gate. This active pullup is present in all NMOS gates except the open-drain configuration. Q_2 and Q_3 are identical low-R_{on} transistors. If both inputs are low, both Q_2 and Q_3 are off and the output is pulled high through the active pullup. If *either* A or B are low and the other is high, the series connection of the two transistors prohibits a path to ground so that the output remains high. Only if both A and B are high do both transistors turn on and connect the output to ground, resulting in a low output. Thus the circuit in Fig. 5-5 is an NMOS NAND gate. Since all MOS circuits are made only from transistors, they are easier to analyze than the bipolar counterparts.

CMOS CIRCUIT TOPOLOGY

Complementary Metal Oxide Semiconductor logic (CMOS) contains both a p-channel and an n-channel transistor in the basic cell. Although even the most basic circuit element requires two transistors, CMOS offers several advantages over pure NMOS or PMOS.

As is the case with the NMOS topology, the basic circuit element in CMOS topology is the INVERTER. Fig. 5-6 shows the simplified schematic of the CMOS INVERTER. Even though there are two transistors, they do not operate in the driver-load configuration of the NMOS topology. Here both operate as drivers, and, in the 4000 series of CMOS as in many other CMOS families, the p-channel and the n-channel transistors have identical parameters (except switching speed); that is, they are *symmetrical*. (RCA calls its CMOS "COS/MOS," for Complementary-Symmetry MOS.)

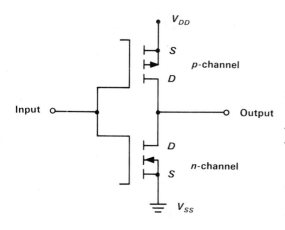

FIGURE 5-6. Basic circuit element for the CMOS logic family. As with the NMOS family, it is an INVERTER. in CMOS, all circuit elements are in complementary pairs—one p-channel and one n-channel transistor. There are never single n- or p-channel transistors.

Let's look at the operation of the INVERTER in Fig. 5-6 as a basis for understanding all CMOS circuits. The first thing to note is that, as in all MOS circuits, *some* input value must be applied at all times; the input cannot be allowed to float. If a low is applied to the input, the p-channel transistor is turned on. The applied low also turns the n-channel transistor off. Thus the n-channel transistor is an open circuit to V_{SS}, but the p-channel is a short circuit to V_{DD}, so the output assumes the value of V_{DD}, or high. (Note that current flows from source to drain in a p-channel MOSFET.) If a high is applied to the input, the p-channel transistor is turned off and the n-channel transistor is turned on. As a result, the output is connected to V_{SS}, and a low occurs at the output. This operation is clearly an INVERTER.

POWER REQUIREMENTS

There are two important facts about the CMOS circuit power requirements that require study:

1. Supply voltage regulation is not critical.
2. Power consumption is very low, falling virtually to zero during steady-state conditions.

Supply Voltage Regulation

Like the NMOS and PMOS circuits, the CMOS circuit contains no resistors, thus eliminating the critical biasing that is necessary in the TTL circuit topology. Consequently, the supply voltage for the CMOS circuit has a very wide range. Operating voltage for the 4000 series CMOS logic family is typically 3–18 V. Thus critical regulation of the voltage to within a small percentage is not a requirement with CMOS. As a result, power supplies are often easier and cheaper to design and build for CMOS than for TTL. This fact, coupled with extremely low power consumption, makes CMOS ideal for battery-powered circuits.

Power Consumption

As the operating description of the CMOS INVERTER revealed, one transistor is on and one is off in both high and low input. Consequently, no current can flow through both transistors from V_{DD} to V_{SS}. Now, assume that the output of the CMOS INVERTER is connected to the input of another CMOS gate, as is usually the case. Because no gate current flows, even though the output transistor may be connected directly to V_{DD}, *no current flows from the output*. The tremendous advantage of CMOS should now be clear: during steady-state conditions (that is, when the gates are not switching), no power is consumed because no current flows. (Actually, a very tiny amount of current flows because of *transistor leakage*, but this current is about 10 pA (picoamp), which for all practical purposes is zero.)

Our discussion of power requirements stressed the importance of steady-state conditions because current does flow during switching, much the same as in the totem-pole output of the TTL gate circuit. Fig. 5-7 shows what happens during the switching operation. Because the transistors require time to switch from the on to the off state, some overlap occurs when both transistors are partially on. During this time, current is allowed to flow directly from V_{DD} to V_{SS} through both transistors. The maximum current flows when the output voltage is equally divided between the two transistors.

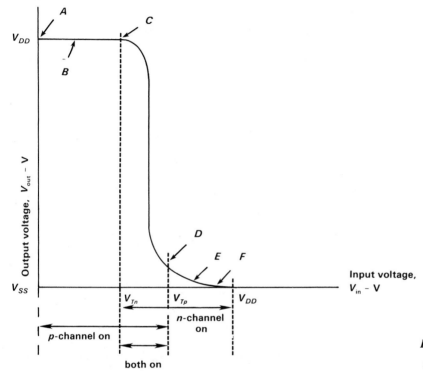

FIGURE 5-7. *Input-output transfer curve for CMOS.*

If the switching operation occurs very infrequently (as in the burglar alarm example in Chapter 3), the average power used would still be virtually zero. In a high-speed switching circuit, however, the average power consumption would rise, proportional to the frequency at which the switching occurs.

Fig. 5-7 details this operation. With the input voltage at 0 (point A), the output voltage assumes V_{DD} since this is the steady-state situation. At point B, the input voltage has risen to the point that the p-channel transistor is beginning to turn off, but the threshold voltage (V_{gs}) of the n-channel transistor has not been reached. At point C, the threshold voltage of the n-channel transistor has been reached but the input voltage is still above the threshold voltage (V_{sg}) of the p-channel transistor, and, as a result, both transistors are solidly on. At point D, the p-channel has now reached the threshold voltage. Current flow decreases dramatically through point E, and at point F the p-channel is solidly off and the n-channel transistor is fully on—again the steady-state situation.

PRECAUTIONS AGAINST STATIC DISCHARGE

One of the major concerns and disadvantages of MOS circuits is their susceptibility to destruction by static discharge. As mentioned earlier, the input impedance of MOS circuits is very high—as a matter of fact, it is on the order of 1 TΩ (Teraohm, 10^{12} or one trillion ohms). A discharge of static electricity, as you experience after walking across a carpet, contains a very small amount of current flow. But, if you solve Ohm's Law using the input impedance of 1 TΩ, you can see that a very large voltage can result from a tiny current flow.

The breakdown voltage of most MOS circuits is on the order of 100 V, and a static discharge can exceed this. Unlike bipolar transistors, which can heal themselves after their breakdown voltage is exceeded, the silicon dioxide layer in a MOS transistor is very thin and can be permanently destroyed.

Special handling precautions must be taken with MOS circuits to avoid destroying them with static discharges. MOS ICs are shipped in special conductive foam to keep charges from building up between V_{DD} and V_{SS}. *Never* use styrofoam in place of this conductive foam, as styrofoam actually accumulates static charges. Also, MOS circuits should never be plugged into a circuit nor removed from it unless power has been removed. Most MOS circuits produced today have input protection circuits, which we will study shortly, that alleviate much of the danger; however, care should always be taken.

THE 4011 QUAD 2-INPUT NAND GATE

Just as the 7400 series denotes the TTL family part numbers, the 4000 series denotes members of the CMOS logic family. Unfortunately, corresponding part numbers do not denote the same function (for example, the Quad 2-input NAND is the 7400 in TTL and the 4011 in CMOS), nor are the pin numbers always the same. (There is, however, another CMOS family, produced by National Semiconductor, whose part and pin numbers do correspond to those of the 7400 family. This is the 74C00 family.)

Fig. 5-8 is the schematic for the CD4011B 2-input NAND gate. The "B" suffix denotes that the gates are buffered to maintain constant loading and provide higher output drive. Most "off the shelf" CMOS gates are of the buffered type.

The two inputs are each connected to a complementary pair of transistors (unshaded area in Fig. 5-9). They operate as follows. If lows are entered at both inputs, the two n-channel transistors are off and the two p-channel transistors are on. Since the two p-channel transistors are in parallel, the state at point I is high. If a low is present at A and a high at B, we find Q_1 on, Q_2 and Q_3 off and Q_4 on. Since Q_2 and Q_3 are in series, the fact that Q_3 is off doesn't matter; Q_1 still makes the state at point I high. If A is high and B is low, the same effect results with the transistors simply reversed.

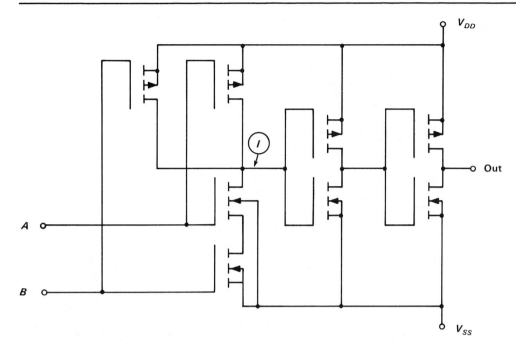

FIGURE 5-8.   *Schematic diagram of the CD4011B 2-input NAND gate. There are four of these circuits in the CD4011B IC package.*

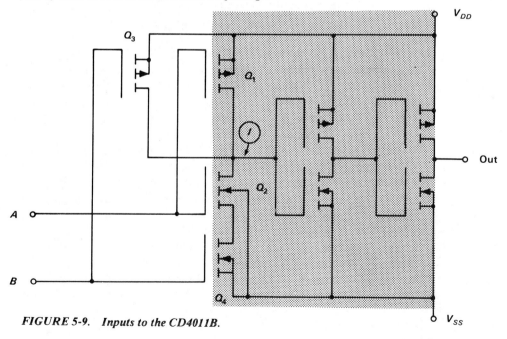

FIGURE 5-9.   *Inputs to the CD4011B.*

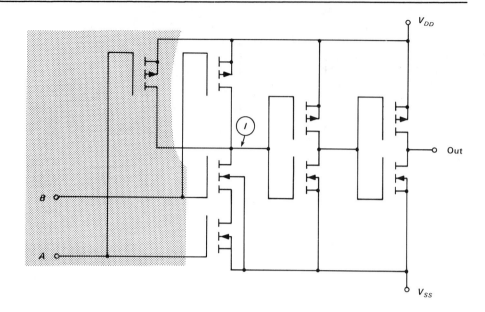

FIGURE 5-10.   Output section of the CD4011B.

If both *A* and *B* have high signals applied, the two *p*-channel transistors (Q_1 and Q_3) are off and the two *n*-channel transistors (Q_2 and Q_4) are on. When both Q_2 and Q_4 are on, the series circuit is completed and point *I* is connected to ground, resulting in the low state. This function is clearly the NAND.

The unshaded area in Fig. 5-10 portrays the output section of the 4011. Notice that this section consists simply of two cascaded CMOS INVERTERs. Recalling from Boolean algebra that a double inversion cancels, we see that the state at point *I* is the same as that at the output. Why, then, waste the transistors necessary to do this double inversion?

The answer is simply that this is the buffer portion of the circuit. The buffering isolates the logical output of the NAND from the output. Buffering enables smaller transistors to be used in the logic circuitry; larger transistors are used only in the output. This allows higher drive capability, while saving space on the silicon chip. In this design, only the two output transistors need to be large enough to carry high current, instead of all four in the NAND. Indeed, the larger NAND gate transistors would take up more additional space than the output transistors.

Also, recall that in the NAND gate two *p*-channel MOSFETs are in parallel. In the parallel resistor combination, the impedance from V_{DD} to the output is twice as much with one transistor turned on as it is with both turned on. If the outputs weren't buffered, the switching speed would vary based on the input combination. This is a highly undesirable trait, which buffering avoids.

The type "B" CMOS gates also incorporate a static discharge protection network, as shown on the inputs in Fig. 5-11. The two diodes and the resistor connected to V_{DD} actually represent a distributed diode-resistor network. If the input exceeds V_{DD}, the diodes conduct and the resistor dissipates power; as a result, no harmful voltage spikes reach the fragile MOS transistors. This protection is good for up to 4 kV (kilovolts). The diode connected to V_{SS} protects against any negative voltage transients.

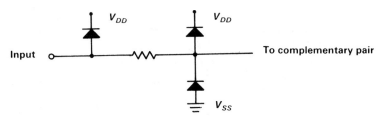

FIGURE 5-11. CMOS input protection circuitry.

Recall from Chapter 4 that diodes have some reverse leakage current. These diodes are no different. Leakage through the diodes results in a somewhat higher power dissipation for the CMOS gate than if the diodes weren't present. We will see this when studying the data sheet.

LOGIC LEVEL VOLTAGES

Fig. 5-12 shows the input and output logic level characteristics for a V_{DD} of 5 V. Compare these levels to TTL logic levels in Figs. 4-14 and 4-16. The CMOS outputs, with no biasing resistors to drop voltage, and almost no current drawn from them, are virtually equal to V_{DD} and V_{SS}—a highly desirable feature. The input levels are quite liberal as far as CMOS interfacing is concerned—also a desirable feature. Note, however, that the CMOS input levels and the TTL output levels are not quite compatible. The interfacing of CMOS and other logic families is detailed in Chapter 8.

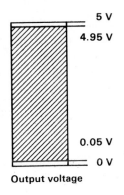

Output voltage

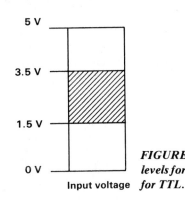

Input voltage

FIGURE 5-12. CMOS input and output voltage levels for V_{DD} of 5 V. Compare to Figs. 4-13 and 4-15 for TTL.

NOISE IMMUNITY

The noise immunity for CMOS circuitry is found in the same way as for TTL. The high-level noise margin is 4.99 V minus 3.50 V, or 1.49 V. The low-level margin is also 1.49 V. Clearly, this level of immunity is considerably better than that of TTL. Also notice that these are the margins with a 5 V supply. Since CMOS is not restricted to 5 V operation, even greater noise margins can be realized by operating CMOS at greater supply voltages.

FANOUT

The fanout of a CMOS gate driving only CMOS circuitry is somewhat more complex than that of TTL. In CMOS, the fanout is limited only by the sum of input capacitances. As these capacitances add up to larger and larger values, the maximum switching speed is reduced, since the capacitors must charge. Consequently, the fanout is at minimum 50. If switching speed is of little consequence, the fanout can include several hundred gates.

PROPAGATION DELAY

As mentioned earlier, the CMOS inputs exhibit some capacitance. But the outputs also have some capacitance, as a result of the MOS transistor geometry. This capacitance is called *parasitic capacitance*. The input protection diodes also, of course, have capacitance. The sum of all these capacitances results in a longer propagation delay in CMOS gates than in TTL gates.

Typical propagation delay in SSI CMOS gates operating at 5 V is 100–150 ns. This is roughly ten times that of TTL. This time can be reduced by increasing the supply voltage, but power consumption also increases with higher supply voltage. Other techniques, which we will study later in this chapter, also decrease the propagation delay of CMOS.

INTERPRETING THE CMOS DATA SHEET

It is, of course, extremely important to be able to interpret the manufacturer's data sheet with proficiency in order to use digital components effectively. CMOS data sheets are very similar to the TTL data sheets covered in Chapter 4, except that some terms carry different names. The following discussion of a typical CMOS data sheet should enable you to make effective use of any CMOS data sheet.

ELECTRICAL CHARACTERISTICS

Fig. 5-13 is an actual manufacturer's data sheet for the "B" series CMOS logic family. The "B" series is the buffered 4000, the most common CMOS logic

	Parameter	Temp. Range	V_{DD} (Vdc)	Conditions	T_{LOW}* Min.	T_{LOW}* Max.	+25°C Min.	+25°C Typ.	+25°C Max.	T_{HIGH}** Min.	T_{HIGH}** Max.	Units
I_{DD}	Quiescent Device Current GATES	Mil.	5			0.25			0.25		7.5	
			10			0.5			0.5		15	µAdc
			15	$V_I = V_{SS}$ or V_{DD}		1.0			1.0		30	
		Comm.	5	All valid input combinations		1.0			1.0		7.5	
			10			2.0			2.0		15	µAdc
			15			4.0			4.0		30	
	BUFFERS, FLIP-FLOPS	Mil.	5			1.0			1.0		30	
			10			2.0			2.0		60	µAdc
			15	$V_I = V_{SS}$ or V_{DD}		4.0			4.0		120	
		Comm.	5	All valid input combinations		4.0			4.0		30	
			10			8.0			8.0		60	µAdc
			15			16			16		120	
	MSI	Mil.	5			5.0			5.0		150	
			10			10			10		300	µAdc
			15	$V_I = V_{SS}$ or V_{DD}		20			20		600	
		Comm.	5	All valid input combinations		20			20		150	
			10			40			40		300	µAdc
			15			80			80		600	
V_{OL}	Low-Level Output Voltage	All	5	$V_I = V_{SS}$ or V_{DD}		0.05			0.05		0.05	
			10	$|I_O| < 1\,\mu A$		0.05			0.05		0.05	Vdc
			15			0.05			0.05		0.05	
V_{OH}	High-Level Output Voltage	All	5	$V_I = V_{SS}$ or V_{DD}	4.95		4.95			4.95		
			10	$|I_O| < 1\,\mu A$	9.95		9.95			9.95		Vdc
			15		14.95		14.95			14.05		
V_{IL}	Input Low Voltage	All	5	$V_O = 0.5\,V$ or $4.5\,V$, $|I_O| < 1\,\mu A$		1.5			1.5		1.5	
			10	$V_O = 1.0\,V$ or $9.0\,V$, $|I_O| < 1\,\mu A$		3.0			3.0		3.0	Vdc
			15	$V_O = 1.5\,V$ or $13.5\,V$, $|I_O| < 1\,\mu A$		4.0			4.0		4.0	
V_{IH}	Input High Voltage	All	5	$V_O = 0.5\,V$ or $4.5\,V$, $|I_O| < 1\,\mu A$	3.5		3.5			3.5		
			10	$V_O = 1.0\,V$ or $9.0\,V$, $|I_O| < 1\,\mu A$	7.0		7.0			7.0		Vdc
			15	$V_O = 1.5\,V$ or $13.5\,V$, $|I_O| < 1\,\mu A$	11		11			11		
I_{OL}	Output Low (Sink) Current	Mil.	5	$V_O = 0.4\,V$, $V_I = 0\,V$ or $5\,V$	0.64		0.51			0.36		
			10	$V_O = 0.5\,V$, $V_I = 0\,V$ or $10\,V$	1.6		1.3			0.9		mAdc
			15	$V_O = 1.5\,V$, $V_I = 0\,V$ or $15\,V$	4.2		3.4			2.4		
		Comm.	5	$V_O = 0.4\,V$, $V_I = 0\,V$ or $5\,V$	0.52		0.44			0.36		
			10	$V_O = 0.5\,V$, $V_I = 0\,V$ or $10\,V$	1.3		1.1			0.9		mAdc
			15	$V_O = 1.5\,V$, $V_I = 0\,V$ or $15\,V$	3.6		3.0			2.4		
I_{OH}	Output High (Source) Current	Mil.	5	$V_O = 4.6\,V$, $V_I = 0\,V$ or $5\,V$	−0.25		−0.2			−0.14		
			10	$V_O = 9.5\,V$, $V_I = 0\,V$ or $10\,V$	−0.62		−0.5			−0.35		mAdc
			15	$V_O = 13.5\,V$, $V_I = 0\,V$ or $15\,V$	−1.8		−1.5			−1.1		
		Comm.	5	$V_O = 4.6\,V$, $V_I = 0\,V$ or $5\,V$	−0.2		−0.16			−0.12		
			10	$V_O = 9.5\,V$, $V_I = 0\,V$ or $10\,V$	−0.5		−0.4			−0.3		mAdc
			15	$V_O = 13.5\,V$, $V_I = 0\,V$ or $15\,V$	−1.4		−1.2			−1.0		
I_I	Input Current	Mil.	15	$V_I = 0\,V$ or $15\,V$		±0.1			±0.1		±1.0	µAdc
		Comm.	15	$V_I = 0\,V$ or $15\,V$		±0.3			±0.3		±1.0	µAdc
C_I	Input Capacitance per Unit Load	All	—	Any Input				7.5				pF

Note: For current flow the convention is positive for current flowing into the device and negative flowing out of the device.
*T_{LOW} = −55°C for Military Temp. Range device, −40°C for Commercial Temp. Range device.
**T_{HIGH} = +125°C for Military Temp. Range device, +85°C for Commercial Temp. Range device.

FIGURE 5-13. Manufacturers data sheet for 4000B series CMOS logic. (Courtesy of National Semiconductor Corporation. Authorization has been given by National Semiconductor Corporation, Santa Clara, California, world leaders in Digital Acquisition Circuitry.)

family. Notice also that these specifications also apply as minimum specifications for the 74C00 family of CMOS logic. The data sheet in Fig. 5-13 is somewhat more complicated than Fig. 4-22 for TTL because the specifications are given for three different operating voltages: 5, 10, and 15 V.

The electrical characteristics are given for a V_{DD} of 5 and 10 V and also for −40 to +85° C (for commercial-grade parts). When designing with any logic family, it is always best to design with the worst case data for the conditions that the circuit will experience. For example, if you are designing a circuit that will always operate indoors, you can use the minimum or maximum (depending on the particular specification) value at 25° C. If the circuit is to operate outdoors, you would use the "worst case" value for either −40 or +85° C. By designing with "worst case" values, you can ensure that the circuit will work under any conditions, with any combination of parts.

Device Currents

The *quiescent device currents*, I_{DD}, compose the current that is required by the entire IC package. For gates, the maximum device current at room temperature at 5 V is 1 μA. Since flip-flops and buffers are more complex, their I_{DD} is somewhat higher, and that of MSI (Medium-Scale Integration) circuits is higher yet, but still only 20 μA. Putting this into perspective, we find that the quiescent currents of four thousand CMOS NAND gates would equal the supply current of *one* TTL NAND.

The output currents, I_{OL} and I_{OH}, specify the output sink current (current into the output when low) and the output source current (current out of the output when high), respectively. For commercial parts operating at 5 V and room temperature, the sink current is 440 μA and the source current is 160 μA. This slight imbalance results from different current-handling capabilities of the *n*-channel and *p*-channel transistors.

The maximum input currents, shown only for a V_{DD} of 15 V, is 0.3 μA. This exceptionally low value is due to the extremely high input impedance.

Device Voltages

The operating voltage for CMOS, as previously mentioned, is 3–18 V. Most operation takes place between 5 and 15 V, and of course the majority of applications use 5 V, since designers are accustomed to working with that voltage as a carryover from TTL circuits.

The output voltages, V_{OL} and V_{OH}, are the low output voltage and the high output voltage, respectively. Again using room temperature, a V_{DD} of 5 V, and commercial parts, the low-level output voltage is 0.05 V maximum (why is only maximum specified?). The minimum (why only minimum?) high output voltage is 4.95 V. Notice that the output voltage does not change with temperature, as do the currents.

Input voltages are represented by the symbols V_{IL} and V_{IH} for the low-level input voltage and the high-level input voltage, respectively. These voltages, also independent of temperature, are 1.5 and 3.5 V, respectively.

The final entry in the data sheet is the input capacitance, which is 7.5 pF per input. Since capacitance adds in parallel, this value increases as the fanout increases.

SWITCHING CHARACTERISTICS

The switching characteristics, or AC electrical characteristics, are shown in Fig. 5-14. The first two, t_{PLH} (rise time) and t_{PHL} (fall time), have the same meaning as for TTL. Notice, however, that these times are roughly ten times longer (at 5 V) than those for TTL.

AC Electrical Characteristics CD4011BC, CD4011BM

$T_A = 25°C$, Input t_r; $t_f = 20$ ns. $C_L = 50$ pF, $R_L = 200$k. Typical Temperature Coefficient is 0.3%/°C.

	PARAMETER	CONDITIONS	TYP	MAX	UNITS
tPHL	Propagation Delay, High-to-Low Level	$V_{DD} = 5V$	120	250	ns
		$V_{DD} = 10V$	50	100	ns
		$V_{DD} = 15V$	35	70	ns
tPLH	Propagation Delay, Low-to-High Level	$V_{DD} = 5V$	85	250	ns
		$V_{DD} = 10V$	40	100	ns
		$V_{DD} = 15V$	30	70	ns
tTHL, tTLH	Transition Time	$V_{DD} = 5V$	90	200	ns
		$V_{DD} = 10V$	50	100	ns
		$V_{DD} = 15V$	40	80	ns
CIN	Average Input Capacitance	Any Input	5	7.5	pF
CPD	Power Dissipation Capacity	Any Gate	14		pF

FIGURE 5-14. AC or switching characteristics for the CD4011B. (Courtesy of National Semiconductor Corporation. Authorization has been given by National Semiconductor Corporation, Santa Clara, California, world leaders in Digital Acquisition Circuitry.)

Because the inputs have relatively large capacitance, an additional time value is given in the AC specifications: transition time (t_{THL}, t_{TLH}). This is the amount of time that it takes an input to change from high to low or low to high due to the charging of the capacitor. Since both times are identical, they are stated on a single line. Fig. 5-15 illustrates both the propagation delay time and the transition time.

The final specifications on the AC electrical characteristic sheet are the average input capacitance (C_{IN}), which we have seen is 7.5 pF, and the power

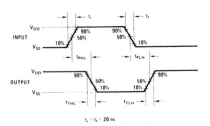

dissipation capacity (C_{PD}), which is 14 pF minimum. The latter capacitance is used to determine the total power dissipation, which is the sum of the quiescent power dissipation (the quiescent current multiplied by the supply voltage) and the dynamic power dissipation. The dynamic power dissipation occurs because of the current surges during switching. The equation for the dynamic power dissipation is:

$$P_{\text{dyn}} = C_{PD} \times V^2_{DD} \times f$$

(f is the switching frequency.) This equation explains why CMOS dissipates more power both at higher voltages and at higher frequencies.

Example 5-1. Calculate the dynamic power dissipation for a 5 V CMOS circuit operating continuously at 5 MHz.

Solution. Typical C_{PD} (from Fig. 5-14) is 14 pF. The equation is:

$$P_{\text{dyn}} = (14 \times 10^{-12}) \times 5^2 \times (5 \times 10^6) = 1.75 \text{ mW}$$

As you can see, the dynamic power dissipation rises dramatically when CMOS is continuously operated at high frequency.

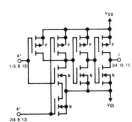

1/4 of device shown

SCHEMATIC FROM THE MANUFACTURER'S DATA SHEET

Fig. 5-16 is the schematic diagram from the manufacturer's data sheet. It is virtually the same schematic as that shown earlier in this chapter. As is typical, only part of the actual IC schematic is shown—in this case, only one of four gates. Also pictured is the input protection circuit.

HIGH-SPEED CMOS

All CMOS discussion so far has referred to the 4000 and the 74C00 families of CMOS logic. The 4000 series was invented in 1968 and the 74C00 series a few years later. Refinements were made thereafter, including the buffering and input protection, but no significant revisions to the CMOS family came about until 1982, when Motorola and National Semiconductor jointly introduced the high-speed CMOS family: the 74HC series.

High-speed CMOS is referred to either as the 74HC series or the 74HCT series. The difference between these two designations will be explained shortly. Most companies manufacturing high-speed CMOS simply refer to it as "High-Speed CMOS Logic," although RCA calls its product line "QMOS" (the "Q" stands for "quick").

The physical structure of the MOS transistors differs in the construction of the gate. In Fig. 5-2, the gate terminal is a metal electrode on top of a thin layer of silicon dioxide. In the high-speed CMOS, the metal is replaced with silicon, forming a *silicon gate* MOSFET. The silicon gate MOSFET reduces the threshold voltage and increases the packing density of the transistors by allowing the transistors to be closer together. This also increases the switching speed of the transistors, giving rise to the increased speed of this CMOS family.

The speed of the 74HC family is equivalent to that of the Low-power Shottky TTL (LSTTL) logic family. The fanout of the 74HC is also increased over that of the 4000 series to ten LSTTL loads. The 74HCT family has an even higher fanout, at 15. Even though the high-speed CMOS is capable of operating at 30 MHz like the LSTTL family, it retains the superb noise immunity of CMOS, and the low power consumption of less than 1 μW. The operating voltage limits of the high-speed CMOS are not as wide as standard CMOS but are still 3–6 V.

74HC DATA SHEET

Fig. 5-17a is the manufacturer's static electrical characteristics data sheet for the 74HC series. The most important specifications to note for the 74HC, as opposed to the 4000 series, are the low-level input voltage (V_{IL}) and the low- and high-level output currents (I_{OL} and I_{OH}). The V_{IL} has been dropped to 0.9 V and the output currents are 4 mA. These changes are necessary to make this family compatible with LSTTL. Note also that the output currents are identical for source and sink.

Also of interest are the facts that the supply current has risen slightly and supply voltage and supply current are called V_{CC} and I_{CC} even though MOS transistors are still used. Additionally, the high-level input voltage has increased 0.1 V over that of the 4000 family.

Fig. 5-17b is the manufacturer's specification sheet for the AC electrical characteristics of the 74HC00, at 25° C and a V_{CC} of 5 V. The t_{PHL} and t_{PLH} are given for both 15 and 50 pF loading capacitance. Notice from this that, to achieve maximum speed, the fanout should be kept small.

Figs. 5-18a and 5-18b show the relationship of the input current requirements and the output current capability of 74HC logic as compared to other logic

DC Electrical Characteristics (Note 4)

Symbol	Parameter	Conditions	V_{CC}	$T_A = 25°C$ Typ	$T_A = 25°C$ ≈	74HC $T_A = -40$ to 85°C Guaranteed Limits	54HC T_A -55 to 125°C Guaranteed Limits	Units
V_{IH}	Minimum High Level Input Voltage		2.0V		1.5	1.5	1.5	V
			4.5V		3.15	3.15	3.15	V
			6.0V		4.2	4.2	4.2	V
V_{IL}	Maximum Low Level Input Voltage		2.0V		0.3	0.3	0.3	V
			4.5V		0.9	0.9	0.9	V
			6.0V		1.2	1.2	1.2	V
V_{OH}	Minimum High Level Output Voltage	$V_{IN} = V_{IH}$ or V_{IL} $|I_{OUT}| \leq 20\ \mu A$	2.0V	2.0	1.9	1.9	1.9	V
			4.5V	4.5	4.4	4.4	4.4	V
			6.0V	6.0	5.9	5.9	5.9	V
		$V_{IN} = V_{IH}$ or V_{IL} $|I_{OUT}| \leq 4.0$ mA	4.5V	4.2	3.98	3.84	3.7	V
		$|I_{OUT}| \leq 5.2$ mA	6.0V	5.7	5.48	5.34	5.2	V
V_{OL}	Maximum Low Level Output Voltage	$V_{IN} = V_{IH}$ $|I_{OUT}| \leq 20\ \mu A$	2.0V	0	0.1	0.1	0.1	V
			4.5V	0	0.1	0.1	0.1	V
			6.0V	0	0.1	0.1	0.1	V
		$V_{IN} = V_{IH}$ $|I_{OUT}| \leq 4.0$ mA	4.5V	0.2	0.26	0.33	0.4	V
		$|I_{OUT}| \leq 5.2$ mA	6.0V	0.2	0.26	0.33	0.4	V
I_{IN}	Maximum Input Current	$V_{IN} = V_{CC}$ or GND	6.0V		± 0.1	± 1.0	± 1.0	μA
I_{CC}	Maximum Quiescent Supply Current	$V_{IN} = V_{CC}$ or GND $I_{OUT} = 0\ \mu A$	6.0V		2.0	20	40	μA

Note 1: Absolute Maximum Ratings are those values beyond which damage to the device may occur.
Note 2: Unless otherwise specified all voltages are referenced to ground.
Note 3: Power Dissipation temperature derating — plastic "N" package: 12 mW/°C from 65°C to 85°C, ceramic "J" package: 12 mW/°C from 100°C to 125°C.

a. DC data.

AC Electrical Characteristics

$V_{CC} = 5V$, $T_A = 25°C$, $C_L = 15$ pF, $t_r = t_f = 6$ ns

Symbol	Parameter	Conditions	Typ	Guaranteed Limit	Units
t_{PHL}, t_{PLH}	Maximum Propagation Delay		8	15	ns

AC Electrical Characteristics

$V_{CC} = 2.0V$ to 6.0V, $C_L = 50$ pF, $t_r = t_f = 6$ ns (unless otherwise specified)

Symbol	Parameter	Conditions	V_{CC}	$T_A = 25°C$ Typ	74HC $T_A = -40$ to 85°C Guaranteed Limits	54HC $T_A = -55$ to 125°C Guaranteed Limits	Units
t_{PHL}, t_{PLH}	Maximum Propagation Delay		2.0V	45	90	113	134	ns
			4.5V	9	18	23	27	ns
			6.0V	8	15	19	23	ns
t_{TLH}, t_{THL}	Maximum Output Rise and Fall Time		2.0V	30	75	95	110	ns
			4.5V	8	15	19	22	ns
			6.0V	7	13	16	19	ns
C_{PD}	Power Dissipation Capacitance (Note 5)	(per gate)		20			pF
C_{IN}	Maximum Input Capacitance			5	10	10	10	pF

b. AC data.

FIGURE 5-17. Manufacturer's static characteristics for the 74HCOO high-speed CMOS gate. (Courtesy of National Semiconductor Corporation. Authorization has been given by National Semiconductor Corporation, world leaders in Digital Acquisition Circuitry.)

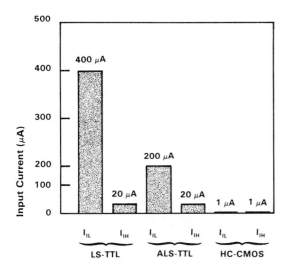

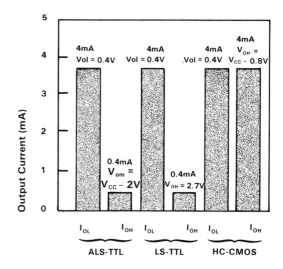

FIGURE 5-18. *Comparison of LSTTL, ALSTTL, and 74HC high-speed CMOS. (Courtesy of National Semiconductor Corporation. Authorization has been given by National Semiconductor Corporation, world leaders in Digital Acquisition Circuitry.)*

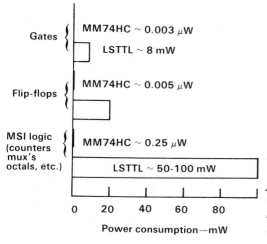

FIGURE 5-19. *Comparison of the power consumption of 75HC high-speed CMOS to LSTTL for simple gates, flip-flops, and MSI circuits. (Courtesy of National Semiconductor Corporation. Authorization has been given by National Semiconductor Corporation, world leaders in Digital Acquisition Circuitry.)*

families. Fig. 5-19 represents the power consumption of the 74HC family and the LSTTL family.

AVAILABILITY

Since production of the 74HC logic family just began in 1982, not all of the functions are available. Most manufacturers anticipate having their entire line in production within three to five years. Many manufacturers also feel that the 74HC family will replace the 4000 family and may even make both the 4000 family and the 74LS family totally obsolete.

SPECIAL CMOS FUNCTIONS

In addition to the features already detailed in this chapter, CMOS logic offers some special functions not found in any other families. Also, CMOS can be used in ways that TTL and other families cannot.

TRANSMISSION GATES

A special logic function found in CMOS is the *transmission gate*. The transmission gate is essentially a CMOS relay, being a switch that is electronically controlled. Fig. 5-20 is a schematic representation of a CMOS transmission gate, and Fig. 5-21 is the logic symbol for the transmission gate. When we studied the CMOS INVERTER, we saw that the p-channel and the n-channel transistors are in series. In the transmission gate, the two transistors are parallel.

The operation of the transmission gate is as follows. The right transistor (Q_1) is turned on when the gate (G_p) is connected to V_{SS}. When G_p is connected to V_{DD}, Q_1 is off. The operation of Q_2, the n-channel transistor, is just the opposite, of course. V_{DD} on G_n turns the transistor on, and V_{SS} on the gate turns it off. As you can see, when either transistor is on, the input is connected directly to the output.

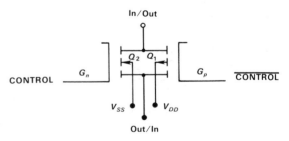

FIGURE 5-20. Schematic of the CMOS transmission gate. As with all CMOS, there is both a p-channel and an n-channel transistor.

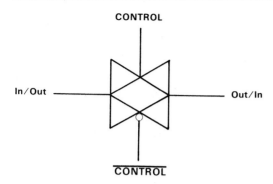

CONTROL

In/Out ———————— Out/In

CONTROL

FIGURE 5-21. Transmission gate logic symbol.

In an actual transmission gate, an INVERTER ensures that G_p and G_n are at opposite states and the transistors are either both on or off. With both transistors on, current can flow either from input to output or vice versa. As a result, transmission gates are sometimes called *bilateral switches* and the terminals are labeled "input/output." Also, because the input and output are directly connected, the output follows the input in a linear fashion. In this application, the transmission gate is often called an *analog switch*.

A quad package transmission gate, or bilateral switch, is available as the 4016. A simplified schematic of one of the switches is shown in Fig. 5-22. Notice how the INVERTER ensures that the two gate signals (called *control signals*) are of opposite polarity. These bilateral switches have a multitude of uses, such as sample and hold circuits and bus transceivers in microprocessor circuits.

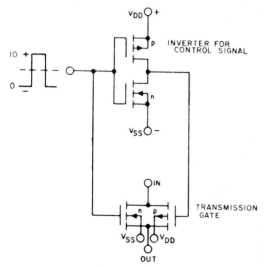

FIGURE 5-22.   Simplified schematic of one of the four bilateral switches in the CD4016 integrated circuit. (Courtesy of RCA.)

Another important use of the transmission gate is in the construction of more complex CMOS logic circuits. Most of the sequential circuits detailed in Chapter 6 are implemented in CMOS with the help of transmission gates.

DUAL COMPLEMENTARY PAIR PLUS INVERTER

Fig. 5-23 is the schematic of the 4007UB. The "UB" suffix denotes an unbuffered logic part. As the schematic reveals, the complementary pairs whose inputs are pins 6 and 3 are not interconnected, and the p-channel source and n-channel drain are connected to pins. This allows complete freedom in connection of the transistors or utilization of a single n- or p-channel transistor.

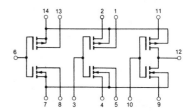

FIGURE 5-23. Schematic of the 4007 dual complementary pair plus INVERTER. (Courtesy of National Semiconductor Corporation. Authorization has been given by National Semiconductor Corporation, world leaders in Digital Acquisition Circuitry.)

The 4007 can often be used when only a single gate or a single MOSFET is necessary to fix a design error. Fig. 5-24a and 5-24b show two possible connections of the 4007, to form either a 3-input NOR or a 3-input NAND. Other combinations are limited only by the designer's imagination.

SUMMARY

1 MOSFET is an acronym for Metal Oxide Semiconductor Field Effect Transistor. MOSFETs are available in both n-channel and p-channel forms, whose operations are digitally equivalent to NPN and PNP transistors, respectively. Fig. 5-1 depicts the schematic symbols, and Fig. 5-2 details the operation of the MOSFET.

2 The MOSFET turns on when the *threshold voltage*, V_t, is reached.

3 When the MOSFET is turned off, the impedance from drain to source is extremely high; it may be as high as 1 TΩ (Teraohms). When the

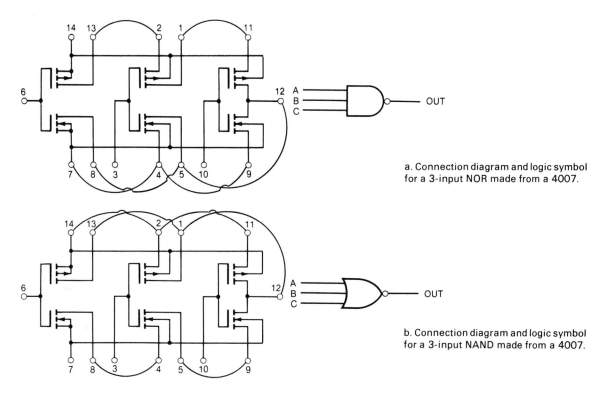

a. Connection diagram and logic symbol for a 3-input NOR made from a 4007.

b. Connection diagram and logic symbol for a 3-input NAND made from a 4007.

FIGURE 5-24. Some possible connections of the 4007. (Courtesy of National Semiconductor Corporation. Authorization has been given by National Semiconductor Corporation, world leaders in Digital Acquisition Circuitry.)

MOSFET is turned on, the drain to source resistance can be as low as a few ohms. Consequently, MOSFETs are nearly ideal switches.

4   MOSFETs take up much less space than bipolar transistors. An NMOS transistor takes only 15 percent of the space of an NPN transistor. This makes the fabrication of extremely complex LSI circuits possible.

5   In an NMOS INVERTER, the load is another MOSFET instead of a resistor. This is called an *active pullup*. There are no circuit elements in any MOS logic families except MOSFETs.

6   If the drive transistors in an NMOS gate are in series, a NAND gate is formed. If the drive transistors are parallel, a NOR gate is formed.

7   CMOS (Complementary Metal Oxide Semiconductor) logic incorporates both a *p*-channel and an *n*-channel MOSFET, connected as a complementary pair. The minimum CMOS circuit therefore has two transistors, and more complex circuits are always constructed using these pairs, never single transistors.

8 In MOS logic, no inputs can be left unconnected; each input must be connected to either V_{DD} or V_{SS}. If not, unpredictable operation or even oscillation can result.

9 In steady-state conditions, CMOS circuits draw virtually no current. During switching, both transistors are on for a brief period and a small amount of current will flow.

10 CMOS logic can operate from power supply voltages ranging from 3 to 15 V and need not be well regulated.

11 Because of their extremely high input impedances, MOS circuits must be protected against static discharge.

12 Buffered CMOS ensures more constant loading and allows higher output drive.

13 Diodes, which have "healable" junctions, are used in the inputs of MOS devices to keep static discharge and overvoltage conditions from destroying the sensitive MOSFET logic.

14 The output levels of CMOS are almost ideal, with $V_{OL} = 0.05$ V and $V_{OH} = 4.99$ V for V_{DD} of 5.0 V.

15 The noise margins for CMOS are quite superior. For a supply voltage of 5 V, both high and low noise margins are 1.49 V.

16 The fanout of CMOS is extraordinarily high, with a conservative minimum of 50.

17 The propagation delay of CMOS is 100–150 ns at 5 V, making it about ten times slower than standard TTL. The propagation delay can be decreased by raising the supply voltage; however, power consumption increases with increased supply voltage.

18 74HC high-speed CMOS is much quicker than standard CMOS, due mainly to its use of silicon for the gate instead of aluminum. 74HC is pin-for-pin interchangeable with LSTTL in virtually all applications.

19 Transmission gates represent another fundamental CMOS building block. In the transmission gate the p- and n-channel transistors are connected in parallel. Transmission gates are also called *analog switches* or *bilateral switches*.

20 In the bilateral switch, both transistors are turned on or off at the same time. When both transistors are on, current can flow in either direction through the switch.

21 Another special CMOS part that can be very useful is the 4007, which has

TABLE 5-1 COMPARISON OF VARIOUS POPULAR LOGIC FAMILIES.

	NMOS*	4000	74HC	7400	74LS
Supply voltage	4-6 V	3-18 V	3-6 V	4.75-5.25 V	4.75-5.25 V
Supply current	—	150 μA	160 μA	4.0 mA	0.8 mA
V_{IH}	2.0 V	3.5 V	3.85 V	2.0 V	2.0 V
V_{OH}	2.4 V	4.95 V	4.95 V	2.4 V	2.4 V
V_{IL}	0.8 V	1.5 V	0.9 V	0.8 V	0.8 V
V_{OL}	0.4 V	0.05 V	0.05 V	0.4 V	0.4 V
I_{IH}	10 μA	0.3 μA	1 μA	40 μA	20 μA
I_{IL}	-10 μA	-0.3 μA	-1 μA	-1.6 mA	-0.4 mA
I_{OH}	-400 μA	-160 μA	-4 mA	-400 μA	-400 μA
I_{OL}	2 mA	440 μA	4 mA	16 mA	4 mA
t_{PHL}	—	250 ns	15 ns	15 ns	15 ns
t_{PLH}	—	250 ns	15 ns	22 ns	15 ns
Fanout	—	50	10 (LS) 50 (CMOS)	10 (TTL)	10 (LS)
Noise immunity	0.4 V	1.45 V	(H) 1.25 V (L) 0.8 V	0.4 V	0.4 V

Note: No NMOS logic family actually exists. These data are averages obtained from LSI data sheets for comparison only.

two unconnected CMOS complementary pairs plus an INVERTER. The complementary pairs can be connected to form a variety of functions.

22 Table 5-1 is a comparison of the MOS logic families with the TTL families.

KEY TERMS

active pullup	LSI	*p*-channel
analog switch	load transistor	PMOS
bilateral switch	MOSFET	silicon gate MOSFET
buffer	MSI	source
complementary pair	*n*-channel	SSI
CMOS	NMOS	threshold voltage
drain	open-drain logic	transmission gate
gate		

QUESTIONS AND PROBLEMS

1   Draw the schematic representation for an enhancement-mode n-channel MOSFET and label the source, drain, gate, and substrate.

2   What significance does the word *enhancement* have in the description of the MOSFET?

3   Does a positive V_{gs} turn an n-channel MOSFET on or off?

4   Are NMOS transistors faster or slower than PMOS transistors?

5   Draw an NMOS INVERTER circuit. Label the driver transistor, load transistor (or active pullup), input, and output.

6   Draw a 3-input NMOS NAND circuit. Label the inputs and output.

7   Draw a 3-input NMOS NOR circuit and label the inputs and output.

8   How many resistors does it take to make a 4-input NMOS NAND gate?

9   Draw a CMOS INVERTER and label the input and output. Which is the p-channel transistor, and which is the n-channel transistor? Draw the logic symbol for this circuit.

10   What is the supply voltage range for standard CMOS logic?

11   Is tight regulation necessary for CMOS power supplies? If so, why? If not, why not?

12   Draw a 3-input buffered CMOS NAND gate. Label the inputs and output.

13   Draw a 2-input buffered CMOS NOR gate and label the inputs and output.

14   Connect the output of the circuit in problem 12 to one of the inputs of the circuit in problem 13. During steady-state conditions, how much current flows out of the first circuit into the input of the second circuit?

15   In the circuit constructed in problem 14, what is the minimum number of additional inputs that could be connected parallel to the input of the circuit from problem 13?

16   At what point in operation does current flow in a CMOS circuit? At what input voltage is this current at its maximum?

17   Draw a typical input protection circuit for CMOS and describe its operation.

18   In a circuit with V_{DD} at 15 V, $V_{OL} = 0.05$ V, and $V_{OH} = 14.95$ V; $V_{IL} = 2.5$ V and $V_{IH} = 12.5$ V (all are "worst case" values). What is the noise immunity of CMOS at 15 V V_{DD}?

19   Which has the greater effect on dynamic power dissipation (P_{dyn}), voltage or frequency?

20 Could you use a standard 4000 series CMOS gate in a circuit designed to operate at 29 MHz? Show calculations to support your answer.

21 How long could the output of a CMOS gate withstand a direct short circuit to ground?

22 Could you substitute a 74HC04 for a 74LS04 in a circuit operating at 27 MHz?

23 For a quad 2-input NAND gate, how much more quiescent supply current is required for a 74HC than for a 4000 series logic gate?

24 Compare the input voltages of the 74HC to the 4000 CMOS logic gate.

Note: You will need a 4000 series CMOS data book for problems 25 through 28.

25 Find the part number of a quad Exclusive OR gate.

26 Find the part number and draw the pin diagram for an 8-input NOR/OR.

27 What two parts (numbers) could you connect to form a 7-input NAND gate?

28 To what pin would you connect V_{DD} on a 4049?

29 Draw the schematic for a transmission gate. How would it be connected to operate as a bilateral switch?

30 Connect the 4007 to operate as a 3-input NOR gate.

6

INTRODUCTION TO SEQUENTIAL LOGIC

The only type of logic system investigated in the text thus far has been combinational logic. But combinational logic has severe limitations because it cannot be readily used to store, or remember, any type of digital information. *Sequential digital logic* offers the ability to remember data and then use these data to perform a useful task.

SEQUENTIAL LOGIC COMPARED WITH COMBINATIONAL LOGIC

The difference between combinational logic and sequential logic is inherent in their individual characteristics. Before proceeding any further, let's look closely at each logical form.

COMBINATIONAL LOGIC

Combinational logic, as demonstrated in Chapters 2 and 3, is capable only of combining present input signals to produce an output. For example, the output of the decoder illustrated in Fig. 6-1 changes in response to the input signals only at the time that they change. In all combinational logic circuitry, the outputs change within a very short period of time after the inputs change.

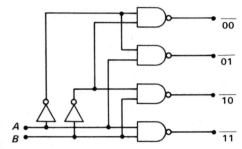

FIGURE 6-1. *Combinational logic circuit: the 2- to 4-line decoder.*

SEQUENTIAL LOGIC

Sequential logic differs from combinational logic because its outputs depend on some *past event*. A digital clock is an example of a sequential system because the present time depends on the past time. The clock advances to the correct time every minute because it has the ability to remember the time of the preceding minute.

The Basic Sequential Logic System

Fig. 6-2 illustrates the basic block diagram of a sequential logic system. Notice that it contains a memory element that is used to remember past events. It also contains a combinational logic circuit that can generate a new output from the

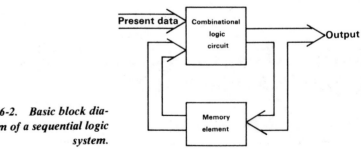

FIGURE 6-2. *Basic block diagram of a sequential logic system.*

past event and present data. For the digital clock, the memory contains the time and the combinational logic circuit adds a "one" to the time once per minute. The present input signal would be a one-minute pulse from an external timing circuit.

THE DIFFERENCE BETWEEN COMBINATIONAL AND SEQUENTIAL LOGIC

Unlike sequential logic circuits, combinational logic circuits analyze only current events. The sequential logic circuit, on the other hand, uses past information along with current information to generate an output. Thus the main difference between these types of logic is *memory* or storage.

This chapter will focus on the most basic type of digital memory: the *flip-flop*. A flip-flop is a device that can store one bit of information. It is extremely important that the operation of the flip-flop be fully understood, because, if it is not, the remainder of this textbook will be unclear.

ASYNCHRONOUS LATCHES

What is an asynchronous latch or, as it is sometimes called, a cross-coupled latch? An *asynchronous latch* is a device that can store a logic 1 or a logic 0 by "latching" or grabbing onto the input information. Because the outputs change at the time that the inputs dictate a change, there can be no synchronizing pulse to time the change. Thus we call the latch "asynchronous," or not synchronized. Later in this chapter, we will discuss *synchronous* flip-flops, which are constructed from asynchronous latches.

THE NAND GATE $\overline{S}\,\overline{R}$ ASYNCHRONOUS LATCH

The NAND gate cross-coupled $\overline{S}\,\overline{R}$ latch is illustrated in Fig. 6-3, along with its logic symbol. The term *cross-coupled* comes from the method of wiring the device. This latch has two active-low inputs: $\overline{R}$ or $\overline{\text{RESET}}$, and $\overline{S}$ or $\overline{\text{SET}}$. A logic 0 on the $\overline{R}$ input will reset the Q output to a 0, and a logic 0 on the $\overline{S}$ input will set the Q output to a 1. Since the logic 0 level activates either the *set Q* or *reset Q* function, these inputs are called *active-low inputs.*

In addition to the $\overline{S}$ and $\overline{R}$ inputs, the NAND $\overline{S}\,\overline{R}$ latch contains two complementary outputs: Q and $\overline{Q}$. Q is an active-high output, and $\overline{Q}$ is an active-low output. If $\overline{R}$ is connected to a logic 0 and $\overline{S}$ is connected to a logic 1, the flip-flop is commanded to *reset Q*, but not to *set Q*. Since Q is the active-high output, it will become a logic 0 and the active-low output, $\overline{Q}$, will become a logic 1—its inactive state.

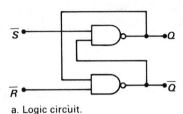

FIGURE 6-3. Logic circuit and symbol of the cross-coupled NAND gate version of the asynchronous $\overline{SR}$ latch.

a. Logic circuit.   b. Logic symbol.

The Function Table for the $\overline{SR}$ NAND Gate Latch

Table 6-1 is the function table for the NAND $\overline{SR}$ latch. It shows all the possible input conditions for $\overline{R}$ and $\overline{S}$, and it indicates the output conditions. Notice that the flip-flop can be set or reset and it can also remember the prior state of Q_n. A 0 on both inputs is not allowed because the outputs would assume the invalid condition of a 1 on both the Q and $\overline{Q}$ outputs.

TABLE 6-1 THE FUNCTION TABLE OF THE NAND $\overline{SR}$ LATCH.

Inputs		Outputs		
$\overline{S}$	$\overline{R}$	Q	$\overline{Q}$	Function
0	0	1	1	Not allowed
0	1	1	0	Set Q
1	0	0	1	Reset Q
1	1	Q_n	$\overline{Q}_n$	Memory

Note: Q_n is the prior state of Q, and $\overline{Q}_n$ is the prior state of $\overline{Q}$.

The Operation of the $\overline{SR}$ NAND Gate Latch

Refer to Fig. 6-4a for the diagram corresponding to the following discussion. If a logic 0 is placed on the $\overline{S}$ input and a logic 1 is placed on the $\overline{R}$ input, the function table indicates that Q will set. Why does Q set? Q sets to a logic 1

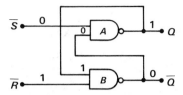

a. Set Q command applied to its inputs.

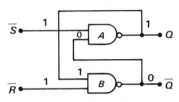

b. After a 1 has been applied to $\overline{S}$. Notice that the outputs do not change following the application of the logic 1 on $\overline{S}$.

FIGURE 6-4. The $\overline{SR}$ latch.

because a 0 on any input of a NAND gate causes the output to become a logic 1. The output of gate A then becomes a logic 1 due to the 0 on $\overline{S}$. Notice that this output has been connected to an input of NAND gate B. The 1 at the output of gate A and the 1 on the $\overline{R}$ input are now connected to NAND gate B. Since 1 NANDed with 1 produces a logic 0, the $\overline{Q}$ output becomes a zero.

Fig. 6-4b shows the effect of placing a logic 1 on both inputs. (Note that this is the memory state.) If the $\overline{S}$ input is raised to a logic 1 from a logic 0, the outputs do not change. NAND gate A now has a logic 0 on one input and a logic 1 on the other input, which causes the output to remain a logic 1. Since the output of NAND gate A has not changed, the inputs to NAND gate B have not changed, and output $\overline{Q}$ will therefore not change.

Fig. 6-5a illustrates the changes that will occur for a series of different inputs to this type of latch. The outputs change exactly as stated in the function table except during time t_d.

To see why this happens, let's stop time at point X. Fig. 6-5b illustrates this change at the time that $\overline{R}$ becomes a logic 0. At this time, the output of NAND gate B becomes a logic 1 after about 10 ns of the propagation delay time. This is the exact point at which time has been stopped. Both of the outputs are now at a logic 1 level. After an additional 10 ns, the output of gate A will become a logic 0. This erroneous output condition, which appears for approximately 10 ns, can in many cases be ignored. This time t_d is often called the *settling time* for the latch because this is the amount of time required for the outputs to settle to their final state.

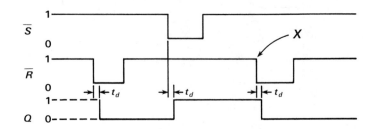

a. Timing diagram.

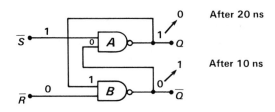

After 20 ns

After 10 ns

b. View of the propagation delay when stopped in the middle of a RESET.

FIGURE 6-5.   The $\overline{S}\overline{R}$ latch.

An Integrated NAND Gate $\overline{SR}$ Latch

Fig. 6-6 illustrates the 74279 quad $\overline{S}\,\overline{R}$ latch integrated circuit along with its function table. This device has two sections that contain more than one $\overline{S}$ input. In these sections, both $\overline{S}$ inputs must be high for the latch to recognize a 1 on the internal $\overline{S}$ input. Both must be high because they are internally ANDed.

QUAD $\overline{S}$-$\overline{R}$ LATCHES

279   DIODE-CLAMPED INPUTS
TOTEM-POLE OUTPUTS

H = high level
L = low level
Q_0 = the level of Q before the indicated input conditions were established.
*This output level is pseudo stable; that is, it may not persist when the $\overline{S}$ and $\overline{R}$ inputs return to their inactive (high) level.
†For latches with double $\overline{S}$ inputs:
    H = both $\overline{S}$ inputs high
    L = one or both $\overline{S}$ inputs low

FUNCTION TABLE

INPUTS		OUTPUT
$\overline{S}^{\dagger}$	$\overline{R}$	Q
H	H	Q_0
L	H	H
H	L	L
L	L	H*

FIGURE 6-6.   The 74279 quad $\overline{SR}$ latch and its function table. (Courtesy of Texas Instruments Incorporated.)

THE NOR GATE SR ASYNCHRONOUS LATCH

The NOR gate SR latch is illustrated in Fig. 6-7. Notice that its inputs are active-high, which means that a logic 1 must be placed on S and a logic 0 on R to set Q to a logic 1.

Table 6-2 illustrates the function table for the NOR gate SR latch. The state that is not allowed is the 1–1 input condition, because it causes both outputs Q and $\overline{Q}$ to become logic 0s. This is exactly the opposite of the invalid input condition of the NAND gate $\overline{SR}$ latch.

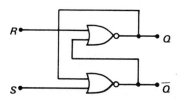

a. Logic circuit.

b. Logic symbol.

FIGURE 6-7.   Cross-coupled NOR gate version of the asynchronous latch.

TABLE 6-2   THE FUNCTION TABLE OF THE NOR GATE SR LATCH.

Inputs		Outputs		
S	R	Q	$\overline{Q}$	Function
0	0	Q_n	$\overline{Q}_n$	Memory
0	1	0	1	Reset Q
1	0	1	0	Set Q
1	1	0	0	Not allowed

Note: Q_n is the prior state of Q and $\overline{Q}_n$ is the prior state of $\overline{Q}$.

a. Mechanical switch.

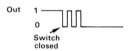

b. Output voltage waveform.

FIGURE 6-8. Effect of bouncing on the output voltage waveform.

CONTACT BOUNCE ELIMINATION

A very common application for the $\overline{S}\,\overline{R}$ latch is *contact bounce* elimination. Whenever a mechanical switch is thrown from one position to another, the contacts physically bounce as the mass of the movable portion of the switch strikes the contact. This contact bounce occurs in the same way that a ball bounces off the floor when it is dropped. When the movable portion of the switch returns to the contact, due to a spring located inside the switch, it might again bounce as when the ball returns to the floor after the first bounce. Fig. 6-8 illustrates a mechanical switch and the voltage waveform found at its output.

Why the big fuss about contact bounce? If bounces were not eliminated from the keyboard on a calculator, for example, each keystroke would produce multiple entries on the display. One push of a 3 key might produce the entry 3333. This would make the calculator totally useless.

The Operation of a Contact Bounce Eliminator

Fig. 6-9 depicts a NAND gate $\overline{S}\overline{R}$ latch connected to an SPDT (single-pole double-throw) toggle switch. This circuit will remove the bounce from the output of the switch via the memory state of the latch. Suppose the switch is moved off the $\overline{A}$ position toward the A position. As soon as the switch leaves the

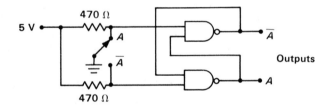

a. Logic circuit.

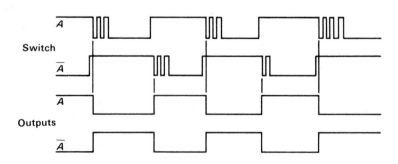

Outputs

b. Voltage waveforms.

FIGURE 6-9. SPDT switch connected to a $\overline{S}\overline{R}$ latch, forming a contact bounce eliminator.

$\overline{A}$ position, both inputs to the latch become logic 1s, causing the latch to remember. When the switch finally makes it to the A position, it grounds the $\overline{R}$ input of the latch, causing the output to become a logic 0. Now the switch will bounce off of the A contact. It might at first appear that the output will change, but in fact, when it is in between $\overline{A}$ and A, both inputs to the latch are high, causing it to remember. This effectively prevents the contact bounces from being passed through the latch.

GATED LATCHES

The *gated latch* is a synchronous latch constructed from either the NAND $\overline{SR}$ latch or the NOR SR latch. The gated latch is used in a system that requires the inputs to be sampled at a particular time. With the asynchronous latch, whatever is presented to the inputs instantly changes the outputs. With the gated latch, the outputs will change only at the time of the GATE pulse.

THE NAND GATE VERSION OF THE GATED LATCH

Fig. 6-10 illustrates the NAND gate version of the gated latch and its logic symbol. Notice that this circuit is basically the $\overline{SR}$ latch with some gating attached to its inputs. This gating allows the data to be connected to the latch only at the time of a gating pulse.

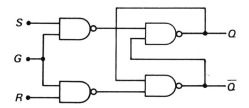

a. Logic circuit.

b. Logic symbol.

FIGURE 6-10. $\overline{\overline{SR}}$ *gated latch.*

Operation of the NAND Gate Version of the Gated Latch

This type of latch is said to have a *level-sensitive* GATE input that passes the S and the R inputs through to the latch. That is, when the GATE is activated, the inputs will affect the latch. Because the GATE of this type of latch is an active-high input, a logic 1 would turn this device on. If a logic 0 is placed on the GATE, the latch will remember the most recent set of inputs. The function table for the gated latch (Table 6-3) shows only the S and R inputs, and these functions only apply when the GATE is active.

TABLE 6-3   THE FUNCTION TABLE OF A GATED NAND GATE SR LATCH WITH THE GATE INPUT ACTIVE.

Inputs		Outputs		
S	R	Q	$\overline{Q}$	Function
0	0	Q_n	$\overline{Q_n}$	Memory
0	1	0	1	Reset Q
1	0	1	0	Set Q
1	1	1	1	Not allowed

Note: If the GATE is inactive, all S and R input conditions cause the flip-flop to remember.

Operation of the GATE Input

Fig. 6-11 shows the input circuit of the gated latch and a timing diagram that illustrates the action of the GATE input. The first thing that should be noticed

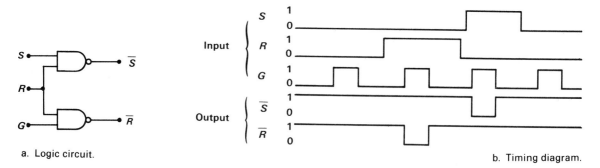

a. Logic circuit.

b. Timing diagram.

FIGURE 6-11.   The $\overline{SR}$ latch. Notice that the only time the outputs can change is during the gating pulse.

from the timing diagram is that inputs are blocked, or inhibited, whenever the GATE input is a logic 0. When the GATE input is a logic 1, the S and R inputs are passed through the NAND gate and are also inverted. This type of gating is called level-sensitive gating because as long as the GATE is active, the outputs will change in response to the S and R inputs.

Operation With the $\overline{SR}$ Latch in Place

Fig. 6-12 shows the operation of the gated latch, including a series of CLOCK pulses that are applied to the GATE input. It is important to note that the outputs of the gated latch will change only *during* the logic 1 level of each CLOCK pulse.

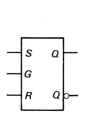

a. Logic circuit.

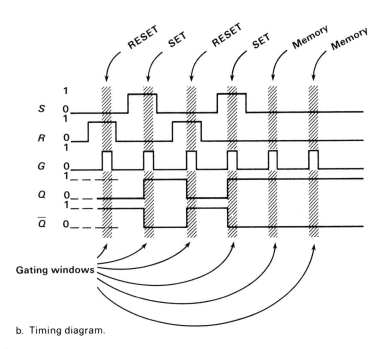

b. Timing diagram.

FIGURE 6-12. The SR gated latch.

THE NOR GATE $\overline{\overline{SR}}$ GATED LATCH

Fig. 6-13 illustrates the NOR gate version of the gated latch. It differs from the NAND gate version of this latch because of its active-low $\overline{S}$, $\overline{R}$, and $\overline{G}$ inputs. Pay particular attention to the fact that the outputs change only during the active (logic 0) level of the GATE input.

Table 6-4 is the function table for the NOR gate gated latch. The main difference between this function table and Table 6-3 is the invalid state. The

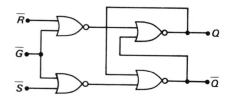

a. Logic circuit.

b. Logic symbol.

FIGURE 6-13. NOR gate version of the
$\overline{SR}$ gated latch.

TABLE 6-4 THE FUNCTION TABLE FOR THE NOR GATE VERSION
OF THE GATED $\overline{SR}$ LATCH.

Inputs		Outputs		
$\overline{S}$	$\overline{R}$	Q	$\overline{Q}$	Function
0	0	0	0	Not allowed
0	1	1	0	Set Q
1	0	0	1	Reset Q
1	1	Q_n	$\overline{Q}_n$	Memory

invalid state in the NOR gate latch occurs when a 0 is placed on both the $\overline{S}$ and $\overline{R}$ inputs during a GATE pulse, while the 1–1 state is invalid for the NAND gate latch.

THE D-TYPE GATED LATCH

Probably one of the most useful forms of gated latch is the D-type latch, or Data latch. The D-type latch is very useful because data stored in a latch often appear on one wire. This latch, shown in Fig. 6-14, has only one signal input besides the GATE. (You will probably recognize the NAND gate version of the SR latch in this illustration.) In fact, the D-type gated latch is nothing more than an SR latch that has an INVERTER connected from the S to the R input. D is connected directly to the S input, and R is connected to $\overline{D}$.

152 **Digital Electronics**

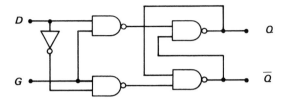

a. Logic circuit.

b. Logic symbol.

FIGURE 6-14.   The D-type gated latch.

Operation of the D-Type Gated Latch

The operation of this device is very simple. If the D input is a logic 1 at the time of the GATE pulse, the Q output becomes a logic 1, and if the D input is a logic 0 at the time of the GATE pulse, the Q output becomes a logic 0. In other words, the D input is *transferred* to the Q output on the GATE pulse.

An Example of an Integrated D-Type Gated Latch

Fig. 6-15 pictures the 74LS373 octal transparent latch. (Sometimes a D-type gated latch is called a "transparent" latch because it looks transparent when the GATE is active.) The 74LS373 is illustrated because it not only includes gated

OCTAL D-TYPE LATCHES

373   3-STATE OUTPUTS
COMMON OUTPUT CONTROL
COMMON ENABLE

*Output control is $\overline{O\,E}$.

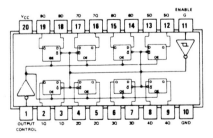

FIGURE 6-15.   The 74LS373 D-type gated latch (Courtesy of Texas Instruments Incorporated.)

latches but also contains Tri-state® output connections. This could be called a state-of-the-art gated latch. Notice that each control input, $\overline{OE}$ and G, are buffered so that they represent one TTL unit load. This is a common practice in most integrated circuits.

AN APPLICATION FOR THE GATED LATCH

Fig. 6-16 shows a simple application for the gated D-type latch. In this application, the latch is used to hold the information from a "mark sense reader." This

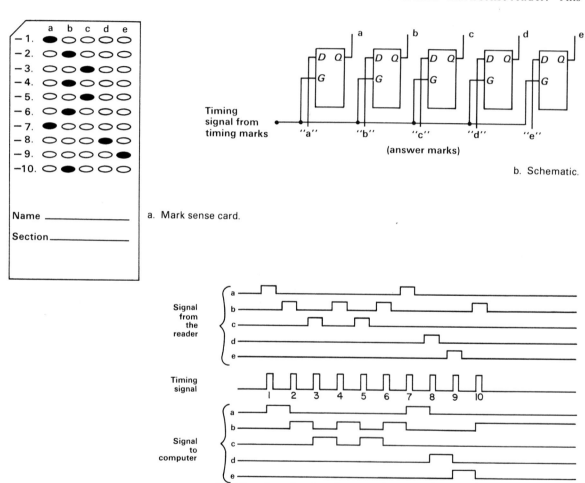

a. Mark sense card.

b. Schematic.

c. Timing diagram.

FIGURE 6-16.   Mark sense reader.

is the same type of reader that an instructor uses to grade multiple-choice tests.

As the multiple-choice answer sheet is passed through the machine, the timing marks on the edge of the page generate the GATE pulses. The GATE pulses allow the marks to be accurately read from the answer sheet. They also provide a signal that allows the reader to accumulate test scores. (The circuit to accumulate scores does not appear in this illustration.) If the timing marks are not present, the machine will not know when to read the answers from the test.

MASTER-SLAVE FLIP-FLOPS

The *master-slave flip-flop* is similar to the gated *SR* latch except with respect to the CLOCK (CLK) or GATE input. While the gated latch uses a level-sensitive GATE or CLOCK input, the master-slave flip-flop uses the *changing state* of the clocking level to modify the output.

THE RACE IN A GATED LATCH

The master-slave flip-flop is more desirable than a gated latch because of a problem called a *race*, which can occur in the gated latch if its output is inadvertently connected to an input. A race condition is illustrated in Fig. 6-17. Whenever the GATE input becomes active, this latch will *oscillate*, or race, to its final state, but it never reaches that final state, as illustrated in the timing diagram of Fig. 6-18. The period of this oscillation is determined by the propagation delay times of the NAND gates.

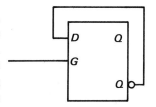

FIGURE 6-17. D-type gated latch wired to race (oscillate).

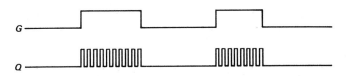

FIGURE 6-18. Waveform present at the output of the circuit in Fig. 6-17.

THE MASTER-SLAVE SR FLIP-FLOP

The master-slave flip-flop is constructed from two gated *SR* latches and an INVERTER, as shown in Fig. 6-19. Table 6-5 illustrates the operation of the *SR* latch at the negative edge ($1-0$ change) of the CLOCK. At all other times, the flip-flop will remember the prior input conditions. The gated latch near the inputs is called the *master,* and the gated latch near the outputs is called the *slave.*

The master will accept and hold data during the time that the CLOCK input, or *T,* is a logic 1. During the time the master is accepting data, the slave will

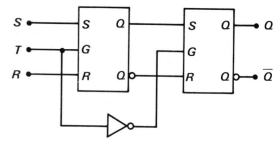

a. Logic circuit.

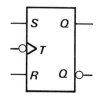

b. Logic symbol.

FIGURE 6-19. *SR master-slave flip-flop.*

TABLE 6-5 THE FUNCTION TABLE OF THE MASTER-SLAVE
SR FLIP-FLOP.

Inputs		Outputs		
S	R	Q	$\overline{Q}$	Function
0	0	Q_n	$\overline{Q_n}$	Memory
0	1	0	1	Reset Q
1	0	1	0	Set Q
1	1	X	X	Invalid

Note: X indicates an unpredictable output condition.

remember the prior state of the master, because its GATE input is at a logic 0, or inactive. When *T* changes from a logic 1 to a logic 0, the contents of the master will be transferred to the slave. This is illustrated in Table 6-5. Note that these functions operate only at the negative edge of the input CLOCK *T*.

This discussion should clearly show that the outputs of the master-slave flip-flop can change only on the 1–0 transition of the CLOCK, or *T*, input. This is often called *negative edge–triggering* because the outputs will only change on the negative edge of the input CLOCK waveform. (Later, however, we will learn that the master-slave flip-flop really isn't an edge-triggered flip-flop.)

Timing for the Master-Slave SR Flip-Flop

Fig. 6-20a illustrates the timing diagram of a negative edge–triggered master-slave SR flip-flop. Notice how the numerous changes on the inputs are transferred to the output only on the negative edge of the CLOCK. For a comparison, Fig. 6-20b shows the outputs that would be present on an SR gated latch. Notice how the timing diagrams differ because of the type of CLOCK input employed by each type of flip-flop.

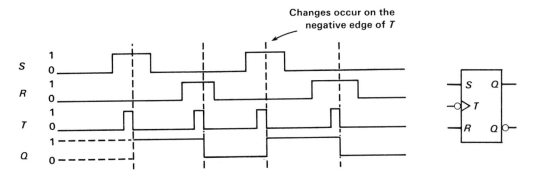

a. For the SR master-slave flip-flop.

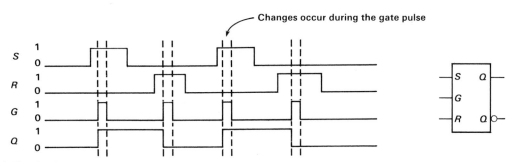

b. For the SR gated latch.

FIGURE 6-20. *Logic symbols and timing diagrams.*

THE MASTER-SLAVE D-TYPE FLIP-FLOP

The master-slave D-type flip-flop is constructed from the master-slave SR flip-flop, as illustrated in Fig. 6-21. The D input is connected to the S input directly and to the R input through an INVERTER. The only difference between the gated and the master-slave D-type flip-flop is the point in the timing diagram at which the output will change.

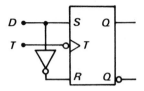

a. Logic circuit.

b. Logic symbol.

FIGURE 6-21.   D-type master-slave flip-flop.

Timing for the Master-Slave D-Type Flip-Flop

Fig. 6-22 shows the timing diagram for the master-slave D-type flip-flop. Notice that the outputs can change only on the negative edge of the input CLOCK waveform.

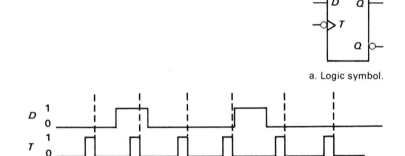

a. Logic symbol.

b. Timing diagram.

FIGURE 6-22.   D-type master-slave flip-flop.

THE MASTER-SLAVE JK FLIP-FLOP

One problem of the master-slave SR flip-flop is that its outputs become invalid when a logic 1 is applied simultaneously to both the S and R inputs. If the negative edge of the CLOCK occurs with the inputs at this level, the outputs will go to an undefined state, which causes problems. The *master-slave JK flip-flop* circumvents this difficulty by defining the 1–1 input condition.

Fig. 6-23 illustrates the master-slave SR flip-flop wired to function as a master-slave JK flip-flop. Notice the additional gating that is added to the inputs of the SR flip-flop. This modification feeds the Q and $\overline{Q}$ outputs back to the inputs. The function table for the master-slave JK flip-flop appears in Table 6-6. The main difference between this table and the function table for the master-slave SR flip-flop (Table 6-5) is the 1–1 input condition. This input condition is invalid in the master-slave SR flip-flop, but it causes Q to be complemented, or *toggled,* in the master-slave JK flip-flop. Also notice that J functions in the same manner as S and K functions in the same manner as R except for the 1–1 input condition.

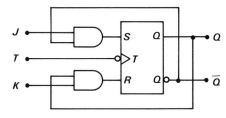

a. Logic circuit.

b. Logic symbol.

*FIGURE 6-23. Master-slave JK
flip-flop.*

TABLE 6-6 THE FUNCTION TABLE FOR THE MASTER-SLAVE
JK FLIP-FLOP.

Inputs		Outputs		
J	K	Q	$\overline{Q}$	Function
0	0	Q_n	$\overline{Q}_n$	Memory
0	1	0	1	Reset Q
1	0	1	0	Set Q
1	1	$\overline{Q}_n$	Q_n	Toggle Q

Note: These functions apply only at the negative edge of the CLOCK input; at all
other times this flip-flop remembers the last input state.

Fig. 6-24a shows the *JK* flip-flop when the output is a logic 0. Notice that
both the *J* and *K* inputs are at their logic 1 levels, causing the output to toggle on
the next CLOCK pulse. Why does this cause the output to toggle? The output of
AND gate *A* is connected to the *S* input of the *SR* flip-flop, and its inputs are
connected to *J* and also to $\overline{Q}$. At the present time, the *S* input is a logic 1 because
the $\overline{Q}$ signal and the *J* signal are both logic 1s. Notice that the output of AND
gate *B* is a logic 0 because of the 0 on output *Q*. On the next CLOCK pulse, the
Q output becomes a logic 1 because of the 1 on *S* and the 0 on *R*. We say that the
output has been "complemented" because 1 is the complement of 0.

Fig. 6-24b shows the same flip-flop with its *Q* output at a logic 1 level. Notice

Introduction to Sequential Logic **159**

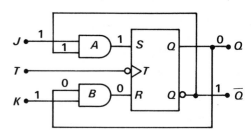

a. Complementing to a logic 1 on the next CLOCK pulse.

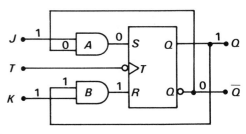

FIGURE 6-24. Master-slave JK flip-flop.

b. Complementing to a logic 0 on the next CLOCK pulse.

a. Negative.

b. Positive.

FIGURE 6-25. The logic symbols for the edge-triggered JK flip-flops.

that the S input to the SR flip-flop is a logic 0 and the R input is a logic 1. On the next CLOCK pulse the Q output will be cleared, or complemented to 0.

Most master-slave flip-flops are negative edge–triggered. Fig. 6-25 depicts the schematic symbols for both negative and positive edge–triggered flip-flops. Fig. 6-26 shows how JK flip-flops, with their inputs at a logic 1 level, will function if one is positive edge–triggered (FFA) and one is negative edge–triggered (FFB). Also notice that both output waveforms are exactly one-half the CLOCK frequency. A JK flip-flop, wired to toggle, will always divide the input frequency by two. This will prove to be very useful in Chapter 7.

THE PRESET AND CLEAR INPUTS

Fig. 6-27 shows the logic symbol of the 7476 dual JK flip-flops with PRESET and CLEAR, along with their function table. The PRESET ($\overline{PR}$) and CLEAR ($\overline{CLR}$) inputs are not applied by the CLOCK as are the J and K inputs. PRESET and CLEAR are *asynchronous inputs* that take effect whenever activated. Notice that Q will become a logic 0 whenever the $\overline{CLR}$ input is placed at a logic 0, and Q will become a logic 1 whenever $\overline{PR}$ is placed at a logic 0.

These inputs are used in many cases to clear or set a flip-flop at the time that power is applied to the flip-flop. A simple circuit that causes a JK flip-flop to clear on the application of power is shown in Fig. 6-28. When power is applied

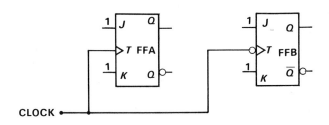

a. Logic circuit.

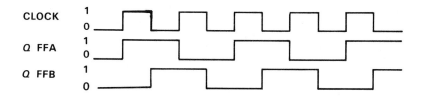

b. Timing diagram.

FIGURE 6-26.
Relationship between the
positive and negative edge-
triggered JK flip-flops.

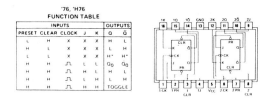

a. Function table.　　　b. Logic symbol.

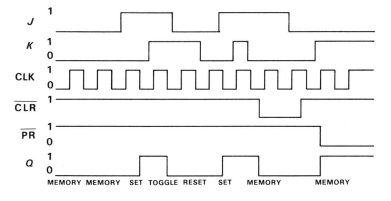

c. Timing diagram.

FIGURE 6-27.   The 7476 dual JK
flip-flops. (Courtesy of Texas
Instruments Incorporated.)

Introduction to Sequential Logic

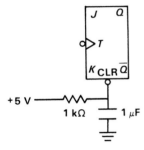

FIGURE 6-28. *Circuit used to clear the JK flip-flop on the application of DC power.*

to this circuit, the capacitor is discharged, placing 0 V on the $\overline{\text{CLR}}$ input, which clears the flip-flop. The capacitor does not remain discharged for very long because of the resistor in series with it. When current flows through this series circuit, it will charge the capacitor to 5 V in a short time. When the capacitor rises above 2.4 V, the $\overline{\text{CLR}}$ input will no longer have an effect on the operation of the flip-flop. This allows the CLOCK and JK inputs of the flip-flop to dictate its operation from this point forward. The principle of *power-on-clearing* (POC) is very important in many digital systems.

IS THE MASTER-SLAVE FLIP-FLOP PERFECT?

The master-slave flip-flop is perfect for many applications, but it does contain a flaw. For most applications, the output will change in response to the inputs—except when the inputs change at the time that the master is receptive to data (i.e., when T is a logic 1). This may cause unreliable output conditions in some applications.

Fig. 6-29 illustrates a master-slave SR flip-flop and its input signals. Notice that some noise has been introduced to waveform S. This causes the output to become set when it should have been cleared, because the inputs are connected to a gated latch that will respond to any change—even noise. In a noisy environment, the master-slave flip-flop might not function properly for this reason. The next section of the text is devoted to the edge-triggered flip-flop, which cures this problem.

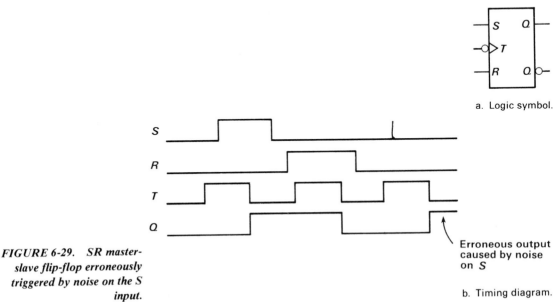

a. Logic symbol.

Erroneous output caused by noise on S

b. Timing diagram.

FIGURE 6-29. *SR master-slave flip-flop erroneously triggered by noise on the S input.*

EDGE-TRIGGERED FLIP-FLOPS

Unlike the master-slave flip-flop, which has a minor flaw, the edge-triggered flip-flop is virtually flawless. Its outputs will change only in response to the input data present at the time of the positive or negative edge of the CLOCK. This makes the edge-triggered flip-flop nearly immune to input noise.

THE BASIC EDGE-TRIGGERED JK FLIP-FLOP

Fig. 6-30 shows the schematic diagram of the negative edge–triggered JK flip-flop. Notice first that none of the previous types of flip-flop circuitry are present. This device contains flip-flops that are similar to the asynchronous latch, but not exactly the same as it. The negative edge–triggered JK flip-flop circuit functions by using the propagation delay time of the gates attached to the J and K inputs.

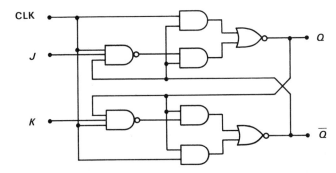

FIGURE 6-30. *Negative edge-triggered JK flip-flop.*

Fig. 6-31a shows the steady-state condition of this flip-flop with the CLOCK at a logic 1 level, and Fig. 6-31b shows the steady state condition with the CLOCK at a logic 0 level. Neither steady-state condition will cause a change in the outputs. The change will occur only at the time that the CLOCK changes from a logic 1 to a logic 0. At the negative transition of the CLOCK, the outputs of the NAND gates are still active even though the CLOCK is a logic 0. This occurs for only 10 ns and is the only place that noise could affect the outputs. This transition is illustrated in Fig. 6-32a. In this diagram the outputs of the NAND gates have not changed but will cause the output state of the latch to change in about 10 ns as illustrated in Fig. 6-32b.

Because the outputs of this device will change only at the negative edge of the clock, it is called a negative edge–triggered flip-flop. This type of flip-flop is found in most modern applications. It is the most reliable type because of its

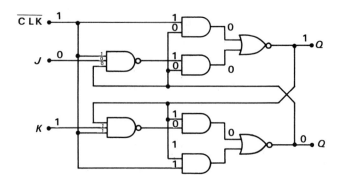

a. With a steady-state logic 1 on the CLK input.

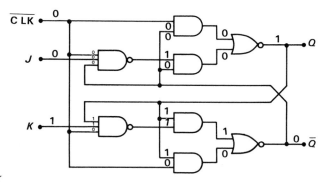

FIGURE 6-31. Negative edge-triggered JK flip-flop.

b. With a steady-state logic 0 on the CLK input.

immunity to noise. In fact, most of the new integrated circuits use this type of triggering in place of the master-slave flip-flop.

The PRESET and CLEAR Inputs

When PRESET and CLEAR inputs are available on the edge-triggered flip-flop, they are connected as shown in Fig. 6-33. If the $\overline{CLR}$ input is grounded, it will instantly cause the Q output to become a logic 0 and the $\overline{Q}$ output to become a logic 1. The PRESET input ($\overline{PR}$) works in the same way, except that Q will be set and $\overline{Q}$ will be cleared.

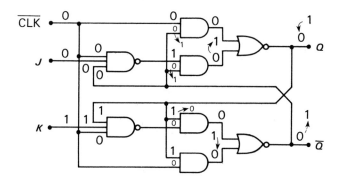

a. At the time of the negative edge of the CLOCK.

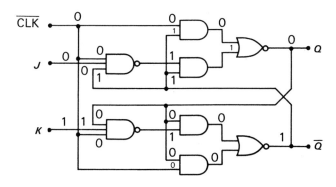

b. After the negative edge of the CLOCK.

FIGURE 6-32. *Negative edge-triggered JK flip-flop.*

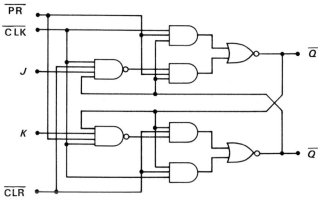

FIGURE 6-33. *Negative edge-triggered JK flip-flop shown with the PR and CLR inputs.*

MONOSTABLE MULTIVIBRATORS

The *monostable multivibrator* is a storage element that will retain data for only a short time. It is sometimes called a *single-shot* or a *one-shot* because of the way that it operates. When triggered, the output of a single-shot will become active for a predetermined amount of time. This time normally depends upon timing components connected to the device.

THE 74121 MONOSTABLE MULTIVIBRATOR

The 74121 monostable multivibrator is shown in Fig. 6-34. This device has three TRIGGER inputs in order to make it as flexible as possible in its application. Notice from the function table that it is possible to trigger this device from a positive- or a negative-going TRIGGER. If A_1 and B are held high, then A_2 is a negative edge-triggered input; if A_1 is grounded, B becomes a positive edge-triggered input.

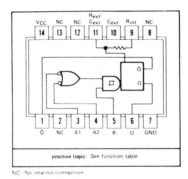

FIGURE 6-34.   The 74121 monostable multivibrator. (Courtesy of Texas Instruments Incorporated.)

a. Pinout

FUNCTION TABLE

INPUTS			OUTPUTS	
A1	A2	B	Q	Q̄
L	X	H	L	H
X	L	H	L	H
X	X	L	L	H
H	H	X	L	H
H	↓	H	⊓	⊔
↓	H	H	⊓	⊔
↓	↓	H	⊓	⊔
L	X	↑	⊓	⊔
X	L	↑	⊓	⊔

b. Function table.

Operation of the 74121

When a TRIGGER input activates the 74121, the Q output will assume a logic 1 level, as illustrated in the timing diagram of Fig. 6-35. Q remains a logic 1 until

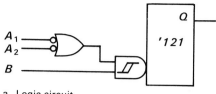

a. Logic circuit.

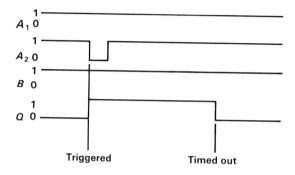

Triggered   Timed out

b. Timing diagram.

FIGURE 6-35. Timing diagram of the 74121 monostable multivibrator showing the effect of the TRIGGER inputs.

the 74121 *times out*. The 74121 times out in the amount of time determined by the timing components R_{ext} and C_{ext}. Typically, this device is capable of developing output pulse widths of from 40 ns to 28 s, depending on the external timing components chosen.

Timing Components for the 74121

Fig. 6-36 illustrates the normal connection of the external timing components to the 74121 monostable multivibrator. Fig. 6-37 illustrates two charts that may prove useful in the proper selection of timing components for a particular time delay. Fig. 6-37a shows the output pulse width versus the timing resistor value, and Fig. 6-37b shows the output pulse width versus the external timing capacitor.

Suppose that an output pulse width of 1 ms (millisecond) is required for a particular application. This is obtained by using an 18 kΩ resistor and a 0.1 μF capacitor. These values are approximations derived from the chart of Fig. 6-37a. In applications where the time is extremely critical, a "trim pot" (precision variable resistor) is used to adjust the value of the resistor to obtain the exact time required. In cases where accuracy is not needed, the approximations from the chart are normally adequate.

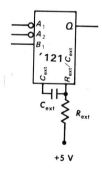

FIGURE 6-36. Normal timing component connections of the 74121 monostable multivibrator.

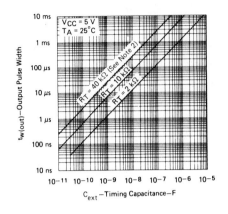

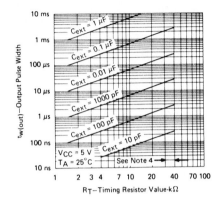

a. Output pulse width versus timing resistor value.

b. Output pulse width versus external capacitance.

FIGURE 6-37. *Timing component selection charts for the 74121 monostable multivibrator. (Courtesy of Texas Instruments Incorporated.)*

Another method for determining the output pulse width is defined by:

$$T_w = C_{ext} \times R_t \times 0.7 \qquad (6\text{-}1)$$

(T_w = output pulse width measured in seconds; C_{ext} = timing capacitance measured in farads; R_t = timing resistor value measured in ohms.) The output pulse width for the 18 kΩ resistor and 0.1 μF capacitor chosen from the graph would actually calculate to 1.26 ms by using Eq. 6-1.

THE 74122 RETRIGGERABLE MONOSTABLE MULTIVIBRATOR

The 74122, shown in Fig. 6-38, functions about the same way as the 74121 except that it can be *retriggered* in the middle of the output pulse. When the 74122 is retriggered, the output pulse width is extended by whatever pulse-width time has been programmed into the 74122 via the external timing components. With the 74121, a TRIGGER in the middle of the output pulse would be ignored completely. Fig. 6-39 shows the relationship between the 74121 and 74122 when each device is retriggered in the middle of the output pulse.

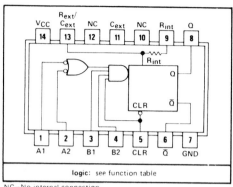

a. Pinout

'122, 'L122, 'LS122
FUNCTION TABLE

INPUTS					OUTPUTS	
CLEAR	A1	A2	B1	B2	Q	Q̄
L	X	X	X	X	L	H
X	H	H	X	X	L	H
X	X	X	L	X	L	H
X	X	X	X	L	L	H
H	L	X	↑	H	⊓	⊔
H	L	X	H	↑	⊓	⊔
H	X	L	↑	H	⊓	⊔
H	X	L	H	↑	⊓	⊔
H	H	↓	H	H	⊓	⊔
H	↓	↓	H	H	⊓	⊔
H	↓	H	H	H	⊓	⊔
↑	L	X	H	H	⊓	⊔
↑	X	L	H	H	⊓	⊔

b. Function table.

FIGURE 6-38. *The 74122 multivibrator. (Courtesy of Texas Instruments Incorporated.)*

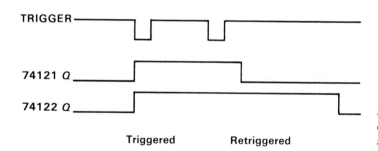

FIGURE 6-39. *Comparison of the effects of triggering the 74121 and 74122 monostable multivibrators.*

Timing Components for the 74122 Monostable Multivibrator

In order to program or set the output pulse width of the 74122, an external capacitor and resistor are connected as shown in Fig. 6-40a. Notice that this is very similar to the timing component connections of the 74121.

Figure 6-40b contains the chart that can be used to select the external timing components for a specific output pulse width when C_{ext} is equal to or less than 1000 pF. If a more precise output pulse width is required, then Eq. 6-2 can be used to calculate the exact component values for a C_{ext} of greater than 1000 pF.

Introduction to Sequential Logic **169**

(In this equation, C_{ext} is measured in picofarads, R_t in kilohms, and T_w in nanoseconds.)

$$T_w = 0.32 \times R_t \times C_{\text{ext}} \times (1 + \frac{0.7}{R_t}) \qquad (6\text{-}2)$$

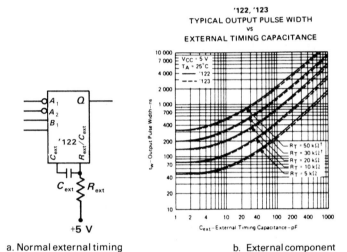

a. Normal external timing component connections.

b. External component selection chart.

FIGURE 6-40. The 74122 monostable multivibrator. (Courtesy of Texas Instruments Incorporated.)

DIGITAL OSCILLATORS

Digital oscillators are used to provide the CLOCK or timing signal in a digital system. A *digital oscillator* is a device that produces a squarewave compatible with digital logic circuits.

USING MONOSTABLE MULTIVIBRATORS TO GENERATE A CLOCK

The monostable multivibrator can be used to generate a digital squarewave, as illustrated in Fig. 6-41a. This circuit will generate the output shown because as each monostable multivibrator times out, it triggers the other. The only problem with this circuit is that, when power is initially applied, the circuit may not begin

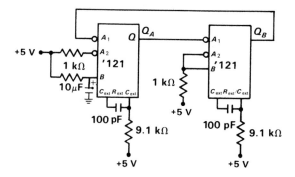

a. An oscillator constructed from a pair of 74121 monostable multivibrators.

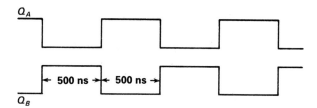

b. Timing diagram.

FIGURE 6-41. The 74121 oscillator.

operating. Operation has been assured by the RC circuit connected to the B input of the first 74121. When power is applied, this circuit provides a positive edge to the B input, which fires the one-shot. The Q output then goes to a logic 1 level for 500 ns. After 500 ns, the Q output drops to a logic 0 and triggers the second 74121, which will then time out in 500 ns just as the first one did. However, its output is tied back to the A_2 input of the first 74121, which causes the first one to trigger again. This process generates the timing diagram illustrated in Fig. 6-41b.

RING OSCILLATORS

The *ring oscillator* is also used occasionally. It is easy to see from Fig. 6-42 where the name comes from. This oscillator will never reach a stable state because of the way it has been connected. If point A is at a logic 1 state, then point B is at a logic 0; point C is at a logic 1; and point D is at a logic 0. Point D is now a logic 0, but point A is connected to point D. This means that the signal at point A is now changed to a logic 0, which will ripple through the INVERTERs and, after a time, again change the state of point A. The period of the output waveform is equal to six times the average propagation delay time of each

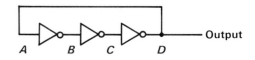

FIGURE 6-42. Three-INVERTER ring oscillator.

INVERTER, since it takes six changes through the INVERTERs to produce one complete waveform at the output.

If a capacitor is placed across one of the INVERTERs, the frequency of the oscillator can be reduced. This is illustrated in Fig. 6-43.

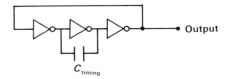

FIGURE 6-43. Ring oscillator with a capacitor that reduces the frequency of oscillation.

Controlling The Ring Oscillator

The ring oscillator can be modified so that it can be started or stopped, as shown in Fig. 6-44. The NAND gate either functions as an INVERTER (when the control input is high), or produces a constant logic 1 at its output (when the control input is low). Thus, if the control input is at a logic 1 level, the circuit will oscillate, and if the control input is a logic 0 level, the output will remain a constant logic 1.

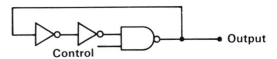

FIGURE 6-44. Ring oscillator that can be started and stopped by a control signal.

Another Form Of Ring Oscillator

Fig. 6-45 depicts a form of the ring oscillator that uses only two INVERTERs. The 330 Ω resistor is critical. In fact, if the value is increased to more than 1000

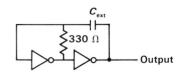

FIGURE 6-45. Modified version of the ring oscillator that uses only two INVERTERs.

Ω or reduced to less than 270 Ω, this circuit will not function. The period of the oscillation is determined by the value of the capacitor (C_{ext}).

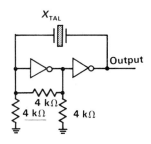

THE CRYSTAL-CONTROLLED OSCILLATOR

In many applications, the frequency of the CLOCK waveform is critical. To obtain a very accurate CLOCK frequency, a *crystal-controlled oscillator* is used. Fig. 6-46 depicts a simple crystal-controlled oscillator that uses two INVERTERs.

FIGURE 6-46. *Crystal-controlled digital oscillator.*

THE 555 TIMER

Another device that is flexible and easy to use is the *555 timer*. It can function as an oscillator or as a monostable multivibrator. The pinout and internal structure of this device is illustrated in Fig. 6-47. As the figure shows, this device is fairly simple. It contains a voltage divider network constructed with three series resistors, two comparators that control an internal *SR* latch, and a transistor switch that is used to discharge an external capacitor.

The Internal Operation of the 555 Timer

Before this device can be used, its internal operation must be understood. The voltage divider divides V_{CC} by a factor of three, so that each resistor has one-third of V_{CC} across it. The bottom comparator compares one-third of V_{CC} with the input labeled TRIGGER. The output of this comparator will set the *SR* latch if voltage on the TRIGGER input drops below one-third of V_{CC}. The other comparator compares two-thirds of V_{CC} with input labeled THRESHOLD. The output of this comparator will reset the *SR* latch if the voltage on the THRESHOLD input exceeds two-thirds of V_{CC}.

The state of the *SR* latch is passed to the output through a buffer stage and also controls the operation of the discharge transistor. The discharge transistor is saturated when $\overline{Q}$ is high and is cut off when $\overline{Q}$ is low.

The 555 Timer Used as an Oscillator

Fig. 6-48 shows the 555 timer wired to function as a digital oscillator, or an *astable multivibrator*. When power is applied to this circuit, the capacitor C is completely discharged to 0 V. As time passes, C begins to charge toward 5 V, or V_{CC}, through R_A and R_B. Since C is initially 0 V, the internal latch is set, causing the output pin to be high. As the capacitor charges toward 5 V, it will eventually reach 3.33 V, or the THRESHOLD voltage. As soon as it exceeds this value, the internal latch is reset, causing the output to become a logic 0. The latch also begins to discharge C through the discharge transistor and R_B. The

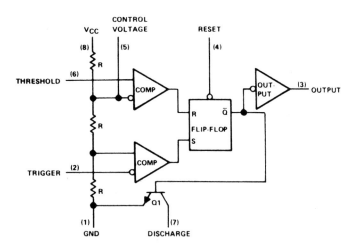

(TOP VIEW)

a. Pinout

b. Functional block diagram.

FIGURE 6-47.   The 72555 timer. (Courtesy of Texas Instruments Incorporated.)

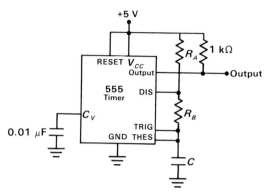

a. Logic circuit.

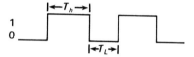

b. Timing diagram.

FIGURE 6-48. *The 555 timer wired to function as an astable multivibrator.*

capacitor will discharge until it reaches 1.67 V (the TRIGGER voltage). When it drops below this value, the internal latch is set, changing the state of the output. The latch also cuts off the discharge transistor, allowing C to begin charging again through R_A and R_B.

In this circuit, the charge time is controlled by C, R_A, and R_B; the discharge time is controlled by C and R_B. The logic 1 output time (T_h) is determined by Eq. 6-3, and the logic 0 output time (T_l) is determined by Eq. 6-4. The duty cycle of the output is determined by Eq. 6-5.

$$T_h = 0.693 \times (R_A + R_B) \times C \qquad (6\text{-}3)$$

$$T_l = 0.693 \times R_B \times C \qquad (6\text{-}4)$$

$$D\% = \frac{R_B}{R_A + R_B} \times 100 \qquad (6\text{-}5)$$

Varying the Control Voltage Input with Astable Operation

If an external voltage is attached to the control voltage input of the 555 timer, as depicted in Fig. 6-49, it is possible to vary the frequency of the timer. Varying the control voltage varies the THRESHOLD and TRIGGER voltage levels.

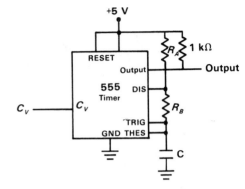

a. Logic circuit.

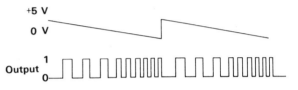

FIGURE 6-49. Effect of varying the control voltage input of a 555 timer wired as an astable multivibrator.

b. Timing diagram.

This, in turn, will vary the output frequency of the 555 timer because it changes the amount of time required to charge and discharge the capacitor.

Monostable Operation Of The 555 Timer

The 555 timer can be operated as a one-shot by connecting it as illustrated in Fig. 6-50. If the TRIGGER input is held at the V_{CC} potential, the internal latch will be reset and the output will be a logic 0. In addition, the discharge transistor will remain saturated and the capacitor C will remain discharged.

When the TRIGGER input is momentarily grounded, the internal latch will be set. This causes the discharge transistor to be cut off and the output to become a logic 1. It also allows C to begin charging toward V_{CC} through R_A. When the voltage of the capacitor reaches the THRESHOLD voltage of two-thirds of V_{CC}, the internal latch is again reset and the discharge transistor is saturated, which then discharges C. This places us back at the starting point. For the cycle to be repeated, another TRIGGER pulse is needed.

The length of time it takes C to charge from 0 V to two-thirds of V_{CC} is determined by the sizes of C and R_A. Eq. 6-6 can be used to calculate the output pulse width (T_p).

$$T_p = 1.1 \times R_A \times C \qquad (6\text{-}6)$$

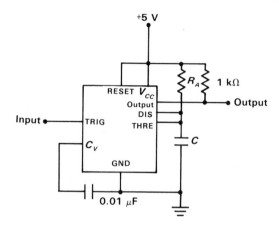

a. Logic circuit.

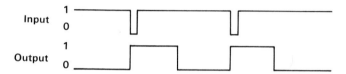

b. Timing diagram.

FIGURE 6-50. *The 555 timer wired to function as a monostable multivibrator.*

SUMMARY

1 The main difference between combinational and sequential logic is the ability of sequential logic to remember past data.

2 The cross-coupled, or asynchronous, latch is the most basic memory element. It contains two logic gates that are cross-coupled.

3 The NAND gate version of the asynchronous latch has active-low inputs, and for this reason it is called an $\overline{S}\,\overline{R}$ asynchronous latch.

4 The NOR gate version of the asynchronous latch has active-high inputs, and for this reason is called an SR asynchronous latch.

5 A common application of the $\overline{S}\,\overline{R}$ latch is the contact bounce eliminator, which removes the mechanical bounce from the electrical signal produced by a mechanical switch.

6 The gated latch is constructed from either the NAND gate $\overline{S}\,\overline{R}$ latch or the NOR gate SR latch.

7 The main advantage of the gated latch is that the application of the input data can be synchronized with an external event or CLOCK.

8 The NAND gate version of the gated latch contains active-high inputs, while the NOR gate version of the gated latch contains active-low inputs.

9 The GATE input of the gated latch is a level-sensitive input. The level of the GATE determines whether the latch will remember or respond to the S and R inputs.

10 Gated D-type latches are often called *transparent* latches because, if the GATE is active, they look like pieces of wire.

11 The main problem with the gated latch and the asynchronous latch is the race condition (oscillation), which occurs if the external circuit provides a feedback path to the inputs.

12 The master-slave flip-flop is constructed with two gated SR latches. The gated latch connected to the inputs is called the *master,* and the gated latch connected to the outputs is called the *slave.*

13 In the negative edge–triggered master-slave flip-flop, the master accepts data when the CLOCK is high and the slave receives data from the master when the CLOCK goes low. This causes the outputs to change on the 1-to-0 transition of the CLOCK.

14 The master-slave JK flip-flop is an SR flip-flop modified to accept a 1–1 input condition. This new input condition causes the JK flip-flop to toggle, or complement, on each CLOCK pulse.

15 The PRESET ($\overline{\text{PR}}$) and CLEAR ($\overline{\text{CLR}}$) inputs on a flip-flop will cause Q to be set or cleared at the instant they are activated. These inputs always override the CLOCK.

16 The main problem with the master-slave flip-flop is that an error can occur if noise affects the inputs while the CLOCK is high. This problem is cured with the edge-triggered flip-flop.

17 The monostable multivibrator is a device that can be triggered to produce an output pulse of a predetermined duration.

18 A retriggerable monostable multivibrator is a device that can be triggered and retriggered at any time during its operation. Each time that it is triggered or retriggered, the output will become active for a predetermined interval of time.

19 Digital oscillators include ring oscillators, crystal-controlled oscillators, and timers.

20 The 555 timer is a device that can be operated as a monostable multivibrator or as an astable multivibrator.

21 The control voltage input on the 555 timer can be used to vary the frequency during astable operation or the pulse width during monostable operation.

KEY TERMS

astable multivibrator	flip-flop	PRESET and CLEAR
asynchronous latch	gated latch	race condition
CLOCK	latch	ring oscillator
contact bounce eliminator	master-slave flip-flop	sequential logic
	memory element	settling time
cross-coupled latch	monostable	single-shot
D-type flip-flop	multivibrator	toggle
edge-triggered flip-flop	one-shot	TRIGGER

QUESTIONS AND PROBLEMS

1 Describe the basic difference between combinational and sequential logic circuitry.

2 Draw the block diagram of a typical sequential logic system.

3 Besides the digital clock discussed in this chapter, what is another example of a digital system that requires past data to generate new data?

4 Draw the logic diagram of an $\overline{S}\,\overline{R}$ asynchronous latch.

5 Draw the logic diagram of an SR asynchronous latch.

6 Describe what occurs when both the $\overline{R}$ and $\overline{S}$ inputs are grounded on the NAND gate version of the $\overline{S}\,\overline{R}$ asynchronous latch.

7 What input conditions are required to place a logic 1 on the Q output of a NOR gate asynchronous latch?

8 Is it possible to use a NOR gate asynchronous latch in a contact bounce eliminator? Explain your answer.

9 Why is it often important that the contact bounce of a mechanical switch be removed from the electrical signal?

10 What is the basic difference between the asynchronous latch and the gated latch?

11 Given the input conditions and CLOCK illustrated in Fig. 6-51, draw the waveform for the Q output.

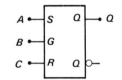

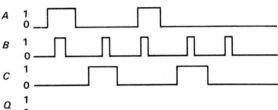

FIGURE 6-51. *Circuit for problem 11.*

12 Given the input conditions and the CLOCK illustrated in Fig. 6-52 draw the waveform for the Q output.

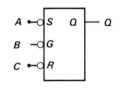

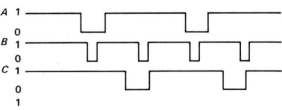

FIGURE 6-52. *Circuit for problem 12.*

13   Given the input conditions and the CLOCK illustrated in Fig. 6-53, draw
the waveform for the Q output.

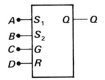

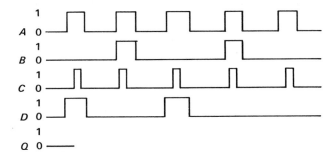

FIGURE 6-53.   *Circuit for problem 13.*

14   Explain why the GATE input of the gated latch is level-sensitive.

15   What is a race?

16   The master-slave flip-flop will accept data during the logic _____
level of the CLOCK input.

17   The master-slave flip-flop will transfer data from the master to the slave
on the logic _____ level of the CLOCK input.

18   Given the input and CLOCK waveforms shown in Fig. 6-54, draw the Q
output waveform.

19   Given the input and CLOCK waveforms shown in Fig. 6-55, draw the Q
output waveform.

20   Draw the function table for the JK flip-flop.

21   If a JK flip-flop has a logic 1 on both its J and K inputs, what logic level
will be present at the Q output after it receives eight CLOCK pulses?
(Assume that the Q output is a logic 0 initially.)

Introduction to Sequential Logic   **181**

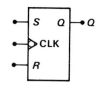

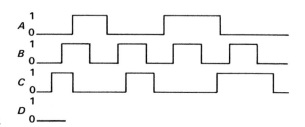

FIGURE 6-54. Circuit for problem 18.

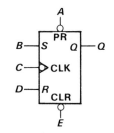

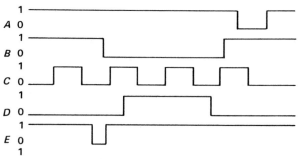

FIGURE 6-55. Circuit for problem 19.

22 Why does the master-slave JK flip-flop toggle for a 1–1 input condition?

23 Are all master-slave flip-flops negative edge–triggered?

24 Given the input conditions and CLOCK illustrated in Fig. 6-56, draw the Q output waveform.

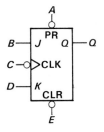

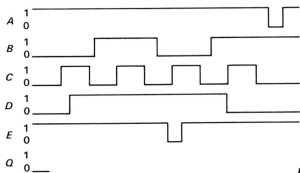

FIGURE 6-56. Circuit for problem 24.

25 What problem can occur when using a master-slave flip-flop?

26 Why is the edge-triggered flip-flop used in place of a master-slave flip-flop?

27 Define the monostable multivibrator.

28 Select timing components for the 74121 so that it produces an output pulse with a duration of 10 ms.

29 Calculate the value of the timing capacitor required to cause the 74121 to produce an output pulse with a duration of 1 second (assume that the value of the resistor is 1 kΩ.

30 Using Eq. 6-2, determine the value of the external timing resistor required to cause an output pulse of 950 ns if the timing capacitor is 1200 pF.

31 Develop a CLOCK circuit that produces a logic 1 at its output for 900 ns and a logic 0 for 1100 ns (refer to Fig. 6-41).

32 If the average propagation delay time for an INVERTER is 12 ns, what is the frequency of the signal generated by Fig. 6-42?

33 Select the proper component values to cause the 555 timer to generate a 1 kHz squarewave with a 50 percent duty cycle.

34 Select the proper component values to cause the 555 timer to generate a pulse with a duration of 5 s each time it is triggered.

7

SEQUENTIAL LOGIC CIRCUITRY

Chapter 6 introduced the flip-flop, which is a major component in most sequential logic systems. This chapter shows how the flip-flop is used in the basic synchronous building blocks of many advanced systems. These building blocks include *counters* and *registers*.

Counters, registers, and some of the combinational logic circuits covered earlier are used in this chapter to illustrate their application in a few simple sequential logic systems.

ASYNCHRONOUS COUNTERS

One of the simplest sequential logic circuits is the *asynchronous counter* or, as it is sometimes called, the *ripple counter*. This section details the operation,

TABLE 7-1 THE COUNTING SEQUENCE OF A 2-BIT BINARY UP COUNTER COMPARED WITH THE DECIMAL COUNTING SEQUENCE.

Decimal	Binary
0	0 0
1	0 1
2	1 0
3	1 1

construction, and application of the following types of binary ripple counters: the up counter, the down counter, and the up/down counter.

BINARY ASYNCHRONOUS UP COUNTERS

One of the most commonly applied counters is the binary up counter. The term *up* describes the counting sequence of the device. Table 7-1 illustrates the counting sequence of a 2-bit binary up counter. Because a 2-bit binary number has four possible values, the 2-bit binary counter has four different counts. A 2-bit binary counter is sometimes called a *modulo-4* counter because it has four output states. (The *modulus* of a counter is equal to the number of different output states.) If a 3-bit binary up counter were used, there would be eight combinations (modulo-8); four bits would yield 16 combinations (modulo-16); five bits would yield 32 combinations (modulo-32); and n bits would yield 2^n combinations (modulo-2^n).

Construction of the 2-bit Asynchronous Binary Up Counter

What type of circuit is used to build a 2-bit binary counter? It would have to be a sequential circuit since the present output state of the counter determines the next output state. This means that the device must be able to store a 2-bit binary number. A flip-flop is used because it is a storage element that can store one bit of information. Thus, two flip-flops will compose one 2-bit counter.

Notice in Table 7-1 that the upward pattern of the counter continues until it reaches a count of 11; it then repeats the sequence from 00. Most counters are *cyclic* in nature and naturally repeat a counting sequence. The counter in this example follows the sequence: 00, 01, 10, 11, 00, 01, etc.

The least significant bit (the bit with the least mathematical weight, or the rightmost bit) changes from a 0 to a 1 or from a 1 to a 0 on each input CLOCK. In this counter, the least significant bit generates the same output as the *JK* flip-flop with a logic 1 on both the *J* and *K* inputs. In this mode of operation, the

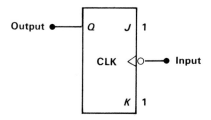

a. Logic symbol.

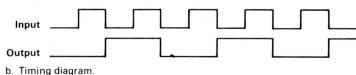

Input

Output

b. Timing diagram.

FIGURE 7-1. JK flip-flop connected to toggle that subsequently divides input by two.

output of the *JK* flip-flop will toggle from 0 to 1 and then from 1 to 0. Fig. 7-1 illustrates the action of this flip-flop and its output waveform.

As shown in Table 7-1, the most significant bit (the leftmost bit) of the count also toggles, but only on every other count. Unfortunately, we don't have a flip-flop that operates in this way, but we do have an output that does. Again referring to the output waveform in Fig. 7-1, notice that the output goes from a 1 to a 0 at the second CLOCK pulse, the fourth, the sixth, etc.

If the output of the least significant flip-flop is used as a CLOCK pulse to the most significant flip-flop, the result is a 2-bit binary up counter, as illustrated in Fig. 7-2.

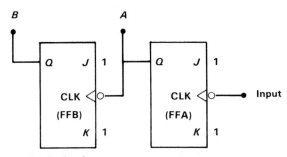

a. Logic circuit.

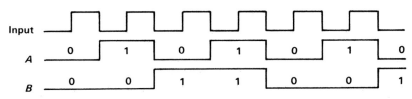

b. Timing diagram.

FIGURE 7-2. Two-bit binary asynchronous counter.

The Operation of the 2-bit
Binary Asynchronous Up Counter

Assuming the count is initially 00, as depicted in Fig. 7-2, the first CLOCK pulse will advance the counter to 01. This happens because the first flip-flop in the series (FFA) receives a negative edge signal from the CLOCK and complements its output (A) to a 1. The second flip-flop (FFB) remains the same because it does not receive a negative edge signal from the output of FFA—its CLOCK. On the second CLOCK pulse, the output of FFA again complements and provides FFB with a negative edge. The overall output now changes from a 01 to a 10. The third CLOCK pulse toggles A to a 1. This does not provide FFB with a negative edge, so the count becomes 11. Finally, on the fourth CLOCK pulse, FFA again toggles and its output becomes a 0. This provides a negative edge to FFB, which also toggles, producing a count of 00. This cycle will repeat itself for as long as a CLOCK is connected to the CLOCK input.

The up counter, as you probably have guessed, can be used for counting CLOCK pulses. Looking at the period of the most significant output (B) in Fig. 7-2, notice that this output is exactly one-fourth of the input CLOCK frequency. The counter has four different counting states, and it divides the CLOCK input by four. It appears from this example that counters can be used to divide a frequency, but let's make sure by looking at counters with more than two bits.

Multiple-bit Binary Asynchronous Up Counters

Keeping in mind that the 2-bit counter used the output of FFA as the CLOCK for FFB, what will happen if the output of FFB is used as the CLOCK for another flip-flop? Fig. 7-3 shows this connection and the waveforms generated by this counter. Notice that the output of FFC is exactly one-eighth of the frequency of the input CLOCK waveform. The number of states of a counter can thus be used to divide the CLOCK frequency. This counter will divide the CLOCK frequency by eight because it goes through eight different counts before returning to zero and repeating the sequence.

Each time that the number of bits of a counter is extended by one, the count doubles. A 4-bit counter will divide the input frequency by 16, as shown in Fig. 7-4. Remember, if a counter divides a frequency, it will also multiply the time (or period). Thus a counter can be used as a frequency divider or as a time multiplier. For example, if a 1 s CLOCK pulse is applied to a modulo-16 counter, it will take 16 s for the count to repeat, and the output waveform will be 16 times longer than the input waveform.

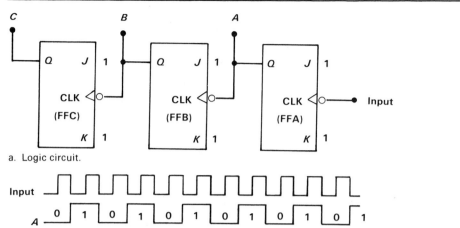

a. Logic circuit.

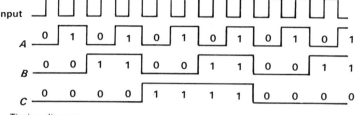

b. Timing diagram.

FIGURE 7-3. Three-bit asynchronous up counter.

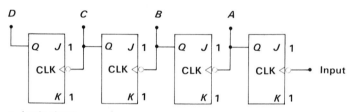

a. Logic circuit.

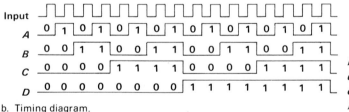

b. Timing diagram.

FIGURE 7-4. Four-bit binary asynchronous counter. Note that this counter will divide the input frequency by 16.

BINARY ASYNCHRONOUS DOWN COUNTERS

The construction of the down counter is almost identical to that of the up counter, as is demonstrated by the 2-bit down counter in Fig. 7-5 and its counting sequence in Table 7-2. There is clearly very little difference between the up and down counter. In fact they only differ in that the CLOCK for FFB comes from $\overline{Q}$ instead of Q.

The Operation of the Binary Down Counter

FFA in Fig. 7-5 functions in the same way as FFA in Fig. 7-2. FFB differs because it receives its CLOCK pulse from the $\overline{Q}$ output of FFA. Every time the Q output of FFA goes from a logic 0 to a 1, FFB will toggle.

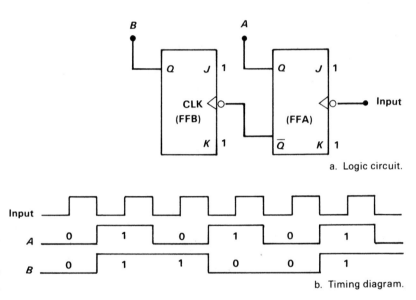

a. Logic circuit.

FIGURE 7-5. Two-bit binary asynchronous down counter.

b. Timing diagram.

Assume that the initial count is 11, as shown in Fig. 7-5b. On the first CLOCK pulse FFA toggles. This will not cause FFB to toggle, since FFA has changed from 1 to 0. The resulting count is a 10. The next CLOCK pulse again causes FFA to toggle. Because its output now changes from a 0 to a 1, FFB will toggle, resulting in a count of 01. It is clear from this description that this counter is counting down.

Fig. 7-6 depicts a 4-bit, or modulo-16 down counter. This counter has been connected in the same way as the counter in Fig. 7-5; the only difference is the

TABLE 7-2 THE COUNTING SEQUENCE OF A 2-BIT BINARY RIPPLE COUNTER COMPARED WITH THE DECIMAL COUNTING SEQUENCE.

Decimal	Binary
3	11
2	10
1	01
0	00

two extra flip-flops. It takes four flip-flops to produce the 16 states of a modulo-16 counter.

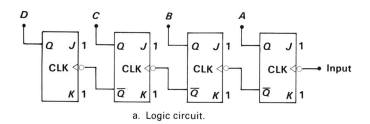

a. Logic circuit.

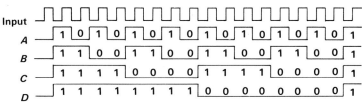

b. Timing diagram.

FIGURE 7-6. Four-bit (modulo-16) binary down counter.

THE BINARY ASYNCHRONOUS UP/DOWN COUNTER

The main difference between the up and the down counter is the CLOCK connection for all flip-flops except the least significant one. Notice that the Q output of the least significant flip-flop is connected to the next flip-flop's CLOCK input for an up counter, while the $\overline{Q}$ output is connected to the input for a down counter. The addition of a switch to select either Q or $\overline{Q}$ will create an up/down counter. Another method of constructing an up/down counter is to insert an INVERTER between the Q output and the CLOCK input for counting down, with the INVERTER left out for counting up.

A Controlled INVERTER

Fig. 7-7 illustrates how the Exclusive OR gate is used as a *controlled INVERTER*. If the control input C is at a logic 0 level, the input waveform passes through to the output without inversion. If C is at a logic 1 level, the input waveform inverts at the output. This simple circuit is often used to control the direction of counting in a binary ripple up/down counter.

a. Logic symbol.

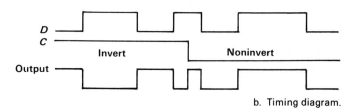

b. Timing diagram.

FIGURE 7-7. *Exclusive OR gate used as a controlled INVERTER.*

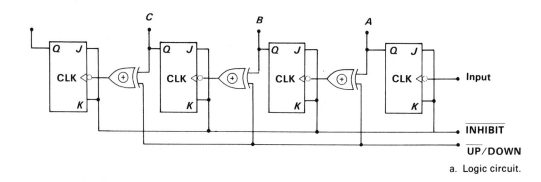

a. Logic circuit.

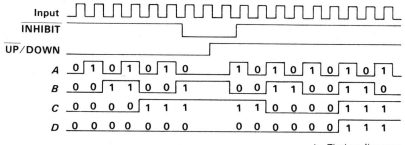

b. Timing diagram.

FIGURE 7-8. *Four-bit binary asynchronous up/down counter. The INHIBIT input is used to stop the counter whenever the direction of counting is changed.*

A 4-bit Binary Up/Down Counter

Fig. 7-8 illustrates how the controlled INVERTER can be built into a counter that will count up or down. The $\overline{\text{UP}}$/DOWN control input will cause this counter to count up if it is at a 0 level and down if it is at a 1 level.

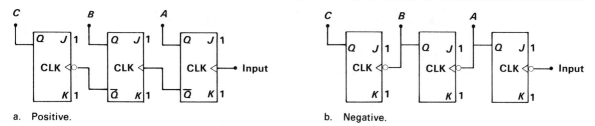

a. Positive.

b. Negative.

FIGURE 7-9.   Three-bit edge-triggered binary asynchronous up counter.

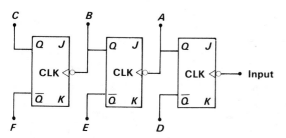

a. Logic circuit.

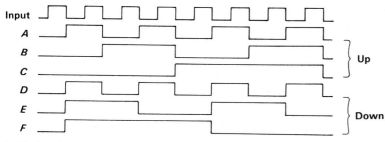

b. Timing diagram.

FIGURE 7-10.   Using both the Q and $\bar{Q}$ outputs of a 3-bit asynchronous counter. The Q outputs will count up and the $\bar{Q}$ outputs will count down in this counter.

The only problem with this connection is that when the direction of counting changes, unwanted CLOCK pulses from the Exclusive OR gates might alter the count. To prevent this problem, all J and K inputs have been tied together and connected to another control input. This additional control input places a logic 0 on all J and K inputs before the direction of counting is changed. This process forces all flip-flops into the memory mode so that, if an unwanted CLOCK pulse does arrive at the CLOCK input, no change in the count occurs.

If positive edge–triggered flip-flops are used in a counter, the direction of the count will be exactly the opposite. Fig. 7-9a shows a 3-bit binary up counter constructed with positive edge–triggered flip-flops. Fig 7-9b shows a 3-bit binary up counter constructed with negative edge–triggered flip-flops.

Fig. 7-10 depicts a simple 3-bit binary counter and the waveforms attainable at the Q and $\bar{Q}$ outputs. The count available at the Q outputs goes up, and the count available at the $\bar{Q}$ outputs goes down.

SYNCHRONOUS COUNTERS

In asynchronous counters, the output of one flip-flop is connected as a CLOCK input to the next flip-flop. This means that the outputs don't change in unison or *synchronization*. The outputs in the synchronous counter, on the other hand, change together.

THE BASIC SYNCHRONOUS UP COUNTER

Fig. 7-11 illustrates a 3-bit binary synchronous up counter. Notice that the CLOCK inputs are tied together, forcing the outputs to change in unison. Also notice that some additional connections exist between the flip-flops.

Let's examine each flip-flop to determine why this counter will count up. FFA has a logic 1 connected to both its J and K inputs so that it toggles on each input CLOCK pulse. This is identical to the connection found on the least significant bit position of the asynchronous counter.

FFB is connected differently from the corresponding flip-flop in an asynchronous counter. Its J and K inputs are tied to the Q output of FFA. Whenever the output of FFA is a logic 0, FFB will not change on the next CLOCK pulse and whenever FFA has a logic 1 output, FFB toggles on the next CLOCK pulse. The counting sequence is 001, 011, 101, 111, etc. See Fig. 7-12 for a view of the effect of this connection.

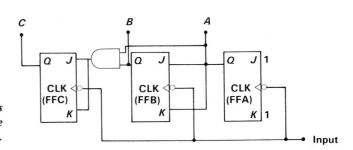

FIGURE 7-11.   Three-bit binary synchronous up counter. Notice how the CLOCK inputs are tied together in a synchronous counter.

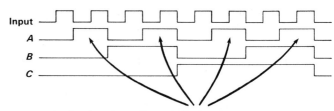

A logic 1 on flip-flop A produces change in flip-flop B on the next clock pulse.

FIGURE 7-12.   The timing diagram of the synchronous counter in Fig. 7-11.

The J and K inputs of FFC are tied through an AND gate to the FFA and FFB Q outputs. As a result, FFC toggles only when both FFA and FFB are logic 1s. This condition occurs for the counts of 011 and 111. On the next CLOCK pulse, FFC toggles to a count of 100 or 000, respectively.

It is important to notice that, in this type of counter, the outputs all change together because all the CLOCKs are tied together. This type of counter is called a synchronous circuit because its next state depends on its present state. In other words, its outputs determine the next state from the present state. This is not exactly the case with the asynchronous counter, because its outputs do not actually determine the next state: they merely provide a CLOCK pulse for the next flip-flop.

THE BASIC SYNCHRONOUS DOWN COUNTER

Just as it is simple to modify an asynchronous up counter into a down counter, it is likewise simple to change the synchronous up counter into a synchronous down counter. It is necessary only to connect the $\overline{Q}$ outputs, instead of the Q outputs, to each subsequent flip-flop.

Fig. 7-13 depicts a 3-bit synchronous down counter before it changes from a 100 to a 011. It also shows the timing diagram for one complete cycle. FFA will toggle from a 0 to a 1 on the next CLOCK pulse, FFB will toggle because its J and K inputs are at a logic 1 level, and FFC will also toggle because both FFA and FFB are presently 0 at their Q outputs. The $\overline{Q}$ outputs of FFA and FFB are both 1s, which cause the J and K inputs of FFC to be logic 1s. This, then, causes

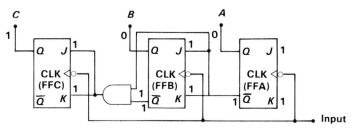

a. Logic circuit.

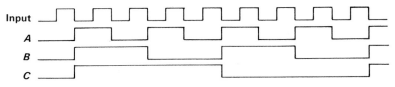

b. Timing diagram.

FIGURE 7-13. Three-bit binary synchronous down counter shown at the count of 100 (binary).

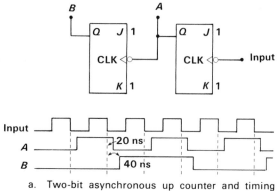

a. Two-bit asynchronous up counter and timing diagram.

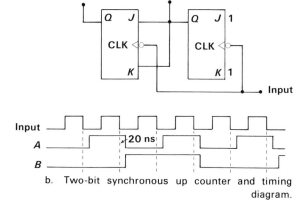

FIGURE 7-14. Effect of propagation delay.

b. Two-bit synchronous up counter and timing diagram.

the output of FFC to toggle on the next CLOCK pulse. In this example, all three flip-flops will toggle at the next pulse, so that the count will change from a 100 to a 011.

A COMPARISON OF ASYNCHRONOUS AND SYNCHRONOUS COUNTERS

Fig. 7-14 depicts the timing diagrams and the counter circuits for a 2-bit synchronous counter and a 2-bit asynchronous counter. A CLOCK frequency of 10 MHz has been chosen to illustrate the difference between these two counters.

Notice that the output waveforms shown in Fig. 7-14a are different from the waveforms shown in Fig. 7-14b. This disparity exists because the outputs of the flip-flops do not change exactly in step with the CLOCK. Most flip-flops require about 20 ns for the output to *settle*, or reach its final state, from the time of the CLOCK input. This is clearly evident in the timing diagrams. The outputs of the asynchronous counter in Fig. 7-14a are skewed by a total of 40 ns from the CLOCK. Output *A* is skewed 20 ns, and output *B* is skewed 40 ns. The outputs of the synchronous counter in Fig. 7-14b are both skewed from the CLOCK a total of 20 ns and occur in unison, or synchronization, with each other.

Maximum Operating Frequency

The maximum operating frequency of a synchronous counter is determined by the *settling time* of the flip-flops used to construct it. If the settling time is 25 ns, then the maximum operating frequency of the counter would be no greater than 40 MHz, the reciprocal of 25 ns. In the asynchronous counter, the maximum operating frequency is also determined by the settling time, but the outputs, in parallel, are not usable at high frequencies because of the skewing.

THE SYNCHRONOUS UP/DOWN COUNTER

Fig. 7-15 illustrates a synchronous 3-bit binary up/down counter and its timing diagram. You may notice that it is similar to the asynchronous up/down

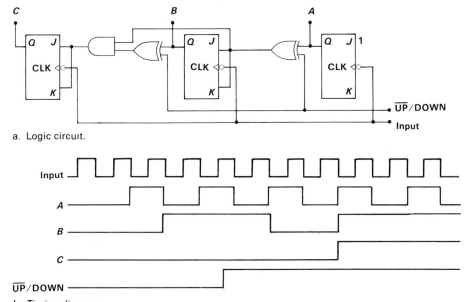

a. Logic circuit.

b. Timing diagram.

FIGURE 7-15. Three-bit synchronous binary up/down counter.

counter. The Exclusive OR gates are used to steer either the Q output or the $\overline{Q}$ output of each flip-flop to each subsequent flip-flop. The asynchronous counter miscounts whenever the $\overline{UP}/DOWN$ control is changed because it provides a false CLOCK to a flip-flop. This cannot occur with the synchronous up/down counter.

MODULUS COUNTERS

The only counters in the text thus far have been *natural binary counters*, that is, counters that have 2^n different states. These counters divide an input frequency by the number of states, or the modulus, of the counter. At times it may be desirable to divide an incoming frequency by a whole number other than a natural binary number. This section describes how modulus counters can be constructed and also how existing MSI counters can be modified to count any number of states.

CREATING A MODULO-X COUNTER BY USING THE CLEAR INPUTS

A modulo-X counter can easily be created by using the asynchronous CLEAR ($\overline{CLR}$) inputs on each flip-flop in a counter. For example, a modulo-6 counter

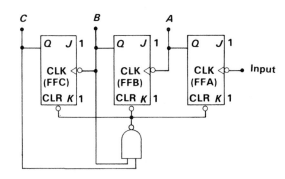

a. Logic circuit.

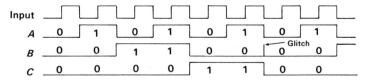

FIGURE 7-16. Modulo-6 counter. Notice that output waveform C requires six CLOCK pulses to repeat.

b. Timing diagram.

would have six different states, 000 through 101. In order to modify a 3-bit binary counter to function as a modulo-6 counter, the CLEAR inputs could be used to clear the counter to 000 whenever it reached a count of 110, or six. Even though the counter would actually reach a count of 110, it would remain there for only about 20 ns.

Fig. 7-16 depicts a modulo-6 counter and its timing diagram. Notice that a NAND gate is used to decode the outputs and generate a $\overline{\text{CLR}}$ pulse for each flip-flop. Each time the FFC and FFB Q outputs become logic 1s, the output of the NAND gate becomes a logic 0. This clears the three flip-flops in a matter of 20 ns. Compare the FFC Q output waveform (C) with the input CLOCK frequency: it is exactly one-sixth of the CLOCK. A modulo-X counter will always divide the input CLOCK frequency by the modulus of the counter.

The Modulo-X Asynchronous Counter Design Algorithm

The algorithm for developing a modulo-X counter is outlined below:

1. Convert X from decimal to binary. The number of bits (N) in this number indicates how many flip-flops will be required to build the modulo-X counter.
2. Design an N-bit binary up counter and connect all the $\overline{\text{CLR}}$ inputs.
3. Connect the 1 bit positions in binary number X to the inputs of a NAND gate. The output of the NAND gate is then connected to the $\overline{\text{CLR}}$ inputs of the flip-flops.

Suppose that you need a modulo-12 counter. The first step is to convert the decimal number 12 to binary form. Because 12 decimal is 1100 binary, four flip-flops are required for this counter. Finally, a 2-input NAND gate is connected to the most significant bit positions of the counter, and its output is tied to all of the $\overline{\text{CLR}}$ inputs. The completed circuit for a modulo-12 counter is illustrated in Fig. 7-17.

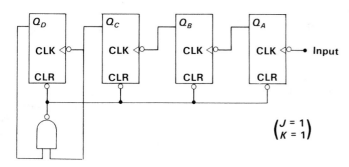

FIGURE 7-17. Modulo-12 counter.

As another example, let's design a modulo-100 counter. This counter would require seven flip-flops and a 3-input NAND gate, as illustrated in Fig. 7-18. It requires seven flip-flops because 100 decimal is 1100100 binary. The 3-input NAND gate is required because there are three 1 bits in this binary number that must be NANDed to produce the $\overline{CLR}$ signal to the flip-flops.

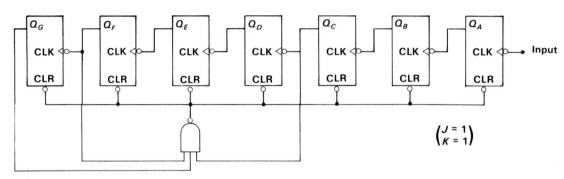

FIGURE 7-18. Modulo-100 counter.

Cascading Modulus Counters

The effect of cascading modulus counters is to multiply the individual moduli of the counters together. For example, if a modulo-10 counter is cascaded with a modulo-12 counter, the resulting modulus is 120. Fig. 7-19 depicts the modulo-10 and modulo-12 counters connected in cascade. Also shown in this illustration is an input CLOCK frequency of 120 kHz and an output frequency of 1 kHz.

FIGURE 7-19. Modulo-10 and modulo-12 counter cascaded to produce a modulo-120 counter.

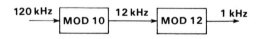

Problems Encountered When Using the $\overline{CLR}$ Inputs for a Modulus Counter

The main problem with using the CLEAR input for a modulo-X counter is that some of the outputs may contain glitches. Fig. 7-20 illustrates the glitches found on the output of the first flip-flop of a modulo-5 counter. The glitch occurs because the counter does reach a count of 101, but only briefly. Here the output of FFA goes to a logic 1 when it should really be a logic 0.

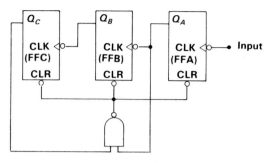

a. Logic circuit.

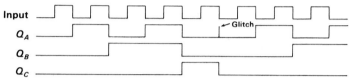

b. Timing diagram.

FIGURE 7-20. Modulo-5 counter. Notice that the glitch occurs during a RESET.

MSI MODULUS COUNTERS

There are a few different types of integrated modulus counters available from various integrated circuit manufacturers. One of the most common types is the 7490 decade counter. Fig. 7-21 illustrates this counter along with the truth table provided by the manufacturer. As its name implies, this is a modulo-10 counter designed to count either from 0000 through 1001, for BCD operation, or in the *biquinary* mode, which generates a perfectly symmetrical squarewave at one-tenth of the input CLOCK frequency.

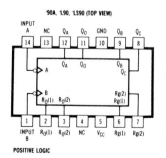

a. Pinout.

'90A, 'L90, 'LS90 BCD COUNT SEQUENCE

COUNT	OUTPUT			
	Q_D	Q_C	Q_B	Q_A
0	L	L	L	L
1	L	L	L	H
2	L	L	H	L
3	L	L	H	H
4	L	H	L	L
5	L	H	L	H
6	L	H	H	L
7	L	H	H	H
8	H	L	L	L
9	H	L	L	H

b. BCD truth table.

'90A, 'L90, 'LS90 BI-QUINARY (5-2)

COUNT	OUTPUT			
	Q_A	Q_D	Q_C	Q_B
0	L	L	L	L
1	L	L	L	H
2	L	L	H	L
3	L	L	H	H
4	L	H	L	L
5	H	L	L	L
6	H	L	L	H
7	H	L	H	L
8	H	L	H	H
9	H	H	L	L

c. Biquinary truth table.

RESET/COUNT FUNCTION TABLE

RESET INPUTS				OUTPUT			
$R_{0(1)}$	$R_{0(2)}$	$R_{9(1)}$	$R_{9(2)}$	Q_D	Q_C	Q_B	Q_A
H	H	L	X	L	L	L	L
H	H	X	L	L	L	L	L
X	X	H	H	H	L	L	H
X	L	X	L	COUNT			
L	X	L	X	COUNT			
L	X	X	L	COUNT			
X	L	L	X	COUNT			

d. Function of the 7490 RESET inputs.

FIGURE 7-21. The 7490 decade counter. (Courtesy of Texas Instruments Incorporated.)

Fig. 7-22 shows the connections for both BCD and biquinary operations of the 7490 decade counter. Notice that this counter is internally divided into two separate portions, a modulo-5 and a modulo-2. In Fig. 7-22a the modulo-2 counter is connected to the CLOCK and the modulo-5 counter is connected to

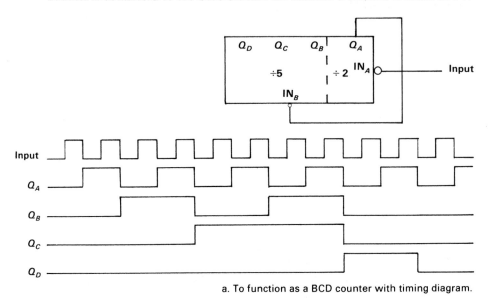

a. To function as a BCD counter with timing diagram.

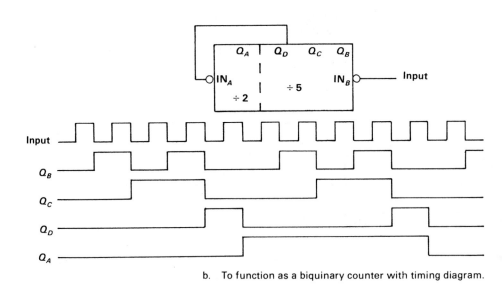

b.   To function as a biquinary counter with timing diagram.

FIGURE 7-22.   Wired 7490 decade counter.

the output of the modulo-2 counter. This generates a BCD counting sequence, which is useful in applications where data must be counted and displayed in decimal. The circuit of Fig. 7-22b connects the modulo-5 counter to the CLOCK and its output to the modulo-2 counter. This configuration will cause the 7490 to count in biquinary, which generates a symmetrical squarewave at the Q_A output.

Returning to Fig. 7-21, notice four additional inputs: $R_{0(1)}$, $R_{0(2)}$, $R_{9(1)}$, and $R_{9(2)}$. The R_0 inputs reset the counter to 0000, and the R_9 inputs reset it to 1001. If both the R_0 inputs are placed at a logic 1, the counter resets to 0000; if both the R_9 inputs are placed at a logic 1, the counter resets to 1001. The R_0 inputs are very useful if this counter is to be used as a modulo-X counter.

Suppose you need a counter that will count from 0 to 59 s in BCD. This requires a 2-digit counter. One counter, for the least significant decimal digit, would count from 0000 through 1001, and the other, for the most significant, would count from 0000 to 0101. Since the 7490 is already a BCD counter that counts from 0000 to 1001, it can be used for the least significant digit. The most significant digit requires a modulo-6 counter. Use of the R_0 inputs can convert the 7490 into a modulo-6 counter. If Q_C and Q_B are connected to $R_{0(1)}$ and $R_{0(2)}$, the 7490 will count from 0000 to 0101. The circuit for a 2-digit BCD 60 s counter is shown in Fig. 7-23. The R_9 inputs on both counters are grounded so that they don't reset to 1001 at all times.

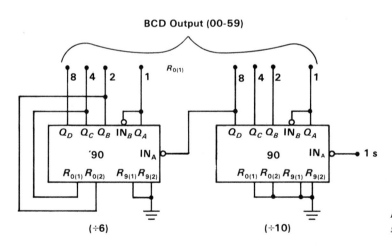

FIGURE 7-23. Two-digit BCD 60-second counter.

PROGRAMMABLE MODULUS COUNTERS

Some counters have LOAD inputs that can be used to program the counter's modulus. An example of this type of counter is the 74163 synchronous 4-bit

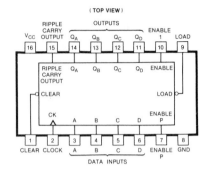

a. Pinout.

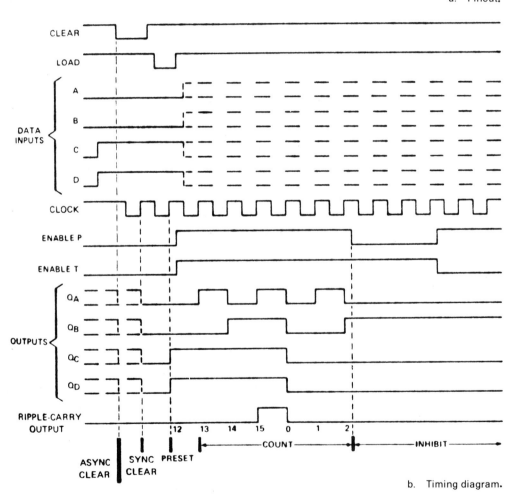

b. Timing diagram.

FIGURE 7-24.   The 74163 synchronous 4-bit counter. (Courtesy of Texas Instruments Incorporated.)

counter illustrated in Fig. 7-24. Whenever this counter functions as a *programmable counter*, its parallel inputs are used to gate a number into the counter. For example, the RIPPLE CARRY output, which goes high only during the fifteenth CLOCK period, can be inverted and connected back to the LOAD input. This connection is illustrated in Fig. 7-25. It is necessary to use the INVERTER because the RIPPLE CARRY output is an active-high signal and the LOAD input is active-low.

With this connection, every time the counter reaches 1111 the RIPPLE CARRY output goes high. The INVERTER then applies a logic 0 to the synchronous LOAD input, which loads the number on the parallel inputs into the counter on the next positive edge of the CLOCK. If a 0000 is attached to the

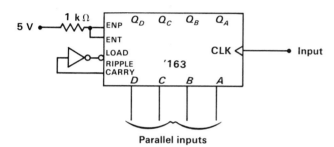

FIGURE 7-25. The 74163 synchronous 4-bit binary counter wired so that the parallel inputs are loaded at the time that the counter's outputs normally return to 0000.

TABLE 7-3 MODULUS OF THE CIRCUIT IN FIG. 7-25.

Data				
D	*C*	*B*	*A*	Modulus
0	0	0	0	16
0	0	0	1	15
0	0	1	0	14
0	0	1	1	13
0	1	0	0	12
0	1	0	1	11
0	1	1	0	10
0	1	1	1	9
1	0	0	0	8
1	0	0	1	7
1	0	1	0	6
1	0	1	1	5
1	1	0	0	4
1	1	0	1	3
1	1	1	0	2
1	1	1	1	1

parallel inputs, the counter will have 16 states because it will count from 0000 through 1111. If a 0001 is attached to the parallel inputs, the counter will count from 0001 through 1111 and have a modulus of 15. The parallel inputs can thus be used to program the modulus of this counter. The resulting modulus from each different combination of the parallel LOAD inputs is depicted in Table 7-3.

Cascading MSI Modulus Counters

The 74163 counters can be cascaded by use of the ENT and ENP inputs. These inputs must be connected to a logic 1 level in order for the counter to count. Cascading is accomplished by connecting the RIPPLE CARRY output of one counter to the ENT and ENP inputs of the next counter. This circuit connection, illustrated in Fig. 7-26, allows the outputs to change in true synchronization. The RIPPLE CARRY output becomes a logic 1 whenever the counter reaches its maximum count of 1111. Because the RIPPLE CARRY output has been attached to the ENT and ENP inputs, the next CLOCK pulse will clear the first counter to 0000 and the second counter will be clocked.

FIGURE 7-26. Using the ENP and ENT inputs when the 74163 is cascaded.

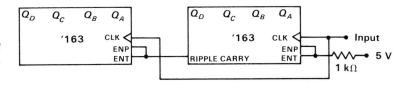

REGISTERS

The register is another very important synchronous building block used in many digital systems. A *register* is a grouping or collection of flip-flops that temporarily holds a multiple-digit binary number. A calculator, for example, can display an answer for as long as the user wishes because the answer is held in a register.

TYPES OF REGISTERS

There are four basic types of registers, and each is suited to a particular form of data transfer. The four types of registers are parallel-to-parallel, parallel-to-serial, serial-to-parallel, and serial-to-serial. Fig. 7-27 shows all types of registers and their data transfer streams.

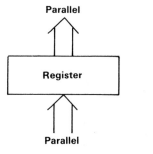

Parallel

Register

Parallel

a. Parallel-to-parallel.

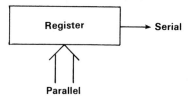

Register → **Serial**

Parallel

b. Parallel-to-serial.

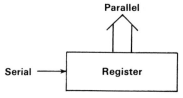

Parallel

Serial → **Register**

c. Serial-to-parallel.

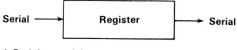

Serial → **Register** → **Serial**

d. Serial-to-serial.

FIGURE 7-27 . The four forms of data transfer.

Parallel-to-parallel Registers

A 4-bit to 4-bit parallel transfer register is illustrated in Fig. 7-28a. This circuit uses four D-type flip-flops that have common CLOCK inputs. If new information is placed on the four parallel inputs and a CLOCK pulse arrives, the data are transferred from these four inputs to the four outputs in parallel. This action is depicted in the timing diagram provided in Fig. 7-28b.

Fig. 7-29 shows the 74175 4-bit parallel register. Notice that this is connected in essentially the same way as the circuit in Fig. 7-28a except for one additional input—the active-low CLEAR input.

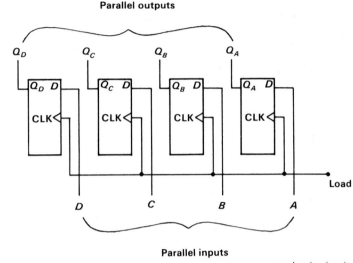

Parallel outputs

Parallel inputs

a. Logic circuit.

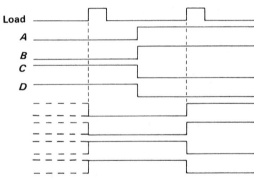

b. Timing diagram.

FIGURE 7-28. Four-bit parallel transfer register and timing diagram. Notice that the CLOCK input is labeled LOAD.

Parallel-to-serial Registers

Parallel-to-serial registers are often called *shift registers* because of the way that information is converted from parallel to serial form. Data are entered in parallel when either the D and the CLOCK inputs or the asynchronous CLEAR and PRESET inputs are used. Fig. 7-28 shows data entered into the D-type flip-flops through the D input by a CLOCK pulse. Fig. 7-30 shows how to enter parallel information through the $\overline{CLR}$ and $\overline{PR}$ inputs of a flip-flop. This technique is called either *jam entry* or *asynchronous entry*. Integrated circuit registers employ both techniques and the choice of one over the other depends

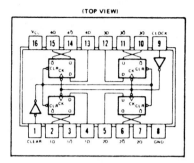

(TOP VIEW)

FIGURE 7-29. The 74175 parallel
register. (Courtesy of Texas
Instruments Incorporated.)

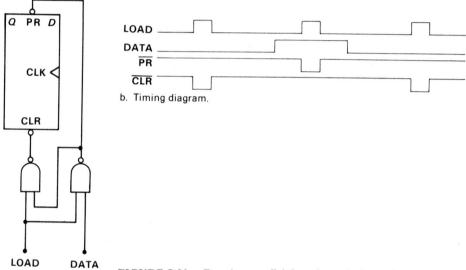

b. Timing diagram.

a. Logic circuit.

FIGURE 7-30. Entering parallel data through the CLEAR
and PRESET inputs of a flip-flop.

on the application. The clocked input technique is used where the outputs must
change in unison with the CLOCK. The asynchronous input technique is used
in less critical applications that do not require the outputs to change in synchronization with the CLOCK.

Data are converted from parallel to serial form by *shifting*, or moving, the
information through each flip-flop in the register either to the right or to the
left. An example of a parallel-loaded serial right-shift register is illustrated in
Fig. 7-31a. The parallel information enters through the $\overline{CLR}$ and $\overline{PR}$ inputs,

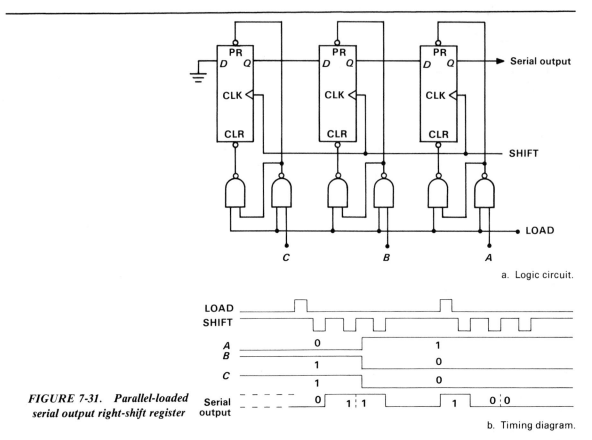

a. Logic circuit.

FIGURE 7-31. *Parallel-loaded serial output right-shift register*

b. Timing diagram.

and the CLOCK shifts the information to the right. The timing diagram provided in Fig. 7-31b clearly demonstrates how the register does the parallel load and the serial shifts. Each time a CLOCK pulse arrives, the information present at each D input passes on to each Q output. Moving through the register, the data appear a bit at a time at the output connection. The timing diagram shows the parallel input and the same information in serial at the output.

If a shift-left register is required, the relative positions of the D and Q pins on the flip-flops can be reversed. A parallel-to-serial shift-left register is illustrated in Fig. 7-32a and its timing is shown in Fig. 7-32b. It will be useful to compare this drawing with Fig. 7-31 to understand the minor change required to convert from right to left.

Serial-to-parallel Registers

The serial-to-parallel register is merely a shift register whose outputs allow the user to view the information in parallel. Data are entered through the D inputs

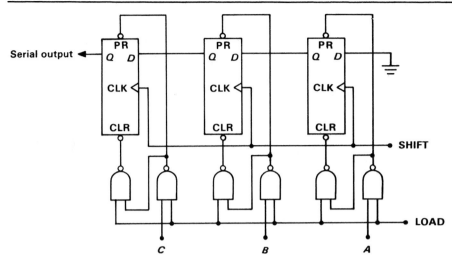

a. Logic circuit.

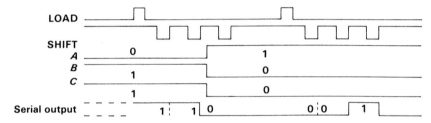

b. Timing diagram.

FIGURE 7-32. Parallel-to-serial shift-left register.

or the $J\overline{K}$ inputs to the shift register. This type of register is illustrated in Fig. 7-33, which shows both the D input method (a) and the $J\overline{K}$ inputs method (b). In the D input method, the serial signal is connected directly to the D input, and in the $J\overline{K}$ method, the J and $\overline{K}$ inputs are tied together and connected to the serial data input.

Data are shifted into this type of register by a periodic CLOCK pulse. The CLOCK shifts the data into the shift register, which makes the data available at the parallel outputs. It takes four clock pulses to shift a serial message completely into a 4-bit serial-to-parallel register.

Serial-to-serial Registers

The serial-to-serial register, shown in Fig. 7-34, is a shift register with one input and one output. The input is connected to the serial data, which then pass through the register to the output labeled "Serial output." This register is basically the same as the serial-to-parallel register except for the point at which the output is taken.

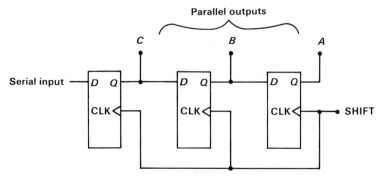

a. *D*-type serial shift-right register.

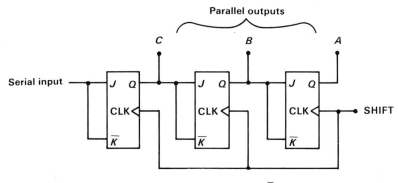

FIGURE 7-33. Serial-to-parallel register.

b. *JK̄* serial shift-right register.

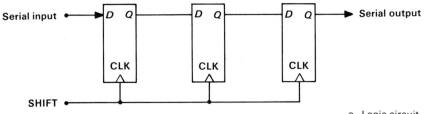

a. Logic circuit.

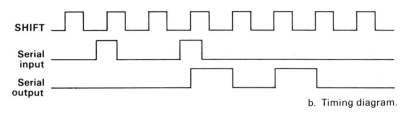

FIGURE 7-34. Serial-to-serial register.

b. Timing diagram.

The 74194 Universal Register

The 74194 is called a *universal register* because it is capable of functioning as any type of register. In order for it to function universally it must contain parallel inputs, serial inputs, serial outputs, parallel outputs, and a method of control. This device is controlled through the S_0 and S_1 inputs, which select parallel loading, right shifting, left shifting, and INHIBIT, or no operation. The 74194 and its timing diagram are illustrated in Fig. 7-35.

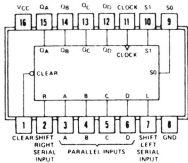

a. Pinout.

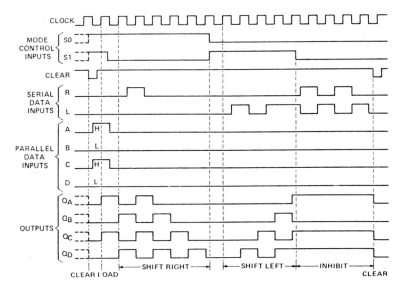

b. Timing diagram.

FIGURE 7-35. The 74194 universal shift register. (Courtesy of Texas Instruments Incorporated.)

RING COUNTERS

The *ring counter* is a special connection of the shift register that allows it to count. A simple 4-bit ring counter and its timing diagram are illustrated in Fig. 7-36. This counter is a serial shift-right register that has its output fed back into its input. As long as at least one bit is present, this type of counter will divide the input CLOCK frequency by four and generate four separate CLOCK phases by continually shifting the logic 1 bit around the ring or loop. The timing diagram illustrates the four phases of the output that are available from this ring counter. This type of counter is sometimes used to generate multiphase CLOCK waveforms. Its big disadvantage is that, if noise changes the logic 1 bit to a 0, the counter will no longer count.

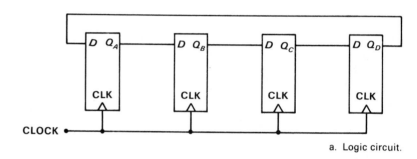

a. Logic circuit.

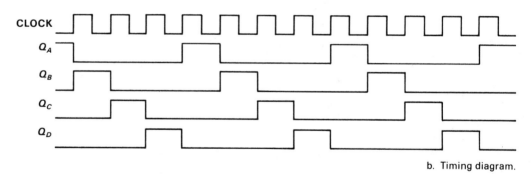

b. Timing diagram.

FIGURE 7-36.   Four-bit ring counter.

THE JOHNSON COUNTER

Another version of the ring counter is the *Johnson counter*, illustrated in Fig. 7-37. The Johnson counter is wired in almost the same manner as the ring counter. Notice that the input is connected to the $\overline{Q}$ output instead of the Q output. This slight modification causes this counter to work quite differently from the ring counter. The basic counting sequence for the 4-bit Johnson

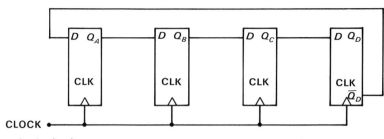

a. Logic circuit.

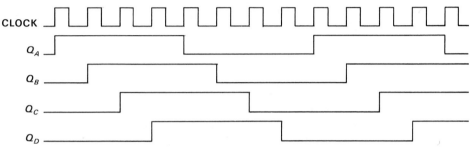

b. Timing diagram.

FIGURE 7-37. *Four-bit Johnson counter.*

counter is 0000, 0001, 0011, 0111, 1111, 1110, 1100, and 1000. Notice that all 1s are shifted through the counter and then all 0s. This 4-bit counter will divide the input frequency by eight, or twice the number of flip-flops; generate four output phases; and generate symmetrical squarewaves at its outputs.

SEQUENTIAL LOGIC DESIGN

Sequential logic design is different from combinational logic design because it starts with a state transition diagram instead of a truth table. The *state transition diagram* is a model of how the sequential system performs.

STATE TRANSITION DIAGRAMS

Fig. 7-38 illustrates the state transition diagram of a modulo-6 counter. Each circle in the diagram represents a unique state in this synchronous machine. Each interconnecting line represents a change from one state to another and

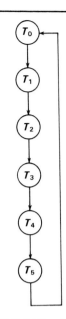

FIGURE 7-38. State transition diagram for a modulo-6 counter.

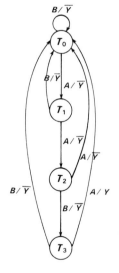

FIGURE 7-39. State transition diagram for a combination lock with combination $AABA$.

also the condition that causes the change in state. This change is most often caused by the CLOCK input to the system.

In order to form a clearer understanding of this type of diagram, let's convert a word problem into a state transition diagram.

Example 7-1. Draw a state transition diagram for a digital combination lock with combination $AABA$.

Solution. A digital combination lock is a good basic example of a sequential system. Our lock has two input switches that are labeled A and B. Because the combination is a sequence of four, this sequential machine requires four unique states. If any button is pressed out of sequence, the system should be able to reset to the initial state. Fig. 7-39 shows how to convert this information into a state transition diagram. States have been labeled T_0 through T_3, and the interconnecting paths have been labeled to indicate the desired flow. In this example, once the system is started at state T_0, the flow can proceed to state T_1 only if the A button is depressed. In addition to defining the inputs, the flow paths also specify the state of the output signal, which would be used to unlock a door or other device. The output signals are shown as the variables Y and $\overline{Y}$. If Y is present, a logic 1 is sent to the unlocking circuit, and if $\overline{Y}$ is present, a logic 0 causes the door to remain locked.

PRESENT STATE–NEXT STATE DIAGRAMS

Once a word problem has been converted to a state transition diagram, each state is assigned a unique binary bit pattern for the synchronous machine. In the digital lock example, the following assignments have been made: $T_0 = 00$, $T_1 = 01$, $T_2 = 10$, and $T_3 = 11$. Two flip-flops are required to specify these four states: these are designated here as flip-flop W and flip-flop X.

The *present state–next state diagram* is a truth table that is constructed a little differently from the type used in combinational logic design. This example, as shown in Table 7-4, has four inputs, or present states, and three outputs, or next states. The diagram illustrates the effect of a given set of inputs (the present state) and the next state of the system. The next state occurs at the time of the system CLOCK pulse, which is not illustrated in the diagram. Once this table has been drawn correctly, the flip-flop circuitry can be designed by developing the logic circuitry required to drive the J and K inputs of the two JK flip-flops. The logic circuitry is designed from a 4-variable Karnaugh map drawn to represent each J and each K input of each flip-flop. This example requires four 4-variable maps. A 4-variable map is required because of the four input signals—A, B, W, and X—and four maps are required because of the two flip-flops.

TABLE 7-4 THE PRESENT STATE–NEXT STATE DIAGRAM FOR THE STATE TRANSITION DIAGRAM OF FIG. 7-39.

Present state				Next state		
A	B	W	X	W	X	Y
0	0	0	0	0	0	0
0	0	0	1	0	1	0
0	0	1	0	1	0	0
0	0	1	1	1	1	0
0	1	0	0	0	0	0
0	1	0	1	0	0	0
0	1	1	0	1	1	0
0	1	1	1	0	0	0
1	0	0	0	0	1	0
1	0	0	1	1	0	0
1	0	1	0	0	0	0
1	0	1	1	0	0	1
1	1	0	0	0	0	0
1	1	0	1	0	0	0
1	1	1	0	0	0	0
1	1	1	1	0	0	0

MAPPING THE J AND K INPUTS FOR THE COMBINATION LOCK

The mapping of J and K inputs should involve only one flip-flop at a time. To single out blocks in the map, we use the inputs for the present state–next state diagram of Table 7-4. This determines the value placed in a block—0, 1, or X *(don't care)*—by the present and next states of the flip-flop. From a present state of 0, it is possible to reach the next state, 1, by either setting or complementing the flip-flop. This is accomplished by placing a 1 in the J map and an X in the K map. Thus a state of 11 would complement it, and a 0 on K and a 1 on J would set it (the effect is the same). Refer to Table 7-5 for a complete list of the possible present states (PS) and the next states (NS) and the values placed in the Karnaugh map. It will be necessary to understand why a 0, 1, or X is placed in a given area of the map before continuing to the next segment of this chapter.

Fig. 7-40a shows the J_W and K_W Karnaugh maps for the W flip-flop of Table 7-4. Starting with the first input condition, plotted in the upper left-hand block, we can use Table 7-4 to determine what W will do when changing from the PS to the NS. In this case W goes from a 0 to a 0 (no change). This can be accomplished by resetting the flip-flop or letting it remember: JK can be 00 or 01. Because K could achieve this result as either a 0 or a 1, we would place an X in this position

TABLE 7-5 VALUES USED IN A KARNAUGH MAP FOR DESIGNING A SEQUENTIAL LOGIC SYSTEM.

PS	NS	J	K	Meaning
0	0	0	X	Don't set, Don't care if reset.
0	1	1	X	Set, Don't care if complemented.
1	0	X	1	Reset, Don't care if complemented.
1	1	X	0	Don't reset, Don't care if set.

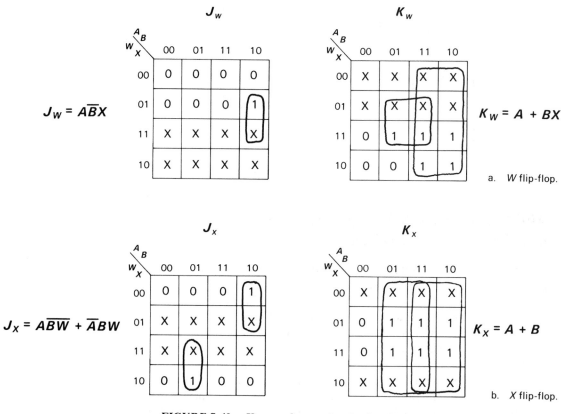

$$J_W = A\overline{B}X$$

$$K_W = A + BX$$

a. W flip-flop.

$$J_X = A\overline{BW} + \overline{A}BW$$

$$K_X = A + B$$

b. X flip-flop.

FIGURE 7-40. Karnaugh maps for the flip-flops used in the combination lock.

of the K_W map and a 0 on the J_W map. The remaining positions of these maps are filled in the same manner. Once the map is completed, the essential prime implicants are selected and the Boolean function for J_W and K_W is written. Fig. 7-40b shows the maps and Boolean functions for the X flip-flop.

THE CIRCUIT FOR THE COMBINATION LOCK

Fig. 7-41 illustrates the two push-button switches, *A* and *B*, which have been "debounced" with INVERTERS. The depression of either switch *A* or switch *B* triggers the 74121 monostable multivibrator, which produces a CLOCK pulse for the combination lock. In practice, the 74121 would be set for an output pulse width of less than 20 ms. Unless the 74121 is triggered for less than 20 ms, the system might miss the next switch *A* or *B* input.

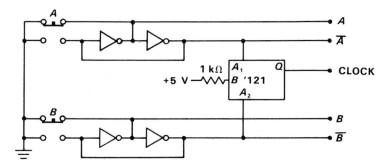

a. Logic circuit.

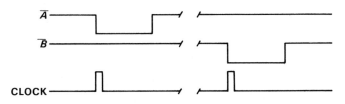

b. Timing diagram.

FIGURE 7-41. Circuit that debounces push-button switches A and B. It also generates the CLOCK for the combination lock.

Fig. 7-42 depicts two *JK* flip-flops driven with the logic circuitry designed in Fig. 7-40. Notice that the flip-flops have a common CLOCK signal since this is a synchronous circuit. The signal *Y*, which is used to unlock the door, must be generated for input state 1011, as illustrated. The 1011 input combination only occurs when switch *A* is pressed during state T_3. Refer to Fig. 7-42 for a detailed view of the operation of this circuit.

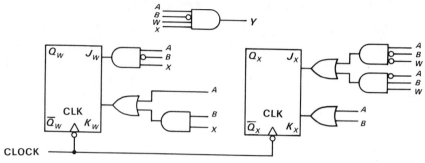

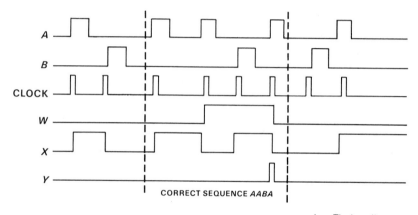

a. Logic circuit.

b. Timing diagram.

FIGURE 7-42. Circuit for
the combination lock and its
timing diagram.

SUMMARY

1 Synchronous counters and registers are examples of synchronous circuits because their next state depends upon their present state.

2 An asynchronous, or ripple, counter is a counter whose outputs do not change in synchronization.

3 An up counter counts in ascending numerical order.

4 A down counter counts in descending numerical order.

5 The modulus of a counter is equal to the number of different states of the counter. A modulo-6 counter has six different states.

6 A counter will divide the input frequency by its modulus or multiply the input time (period) by its modulus.

7   An Exclusive OR gate can be used as a controlled INVERTER.

8   An asynchronous counter constructed from positive edge–triggered flip-flops will count in one direction, while an asynchronous counter constructed with negative edge–triggered flip-flops will count in the opposite direction. This is true only if both counters are wired in exactly the same way.

9   The synchronous counter is a device that will count and produce outputs that change together or in synchronization with each other.

10   The maximum operating frequency of a counter is determined by the settling time of the flip-flops used in its construction.

11   When modulus counters are cascaded, their individual moduli are multiplied.

12   A decade counter is a modulo-10 counter.

13   A programmable modulus counter is a counter whose modulus can be changed by varying the number loaded at the end of each counting sequence.

14   There are four basic types of registers: parallel-to-parallel, parallel-to-serial, serial-to-parallel, and serial-to-serial.

15   A ring counter is a special form of register that will divide the input frequency by the number of flip-flops in the counter.

16   The Johnson counter is a special form of register that will divide the input frequency by a factor equal to two times the number of flip-flops in the counter.

17   Both ring and Johnson counters can be used to generate multiple-phase

KEY TERMS

asynchronous counter	modulus counter	ring counter
BCD counter	present state–next	ripple counter
biquinary counter	state diagram	shift register
decade counter	programmable counter	state transition diagram
down counter	programmable	synchronous counter
Johnson counter	modulus counter	up counter
modulus	register	

CLOCK waveforms. The outputs of the ring counter are asymmetrical, and the outputs of the Johnson counter are symmetrical.

18 A state transition diagram is used in sequential logic design for defining the problem in a logical manner.

19 The present state–next state diagram is used to develop the sequential logic circuit.

20 Each J and K input of each flip-flop is mapped in sequential logic design.

QUESTIONS AND PROBLEMS

1 How many different counting states do each of the following binary counters contain: 2-bit, 3-bit, 7-bit, and 9-bit?

2 Why is the JK flip-flop connected to toggle when it is used in the construction of binary ripple counters?

3 Most counters are cyclic in nature. What is meant by this statement?

4 A 3-bit binary counter divides the input frequency by what factor?

5 A 6-bit binary counter divides the input frequency by what factor?

6 Draw a 5-bit binary asynchronous up counter and its input and output waveforms.

7 Draw a 5-bit binary asynchronous down counter and its input and output waveforms.

8 What is the modulus of a counter?

9 A modulo-32 counter requires how many flip-flops?

10 The circuitry often used to cause an up/down counter to count up or down is the controlled INVERTER (see Fig. 7-7). Explain the operation of this circuit.

11 Why do the J and K inputs of Fig. 7-8 need to be controlled?

12 If a binary up counter is constructed, but the outputs are taken from $\overline{Q}$, what will be the effect on the counting sequence?

13 What is the main difference in the outputs of a synchronous counter when compared to an asynchronous counter?

14 Draw a 5-bit binary synchronous up counter and its input and output waveforms.

15 What determines the maximum operating frequency of a counter?

16 Is it possible for a binary counter to count from 000 to 001 to 010 and back to 000 again?

17 List the steps required to develop a modulo-X counter.

18 Draw the circuitry required to construct a modulo-7 counter, and also draw the output waveforms.

19 Draw the circuitry required to construct a modulo-14 counter, and also draw the output waveforms.

20 Draw the circuitry required to construct a modulo-100 counter, and also draw the output waveforms.

21 If a modulo-10 counter is cascaded with a modulo-5 counter, what would be the modulus of the resulting combination?

22 Glitches may occur in the outputs of some modulus counters. Why?

23 What is a biquinary counter?

24 Connect a 74163 so that it counts from 0001 to 0111 (a modulus of seven) and draw its output waveforms.

25 List the four different types of registers.

26 Draw four 3-bit circuits that represent each of the four different types of registers.

27 Explain how a shift register shifts.

28 A 5-bit ring counter will divide the input frequency by what factor?

29 A 5-bit Johnson counter will divide the input frequency by what factor?

30 Develop a state transition diagram for a 2-bit binary up counter that can be stopped at any count. (Use signal A to stop the counter when high.)

31 Develop the present state–next state diagram from the state transition diagram of problem 30.

32 Using Karnaugh maps, write the J and K input functions from the present state–next state diagram of problem 31.

33 Draw the logic circuit from the J and K Boolean expression developed in problem 32.

34 Repeat problems 30 through 33 for a combination lock with three inputs (A, B, and C). The combination is up to you.

INTERFACING LOGIC ELEMENTS

In previous chapters we have studied the interconnection of logic elements to make useful circuits (Chapters 3, 6, and 7) and several different logic families (Chapters 4 and 5). Now we turn to the interconnection of elements of *different* families and the connection of the logic circuits to external, or "real world," devices. Without these interconnections, most circuits would be of little use.

INTERFACING TTL WITH MOSFET LOGIC

Although it is generally desirable to design an entire system from a single logic family, it is sometimes necessary to connect members of different families. In this section, we will investigate the connection of TTL outputs to MOSFET inputs.

225

REVIEW OF TTL OUTPUT AND
MOS INPUT CIRCUITS

As Fig. 4-20 illustrated, the output voltages for TTL (worst case) are 0.4 V (V_{OL}) and 2.4 V (V_{OH}). The output current capability for TTL (see Table 4-4) is from 16 mA (I_{OL}) to −400 μA (I_{OH}).

The input parameters for NMOS and CMOS logic (at V_{DD} of 5 V) are taken from Table 5-1. For NMOS, V_{IH} is 2.0 V and V_{IL} is 0.8 V; I_{IH} and I_{IL} are negligible. For CMOS, V_{IH} is 3.5 V and V_{IL} is 1.5 V; and I_{IH} and I_{IL} here are also negligible, at less than a microamp each. (The parameters for 74HC can also be found in Table 5-1.)

A comparison of these input and output characteristics will reveal that the output currents of the TTL family are more than adequate to supply the requirements of either NMOS or CMOS, as is the low-level output voltage. The high-level output voltage is high enough for NMOS circuits, but it is not sufficient for 4000 series CMOS. Fig. 8-1 depicts the output stage of a TTL circuit and the input stage of a CMOS circuit. The two are not connected because some additional circuitry is necessary to make them compatible.

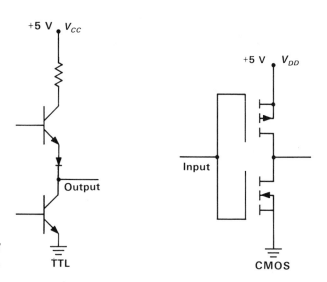

FIGURE 8-1. TTL output circuitry and CMOS input circuitry.

Since the TTL output voltage will not swing high enough to provide a logic 1 level for CMOS, some "help" must be provided for the TTL. This takes the form of a pullup resistor. Fig. 8-2 shows the connection of the pullup resistor. Since the input impedance of CMOS is so high, the pullup resistor will ensure that the high-level output voltage of TTL is virtually 5 V. However, when the

TTL gate is in the low state, the pullup resistor must not be able to pull the output voltage over 1.5 V, which is the "worst case" V_{IL} for CMOS. Also, the output circuits for different subfamilies of TTL (74LS, 74H, etc.) have slightly different characteristics. Values for the pullup resistor of between 1.5 and 4.7 kΩ provide enough pullup but will not sink the low output below the minimum for CMOS. A pullup value of 2.2 kΩ is probably the most common, as it falls in the middle of the range.

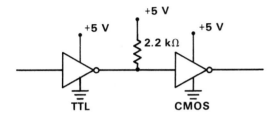

FIGURE 8-2. Pullup resistor used for interfacing TTL to CMOS.

INTERFACING TO CMOS WHEN V_{DD} IS NOT 5 V

As we have illustrated, interfacing TTL outputs to CMOS inputs is not too difficult—as long as the supply voltage for both circuits is the same. However, in some systems the CMOS circuitry must operate at voltages other than 5 V, and this requires some additional interfacing circuitry.

CMOS with V_{DD} Greater Than 5 V

When the supply voltage for CMOS is greater than 5 V, the interface can take either of two forms. If the TTL outputs are open-collector, the solution is easy. The same type of interface just described is used, except that the pullup resistor is connected to the higher supply voltage, as shown in Fig. 8-3. The value of this resistor is determined by the higher supply voltage. For V_{DD} of 10 V, 33 or 39 kΩ is appropriate. For V_{DD} of 15 V, 68 kΩ is a good value.

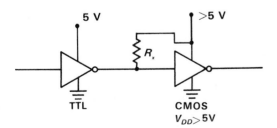

FIGURE 8-3. TTL interface when CMOS is powered by a supply voltage greater than 5 V.

If TTL outputs are not open-collector, a level-shifting circuit becomes necessary. A schematic of the CD40109B quad low-to-high voltage level shifter is shown in Fig. 8-4. The 40109 has *two* power supply connections with a common ground. The TTL supply is connected to the V_{CC} pin, and the higher CMOS supply is connected to the V_{DD} pin. The TTL output is thus connected to the A input, and the output is completely compatible with the high-voltage CMOS. The ENABLE pin in most circuits is connected to V_{CC}. However, it can be used to enable the CMOS inputs by connecting an appropriate TTL signal to it.

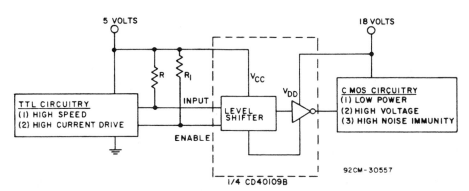

FIGURE 8-4. CD40109B quad voltage level shifter. (Courtesy of RCA.)

CMOS With V_{DD} Less Than 5 V

Some circuits must interface 5 V TTL with CMOS at some lower voltage (usually 3 V), although this is not common. This interfacing is quite simple, but it requires an additional IC package. The CD4049UB and the CD4050B are inverting and noninverting buffers, respectively. Unlike most CMOS gates, whose maximum input voltage is V_{DD} +0.5 V, these two ICs have an input range of up to +20.5 V, regardless of the supply voltage (V_{DD}). The abbreviated data sheet for these ICs is shown in Fig. 8-5.

Fig. 8-6 is the completed interface between 5 V TTL and 3 V CMOS using the 4049 as a buffering level translator. Since the 4050 is connected to the 3 V supply, its output will swing between 2.95 V and 0.05 V. To invert the TTL signal, the 4049 can substitute for the 4050.

NMOS to CMOS Interfacing

You may have noticed from Table 5-1 that the output voltages for NMOS are identical to those for TTL and suspected that a problem might exist between an NMOS output and a CMOS input. Notice, however, that those output voltages

STATIC ELECTRICAL CHARACTERISTICS

CHARAC-TERISTIC	CONDITIONS			Limits At Indicated Temperatures (°C) Values at −55,+25,+125 Apply to D,F,H Pkgs. Values at −40,+25,+85 Apply to E Package							UNITS
	V_O (V)	V_{IN} (V)	V_{CC} (V)	−55	−40	+85	+125	+25			
								Min.	Typ.	Max.	
Quiescent Device Current, I_{DD} Max.	−	0.5	5	1	1	30	30	−	0.02	1	μA
	−	0.10	10	2	2	60	60	−	0.02	2	
	−	0.15	15	4	4	120	120	−	0.02	4	
	−	0.20	20	20	20	600	600	−	0.04	20	
Output Low (Sink) Current I_{OL} Min.	0.4	0.5	4.5	3.3	3.1	2.1	1.8	2.6	5.2	−	mA
	0.4	0.5	5	4	3.8	2.9	2.4	3.2	6.4	−	
	0.5	0.10	10	10	9.6	6.6	5.6	8	16	−	
	1.5	0.15	15	26	25	20	18	24	48	−	
Output High (Source) Current I_{OH} Min.	4.6	0.5	5	−0.81	−0.73	−0.58	−0.48	−0.65	−1.2	−	
	2.5	0.5	5	−2.6	−2.4	−1.9	−1.55	−2.1	−3.9	−	
	9.5	0.10	10	−2.0	−1.8	−1.35	−1.18	−1.65	−3.0	−	
	13.5	0.15	15	−5.2	−4.8	−3.5	−3.1	−4.3	−8.0	−	
Output Voltage: Low-Level, V_{OL} Max.	−	0.5	5	0.05				−	0	0.05	V
	−	0.10	10	0.05					0	0.05	
	−	0.15	15	0.05				−	0	0.05	
Output Voltage: High-Level, V_{OH} Min.	−	0.5	5	4.95				4.95	5	−	
	−	0.10	10	9.95				9.95	10	−	
	−	0.15	15	14.95				14.95	15	−	
Input Low Voltage V_{IL} Max. CD4049UB	4.5	−	5	1				−	−	1	V
	9	−	10	2				−	−	2	
	13.5	−	15	2.5				−	−	2.5	
Input Low Voltage V_{IL} Max. CD4050B	0.5	−	5	1.5				−	−	1.5	V
	1	−	10	3				−	−	3	
	1.5	−	15	4				−	−	4	
Input High Voltage V_{IH} Min. CD4049UB	0.5	−	5	4				4	−	−	
	1	−	10	8				8	−	−	
	1.5	−	15	12.5				12.5	−	−	
Input High Voltage V_{IH} Min. CD4050B	4.5	−	5	3.5				3.5	−	−	
	9	−	10	7				7	−	−	
	13.5	−	15	11				11	−	−	
Input Current, I_{IN} Max.	−	0.18	18	±0.1	±0.1	±1	±1		$±10^{-5}$	±0.1	μA

RECOMMENDED OPERATING CONDITIONS at T_A=25°C, Except as Noted.
For maximum reliability, nominal operating conditions should be selected so that operation is always within the following ranges:

CHARACTERISTIC	LIMITS		UNITS
	Min.	Max.	
Supply-Voltage Range (V_{CC}) (For T_A=Full Package-Temperature Range)	3	18	V
Input Voltage Range (V_{IN})	V_{CC}*	18	V

*The CD4049 and CD4050 have high-to-low-level voltage conversion capability but not low-to-high-level; therefore it is recommended that $V_{IN} \geqslant V_{CC}$.

FIGURE 8-5.   Abbreviated data sheet for the 4049 and the 4050. (Courtesy of RCA.)

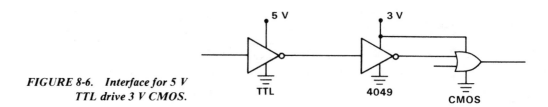

FIGURE 8-6. Interface for 5 V
TTL drive 3 V CMOS.

are driving TTL. When driving a CMOS input, with its very high impedance, the NMOS output will swing almost from rail-to-rail (V_{DD} to V_{SS}), so no interfacing problem exists.

INTERFACING MOSFET LOGIC WITH TTL

While the previous section covered the interfacing of TTL outputs to MOS inputs, this section deals with the converse problem: interfacing MOS outputs to TTL inputs. This type of interfacing is probably more common than TTL to MOS interfacing.

In the interfacing of CMOS logic to TTL logic, the primary concerns are the output voltages and currents of the CMOS and the input voltage and current requirements for TTL. Again referring to Table 5-1, we can determine these values to see if a problem exists. The output voltages of the CMOS gates easily surpass the necessary requirements for TTL inputs; however, the output sink current I_{OL} does not. What actually happens is that the CMOS output cannot sink enough current to bring its output voltage below the 0.4 V necessary for a TTL low input. (Remember, Table 5-1 gives the values for connecting CMOS to CMOS.) When sinking a TTL load, the output voltage will not be able to meet the specified value. Notice, however, that standard (buffered) CMOS does meet all criteria for LSTTL—but just barely. In practice, the fanout for CMOS to LSTTL is only one.

Therefore, if a CMOS output is driving an LSTTL load of one, no interfacing problems exist and the circuitry can be connected as if it were all of the same family. However, if the circuit were modified at a later time, it would be important not to connect an extra LSTTL input to a CMOS input.

For interfacing with standard TTL, another method must be used. This method is the same as that used in interfacing 3 V CMOS with TTL. In addition to a tolerance of high voltage swings on their input, the 4049 and 4050 have greater current output source and sink capabilities. As Fig 8-5 shows, the low-level output current (at 25° C) is 4.6 mA minimum, and the high-level output current is -1.1 mA minimum. These currents easily drive standard TTL loads. Table 8-1 shows the fanout of a 4049 or 4050 when interfaced with

TABLE 8-1 FANOUT FOR THE 4049/4050 INTERFACED WITH VARIOUS TTL FAMILY MEMBERS. (Courtesy of RCA.)

4049/4050	TTL Family				
Fanout	74	74H	74L	74LS	74S
Minimum	1	1	14	7	1
Typical	3	2	23	14	2

various members of the TTL logic family. Notice that the minimum for any family is one. Therefore, the same caution must be used in modifying existing systems as that used in modifying standard CMOS and LSTTL.

If a fanout of more than one standard TTL input is necessary, one of two methods can be employed. First, the 4049 or 4050 can drive a TTL buffer, which in turn has a standard TTL fanout of ten. Also, several buffers from the 4049 or 4050 package can be connected in parallel to give extra drive capability. Paralleling outputs in CMOS is perfectly acceptable, because the process tolerances are so similar that one gate will not become a current hog and burn itself up. Even better matching can be achieved by paralleling gates or buffers on the same IC.

As we have seen, CMOS circuits don't always operate at 5 V. The interfacing of a CMOS circuit operating at a voltage other than 5 V to TTL requires the application of the 4049 or 4050. The circuit is nearly the same as that used for interfacing to 3 V CMOS except that the input and output families are reversed. Fig. 8-7 depicts a typical circuit for interfacing high-voltage CMOS to TTL. The use of high-voltage CMOS in this type of circuit is likely to gain additional noise immunity.

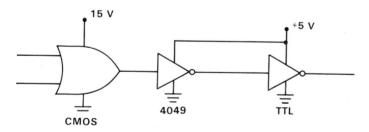

FIGURE 8-7. Interface for high-voltage CMOS to TTL.

INTERFACING SWITCHES WITH LOGIC

One of the most common input devices to a digital circuit is a simple switch. Switches take many forms: normally open push-button, normally closed push-button, toggle switches, and so on. This section investigates the methods and problems associated with interfacing switches to logic circuitry.

INTERFACING TTL TO SWITCHES

A typical switch-to-TTL circuit is shown in Fig. 8-8. This configuration is for an active-high switch; that is, closing the switch will force a high state on the input (and in this case the output) of the gate. Resistor R_1 forces a low state on the gate input when the switch is open. We calculate its resistance value by using the maximum low-input current value of 1.6 mA. To sink 1.6 mA and still have a maximum voltage of 0.8 V, the maximum value of resistance is 500 Ω for standard TTL. It is desirable to choose a value near the maximum so as not to draw excessive current through the resistor when the switch is closed. Also, remember that the resistance value has a tolerance range of 10 percent. The closest standard value is 470 Ω; however, we can't use this value with reliance, because at 10 percent tolerance the range of values is 423–517 Ω. Therefore, we must choose the closest lower value, which is 390 Ω.

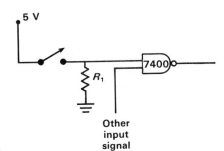

FIGURE 8-8. *Switch-to-TTL interface.*

The circuit in Fig. 8-9 is similar to that in Fig. 8-8, but this time the switch is active-low. In other words, when closed, it connects the gate input to ground. The resistor in this circuit pulls the input high when the switch is open. As mentioned before, a TTL input will float high; however, this is not universally reliable, so the insurance of a pullup resistor is well worth the trouble. For the pullup, only 40 μA must be sourced, and the minimum high-voltage level is 2.0 V. Application of Ohm's Law yields a value of 75,000 Ω. We could use a 68

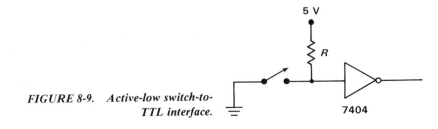

FIGURE 8-9. *Active-low switch-to-*
TTL interface.

kΩ resistor and be within tolerance. However, a lower value will increase noise tolerance and not appreciably increase power supply current. A typical range for this resistor, then, is from 10 to 68 kΩ.

SWITCH CONTACT RESISTANCE

Most of the common mechanical switches, such as toggle and push-button switches, have very small resistances from input to output (called *contact resistance*). When using these switches with TTL logic, we can assume the contact resistance to be zero and ignore it. However, many modern circuits utilize mylar switch overlays and keyboards like the one shown in Fig. 8-10. Instead of mechanical contacts, these use a conductive ink as the switch contact. Depending on the process used, the contact resistance can be as high as several hundred ohms. When resistance is this high, the circuit of Fig. 8-8 must be converted to that of Fig. 8-11. When the switch is closed, there is a voltage divider across the input of the gate, and the resistors must be resized to compensate for this.

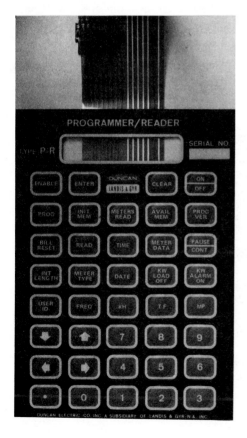

FIGURE 8-10. Mylar keyboard assembly.

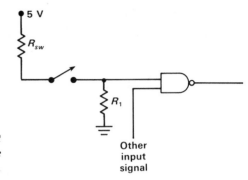

*FIGURE 8-11.   Equivalent circuit
to Fig. 8-8 when switch resistance
is high.*

CONTACT BOUNCE

Switches exhibit a phenomenon that can disrupt circuit operation significantly if not compensated for. When a switch is closed, contact bounce (or switch bounce) is generated. The effect of closing a switch is shown in the oscilloscope photo in Fig. 8-12. Notice how the output oscillates for a time before assuming the low state. In mechanical switches, this is caused by the contacts actually bouncing against each other. In conductive ink switches, it is caused by a combination of bouncing and grinding of the two contact surfaces against one another, since they are not as smooth as the metal contacts used in mechanical switches.

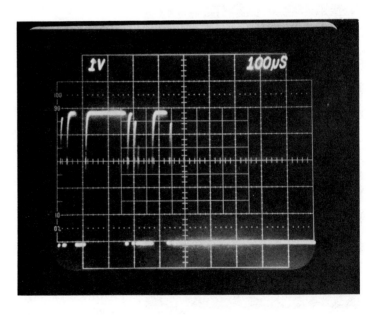

*FIGURE 8-12.   Oscilloscope trace
showing contact bounce.*

The oscillations associated with contact bounce damp out after a few milliseconds. For some circuits this oscillation has no effect on the operation. Most circuits, however, either switch in response to the oscillations or go into an undefined state. Either of these results will cause the circuit to operate improperly or not at all.

Fortunately, we can eliminate contact bounce with a logic circuit. Fig. 8-13 is a schematic representation of a switch debounce circuit. When the switch closes, the capacitor begins to discharge. When a positive bounce occurs (as in Fig. 8-12), the capacitor begins to charge again. The RC time constant is selected so that the capacitor takes several milliseconds of constant low-level current to discharge it enough for a logic 0 state. The feedback resistor helps provide a snap action by feeding back some of the stable state from the isolated second INVERTER. This hysteresis effect of the feedback resistor also helps eliminate undefined states on the input of the first INVERTER. Fig. 8-14 is an

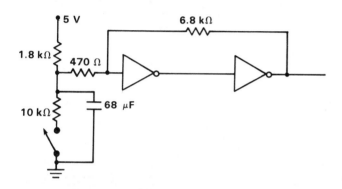

FIGURE 8-13.   Switch debounce circuit.

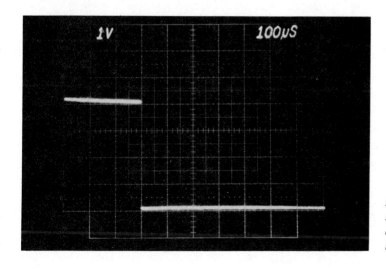

FIGURE 8-14.   Oscilloscope trace showing the action of the same switch as in Figure 8-12 after the installation of the contact debounce circuit.

oscilloscope photograph of the same switching as Fig. 8-12, after passing through the debounce circuit. A similar circuit can be used for an active-high switch.

INTERFACING SWITCHES TO MOS LOGIC

Interfacing switches to MOS logic is essentially the same as interfacing to TTL. With MOS logic, there is no choice about floating inputs; they must be avoided. Figs. 8-15 and 8-16 show circuits for interfacing active-high and active-low switching configurations. The only difference from TTL interfacing is the value of the pullup or pulldown resistor. The calculated value for the pulldown resistor, based on the values in Table 5-1 for CMOS, is 5 MΩ (for 5 V). However, due to the noise considerations, which are the same as those for the TTL circuit, a better choice would be a value in the range of 100 kΩ to 1 MΩ. Even for a 100 kΩ resistor, the leakage current through the resistor when the switch is on is only 50 μA. For battery circuits, it is usually better to choose the 1 MΩ resistor to cut current consumption to the bare minimum.

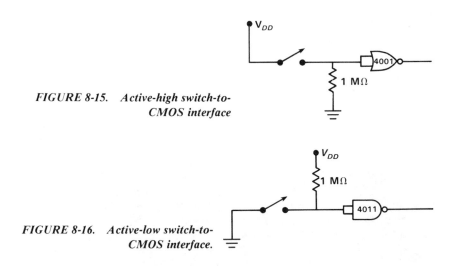

FIGURE 8-15. *Active-high switch-to-CMOS interface*

FIGURE 8-16. *Active-low switch-to-CMOS interface.*

For the pullup resistor in Fig. 8-16, we again calculate a value, based on 5 V parameters, of 5 MΩ (since we can drop a maximum of 1.5 V across the pullup and still maintain the minimum 3.5 V on the input). According to the same reasoning a range of 100 kΩ to 1 MΩ is a more practical choice.

Mylar Switches

Interfacing mylar/conductive ink switches to CMOS is no problem since the gate inputs draw virtually no current. Thus the resistance through the switch does not drop, even if it is several thousand ohms. This fact can be exploited by the designer. In most instances, the price of the conductive ink switch falls as resistance is allowed to rise. So, if a mylar/conductive ink switch is designed into a CMOS circuit, the price of the switch can be reduced with a noncritical ink resistance specification.

The debouncing problem is the same with CMOS as it is for TTL. The circuit in Fig. 8-13 will work for CMOS, but the circuit in Fig. 8-17 takes advantage of the built in hysteresis of the Schmitt trigger.

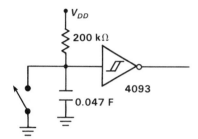

FIGURE 8-17.   Switch debounce circuit for CMOS.

MULTIPLEXERS AND DEMULTIPLEXERS

Sometimes an interface is necessary between different parts of a circuit, which may even be of the same logic family but which operate in different modes. Multiplexing is a scheme of putting multiple signals on the same channel. To interface the multiplexer to a part of the circuit that can only handle one signal at a time, a *demultiplexer* must be used.

The circuit in Fig. 8-18 is a 16-input multiplexer: a 74150. The truth table for the circuit appears in Table 8-2. The operation of the multiplexer is simple. Any of the 16 inputs can be presented to the output (in the inverted state) when the address of that input is presented at the four SELECT inputs. An additional input, the ENABLE line, is used to disable the device. One useful application of a multiplexer is a 16-digit keypad. Each of the keys would be one of the 16 inputs, and they could be scanned by sequentially addressing the 74150.

The advantage of multiplexing is in reducing the number of wires or printed circuit traces that must be routed. With microprocessor circuits, it is generally desirable to minimize the number of "wires," called input/output or I/O, that any part of a circuit sends to the microprocessor. With multiplexing, the microprocessor can use its limited I/O for more tasks. The microprocessor can be internally programmed to handle the multiplexed data as easily as it handles 16 individual inputs.

Sometimes there is a part of the circuit that cannot handle multiplexed information. In this case, a demultiplexer circuit is necessary. For example, a

TABLE 8-2 TRUTH TABLE FOR 74150 16-BIT MULTIPLEXER.
(Courtesy of National Semiconductor Corporation.
Authorization has been given by National Semiconductor
Corporation, Santa Clara, California, world leaders in
Digital Acquisition Circuitry.)

| | Inputs | | | | |
| | Select | | | Strobe | Output |
D	C	B	A	S	W
X	X	X	X	H	H
L	L	L	L	L	$\overline{E_0}$
L	L	L	H	L	$\overline{E_1}$
L	L	H	L	L	$\overline{E_2}$
L	L	H	H	L	$\overline{E_3}$
L	H	L	L	L	$\overline{E_4}$
L	H	L	H	L	$\overline{E_5}$
L	H	H	L	L	$\overline{E_6}$
L	H	H	H	L	$\overline{E_7}$
H	L	L	L	L	$\overline{E_8}$
H	L	L	H	L	$\overline{E_9}$
H	L	H	L	L	$\overline{E_{10}}$
H	L	H	H	L	$\overline{E_{11}}$
H	H	L	L	L	$\overline{E_{12}}$
H	H	L	H	L	$\overline{E_{13}}$
H	H	H	L	L	$\overline{E_{14}}$
H	H	H	H	L	$\overline{E_{15}}$

microprocessor might be programmed to turn on one of 16 LEDs to indicate some state. Rather than use up 16 output lines on the microprocessor, the multiplexed data can be sent out on four outputs as a binary number, or address, and then demultiplexed at the LED. The circuit in Fig. 8-19 uses a 74154 for this purpose. With the ENABLE inputs low, the output line defined by the address goes low. This then turns on the LED specified. The problems at the end of this chapter will provide some additional experience with multiplexers and demultiplexers.

There are other types of multiplexers and demultiplexers (often called "mux" and "demux" circuits). These are beyond the scope of this book in terms of detailed study, but will be mentioned. One such multiplexer is the time-division multiplexer.

A time-division multiplexer places several digital signals on a channel; however, a given input is placed on the channel only for a brief time, and then other inputs are placed on the channel. After the other inputs have been placed,

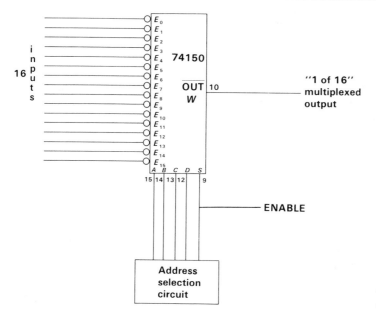

FIGURE 8-18.   A 74150 16-input multiplexer.

the first input is again put onto the channel for a brief time, and the cycle is repeated. The signals are demultiplexed at the receiving end with synchronous techniques. This scheme is often used in communications systems.

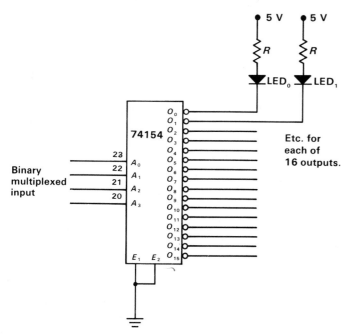

FIGURE 8-19.   74154 4- to 16-line demultiplexer.

DISPLAY DEVICES

One of the most common interfaces to a digital circuit is a *display device.* LEDs, LED displays, LCDs, vacuum fluorescent displays, and CRTs (TV screens) are the most common display interfaces. Fortunately for the digital designer, the interface circuitry for most of these interfaces has been incorporated into monolithic IC packages. But the interface must be understood before these devices can be used properly.

LIGHT-EMITTING DIODE INTERFACING

Perhaps the most common interface to a digital circuit is the light-emitting diode, or LED. LEDs are available as a single "bulb" or "discrete LED," as individual- or multiple-digit displays, or as bar graph indicators. Regardless of the configuration, the interfacing is mostly the same.

The voltage drop from anode to cathode across an LED is different from that of the silicon diode since the semiconductor used in most LEDs is made of gallium. The typical forward voltage drop across an LED is 1.5–2.0 V, depending on the LED used. Most LEDs can handle a fairly large reverse voltage, on the order of 5–13 V (i.e., large with respect to logic voltages). Probably the most critical parameter in interfacing to LEDs is the forward current. The light intensity is proportional to the current flowing through the LED, so in many cases the designer wants to put in as much current as he safely can to get maximum brightness.

A typical LED data sheet, for the National Semiconductor NSL5020, is shown in Fig. 8-20. The specified absolute maximum forward current (I_f) is 70 mA. In practical use, we would never want to put 70 mA through this LED continuously. It is always desirable to design somewhat conservatively. The practical maximum for this LED is about one-half of the absolute maximum, or 35 mA. Notice that the parameters are stated for a forward current of 20 mA. The value given here is always a good place to start when designing with an LED. If more brightness is needed, some additional current can be drawn, or if the LED is too bright, the current can be reduced.

Table 5-1 will show that for both TTL and CMOS the current-sinking capability is much greater than the current-sourcing capability. Only high-speed CMOS has symmetrical current sink/source characteristics. Notice, also, that even TTL, which can sink the greatest current, can only sink 16 mA.

TTL to LEDs

Because of the large difference in sink and source capabilities of TTL, it is usually easiest to interface TTL so that it turns the LED on with a low and off

Electrical and Optical Characteristics (25°C)

PARAMETER	CONDITIONS	5020
Forward Voltage (V_F) Typ Max	$I_F = 20$ mA	 1.8 2.0
Reverse Breakdown Voltage (BV_R) Min	100 µA	 5.0
Light Intensity (I) Min Typ	$I_F = 20$ mA	 0.5 1.0
Peak Wavelength Typ	$I_F = 20$ mA	 660
Spectral Width, Half-Intensity Typ	$I_F = 20$ mA	 40
Light Rise and Fall Time, 10%–90% Typ	Step Change of I_F, 50 Ω System	 50
Angle of Half-Intensity Off Axis Typ		 40
Capacitance Typ	V = 0, 1 MHz	 75

FIGURE 8-20. Data sheet for National NSL5020 LED display. (Courtesy of National Semiconductor Corporation. Authorization has been given by National Semiconductor Corporation, Santa Clara, California, world leaders in Digital Acquisition Circuitry.)

with a high. A typical current for interfacing a TTL gate to an LED is depicted in Fig. 8-21. The LED anode is connected to V_{CC}, and the cathode is connected to the gate output. A series resistor limits the amount of current passing through the LED and into the gate.

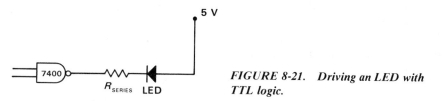

FIGURE 8-21. Driving an LED with TTL logic.

If the LED is to be turned on with a logic high, or if LSTTL, which can only sink 4 mA, is used, an external transistor can be used as a DC current amplifier; this configuration is drawn schematically in Figs. 8-22a and 8-22b. The resistor in the base of the transistor limits the current going out of the gate and into the transistor when it is on. Since the DC gain of a small-signal transistor is typically around 100, it is important that both the base resistor and the series LED resistor be properly sized.

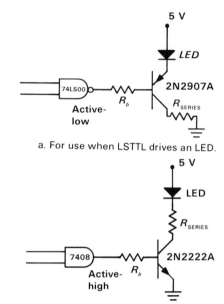

a. For use when LSTTL drives an LED.

b. For use when a logic high is needed to drive an LED.

FIGURE 8-22.   Transistor amplifier.

CMOS to LEDs

Interfacing CMOS to LEDs has the same problem as interfacing a logic high or LSTTL: sufficient drive current is not available out of the CMOS gate. Therefore, the interfacing of CMOS to an LED requires an amplifier. In Fig. 8-23, a PNP transistor is used, which will turn on the LED when a low appears on the output of the gate. A PNP transistor is used for the same reason an active-low input is used to turn on the LED with TTL: the sink current of CMOS is greater than the source current. With a transistor gain of 100, the 160 μA of source current might not be enough to illuminate the LED fully when amplified to 16 mA.

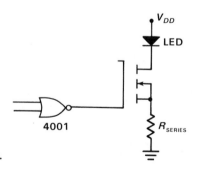

FIGURE 8-23.   CMOS-to-LED interface.

If it is necessary to use a logic 1 to turn on the LED, it is best to use a discrete MOSFET as an amplifier, as shown in Fig. 8-23. The resistor from gate to ground ensures predictable V_{gs} to turn the MOSFET on. As usual, the series LED resistor limits the current.

LED Displays

The most common LED display is a *seven-segment display*. The segments are labeled as shown in Fig. 8-24. To show a 3 on the display, for example, we would illuminate segments a, b, c, d, and g. Each segment is interfaced as a discrete LED—because that is exactly what it is.

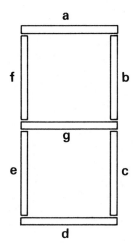

FIGURE 8-24. *Labeling of the segments of a seven-segment display.*

One problem with LED displays occurs when many digits are needed in an application. If every digit requires seven drivers, it is evident that a 6-digit display would be quite complex. One solution to this problem is to use an integrated display driver such as the 7447. In addition to supplying the proper seven-segment outputs, it accepts BCD inputs, which are more easy to generate than the seven-segment codes. Thus, in addition to being a display driver, the 7447 is also a decoder or demultiplexer. Fig. 8-25 shows a typical circuit for a 3-digit LED display.

In the 3-digit display of Fig. 8-25, each digit is driven directly by a 7447. When many digits are needed, the solution is to multiplex the display, using a form of time-division multiplexing. Fig. 8-26 is a block diagram of a 3-digit display multiplexer. The switch is actually a sequential logic circuit, the operation of which is quite simple. The amount of time that any digit is actually illuminated is one-third (since there are three digits) of the total time. This is called the *duty*

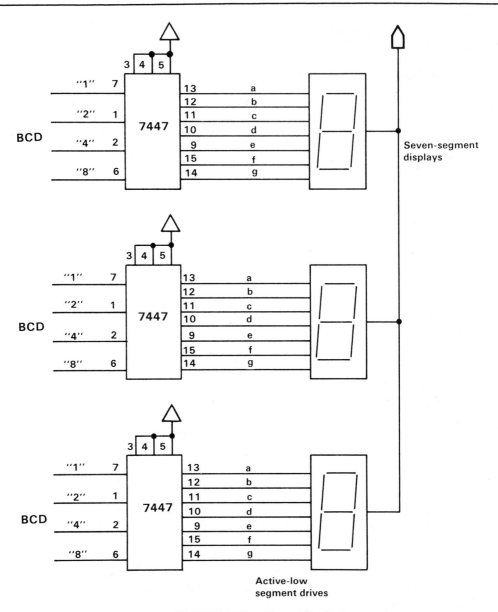

FIGURE 8-25. Three-digit display circuit.

cycle. Let's say that we want to display the number 375. When the switch is in the first-digit position, segments a, b, c, d, and g are turned on. But, since only the first digit has voltage applied to it, it is the only one that illuminates. Similarly, when the switch is in the second-digit position, segments a, b, and c are illuminated; since only the second digit has power applied to it, only it will

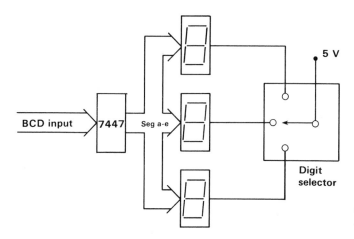

FIGURE 8-26. *Block diagram of a 3-digit multiplexer.*

illuminate. The same procedure occurs with the third digit. Most multiplexer circuits usually have a short interdigit blanking period, during which time all segments are turned off. This eliminates "ghosting"; for our example, it would eliminate the faint image of a 3 appearing on the position of the second digit.

Because each digit is on only one-third of the time, the display will not be as bright as a continuously driven LED using the same current. When multiplexing, therefore, we use greater currents. Instead of stating the maximum forward current, specifications for displays intended for multiplexing state the maximum average current and the maximum peak current. We determine the average current by multiplying the on current by the duty cycle in decimal form. For the example above, suppose the on current is 15 mA. If we multiply this by 0.33 (one-third duty cycle), we find that the average current is 5 mA. The LED will have about the same brightness at 15 mA and 33 percent duty cycle as it will with 5 mA and continuous duty cycle. The intensity will always be somewhat less when multiplexing is used.

INTERFACING LIQUID CRYSTAL DISPLAYS

Liquid crystal displays, or LCDs, are used for a variety of applications. Modern LCDs can be custom-designed to fit the user's exact application and can even be colored. LCDs have two major advantages over LED displays: (1) The current drawn by an LCD is very small; hence power consumption is quite low, making it a good match with CMOS; and (2) LCDs are easily seen in daylight, whereas LEDs are usually difficult to see in direct sunlight. LCDs also have two major disadvantages when compared to LEDs: (1) Unless lit by an external means, LCDs cannot be viewed at low light levels; and (2) The temperature range of LCDs is generally not as great as that of LEDs.

LCDs must be driven with AC voltage instead of DC, which is used for LED displays. If driven with a DC component, the LCD will soon be ruined because the DC will cause plating to occur on the conductive electrodes through electrolysis. Because of this, virtually all LCD drive circuits use integrated circuit drivers.

One of the most common drivers is the 4055, which is a standard CMOS part. The 4055 is the CMOS-LCD equivalent of the 7447 TTL-LED driver, since it is a BCD-to-seven-segment LCD decoder/driver. Also available is the 4056, which can latch the inputs.

The 4055 operates over the standard CMOS power supply range of 3–18 V, and from this supply it generates the necessary AC waveforms for the LCD. The AC frequencies available range from 30 Hz to 200 kHz. Below 30 Hz, flicker becomes noticeable on the display, and higher frequencies cause higher power consumption. Fig. 8-27 is a schematic of a typical LCD circuit.

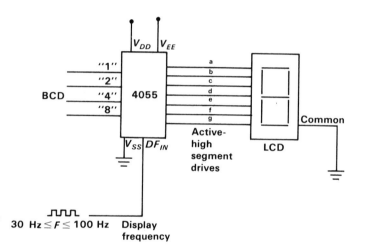

FIGURE 8-27. *Typical LCD interface circuit.*

The inputs to the 4055 are CMOS or TTL (with the 4055 at 5 V) levels with the information in BCD format. The 4055 then decodes the information and energizes the proper segments of the liquid crystal display. Like the 7447, the 4055 is limited when many digits are needed, since one 4055 would be required for each. Most displays, therefore, use a multiplexed scheme.

Fig. 8-28 shows a circuit using the NEC 7225 intelligent LCD controller/driver. This is an extremely powerful display interface IC, of which we will be able to study only the highlights.

In the construction of very large LCDs, the elements (either seven-segment characters, alphanumeric characters, or other symbols) are constructed in layers; each layer is called a *backplane*. Multiple backplanes can be multiplexed to minimize the number of connections to the LCD. This is the same basic idea that we studied in the section on multiplexing. The 7225 is capable of driving

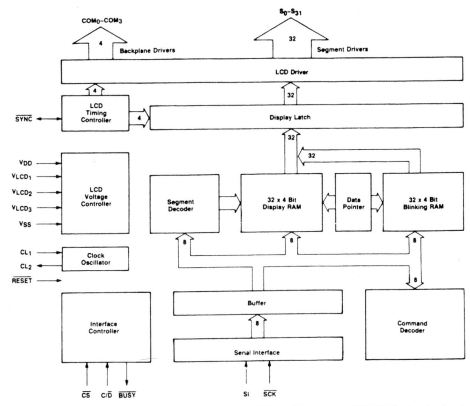

FIGURE 8-28. *The NEC 7225 intelligent LCD driver. (Courtesy of NEC Electronics Inc.)*

four backplanes with 32 segments each. Obviously, this would be a very complex LCD!

The input to the 7225 is serial. Because of the complexity of its use of four backplanes, 32 segments, and serial input, the 7225 is intended for use with a microprocessor. Basically, the 7225 receives a backplane and segment location (address) through the serial input. In turn, it generates and applies the correct waveform to turn on that particular segment. With large-scale integration (LSI), an extremely complex operation is reduced to relatively simply design procedures—in this case a computer program for the microprocessor.

VACUUM FLUORESCENT DISPLAYS

The third common display device is the vacuum fluorescent, or VF. The VF display has some of the advantages of LED, some of LCD, and few disadvantages. One of the advantages is that the VF has low power consumption (although not quite as low as the LCD). It is bright enough to be read in sunlight but can still be read at night; it can be somewhat customized; and it has a wide temperature range. The main disadvantages are that the VF displays cannot be customized to the extent that LCDs can, and most VFs require a high-voltage drive, typically 12–30 V. The second drawback has minimized the popularity of the VF, except in the automotive industry, where a 12 V supply is easily obtained. New VFs are available for 5 V operation, but the models produced so far are not as bright as the higher-voltage versions.

The vacuum fluorescent display is much like an old vacuum tube triode in operation. A constant voltage applied to the cathode (or filament) causes electrons to be emitted from the oxidation layer. If the other two parts of the VF—the grid and the anode—are sufficiently positive with respect to the cathode, the electrons accelerate through the grid and strike the anode. The anode has been treated with a special fluorescent material that will luminesce when excited by impinging electrons. Bringing the grid and anode to a lower potential than the cathode (usually ground) turns the display off. The entire configuration is inside an evacuated glass envelope—hence the name *vacuum fluorescent*. In the vacuum fluorescent display, the anode and grid draw virtually no current. Only the cathode draws current, which is relatively small, so the VF maintains reasonably low power consumption.

VF displays can be purchased for either AC or DC drive of the cathode and for either constant or multiplexed digit drive. In the constant drive version, the grids are all common, and each segment (anode) is individually terminated. As with the other displays, when several digits are needed, the interface circuitry becomes quite complex and we must turn to multiplexing. In multiplexing, each of the seven segments is common for all digits, and the grids are terminated individually for each digit. By placing the segment information on the anode lines and bringing the appropriate digit grid to high voltage potential, the display is multiplexed.

Integrated driver/decoders are available for the VF displays as they are for LED and LCD. These ICs operate very similarly to those we have already studied, so detail is unnecessary here.

OTHER DISPLAY DEVICES

Three other display devices that bear mentioning are the incandescent display, the electrochemical display, and the cathode-ray tube (CRT). The incandescent display uses incandescent bulbs to illuminate segments on a display and is similar in appearance to an LED or VF display. It has good brightness but requires a great deal of power. Because of this shortcoming and the availability of LED and VF at reasonable prices, the incandescent display is not very popular.

Another display that is not popular at this time is the electrochemical display. This display has good resolution and power consumption, but its cost is

TABLE 8-3 SOME OF THE AVAILABLE CRT CONTROLLERS.

Part number	Manufacturer	Microprocessor family
6845	Motorola	6800 family
8275	Intel	8051, 8085, 8086, 8088
8276	Intel	8051, 8085, 8086, 8088
6545	Synertek	6502 family
1861	RCA	1800 family
1869/1870	RCA	1800 family

very high. Advances in production technology could, however, make this a viable display option.

The CRT is one of the most popular display devices, but its interfacing is too complex to be covered in detail in this text. Almost every computer has, or can have, a CRT terminal attached. The CRT shoots a beam of electrons at a phosphorescent screen (the front of the "picture tube"). Generally, the screen is divided into a row-column matrix of dots. A dot is illuminated if the electron beam is turned on as it hits that point and is not illuminated if the electron beam is off at that point. Strict control of the exact time during the scan that the electron beam is turned on allows letters, numbers, characters, and even pictures to be displayed on the screen.

The interfacing problem exists because the beam is scanning the whole screen many times per second and is controlled by high voltage. Fortunately, several LSI CRT controllers are available to alleviate some of the interface problems. These are generally used in conjunction with microprocessors, and most are tailored to operate with a specific family of microprocessors. Some of the available CRT controllers are listed in Table 8-3.

INTERFACING TO RELAYS, MOTORS, AND SOLENOIDS

Digital circuits frequently interface to other types of mechanical devices besides switches, including relays, motors, and solenoids. Electromechanical devices have different interfacing problems from many of those we have studied so far.

RELAYS

Relays have several interfacing problems. First, the coil currents required to switch most relays are too large for CMOS, and many are too large for TTL.

Second, the presence of a coil introduces a back-EMF, or reverse electromotive force, which can destroy logic circuits. Third, the switching of the relay contacts causes electromagnetic interference (EMI) or "noise," which can disrupt many logic systems. Finally, relays take a long time, with respect to logic systems, to operate. This can lead to subtle problems that we will discuss shortly.

An Interface Circuit

Fig. 8-29 is a schematic of a commonly used relay interface circuit. As with LED driver circuits, a transistor current amplifier is the main component. Also, a diode is placed across the relay coil. While this does not adversely affect the operation of the relay, it does snub the reverse EMF spike, thus sparing the logic circuit. This cuts down on some of the noise, too. However, a bypass capacitor across the contacts is desirable if it will not disrupt the external part of the circuitry.

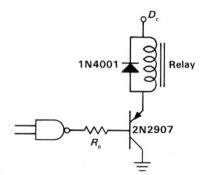

FIGURE 8-29. Relay interface circuit.

As a general rule, the larger the currents that are switched and the faster the switching frequency, the more the EMI that will be generated. Controlling EMI is a subject in itself, but bypassing capacitors and sometimes shielding are usually enough to keep the noise from disrupting the logic circuit's operation.

Switching Speed

As mentioned, the switching speed of relays is very slow compared to that of logic circuits. The switching time, which includes bounce (yes, relays bounce, too), can be milliseconds in length. This can lead to some problems with interfacing. For example, one particular system used a logic signal to close a relay. Then the digital system would check to make sure the contacts had actually closed. If they had not, the relay would be assumed to have failed and would be powered down. Unfortunately, the digital circuit checked so quickly that the relay didn't have time to close and was always assumed to have failed! This illustrates that in practical circuits, a great deal of thought needs to be taken before the circuitry is actually designed.

Solid State Relays

As an alternative to the electromechanical relay, *solid state relays* (SSRs) are available. These aren't actually relays in the traditional sense but are optical switches. Solid state relays are used only in low-current applications. There are basically two differences between electromechanical and solid state relays: (1) The solid state relay does not offer total isolation from the control circuitry; however, the isolation can be as large as tens of megohms; and (2) There is no mechanical bounce; hence, the SSR has much less EMI.

The connection of the SSR is exactly the same as for the electromechanical relay. One side (corresponding to the coil in an electromechanical relay) is the control side, and the other side of the relay is the contact side.

SOLENOIDS

A solenoid is simply a coil that sets up a magnetic field that causes a plunger to move in or out of a cylinder. Since the solenoid has a coil, the same precautions must be taken as with relays. Basically, the same interface circuit can be used with a solenoid as with the relay.

MOTORS

There are two types of motors that a designer might want to interface to: linear, or continuous, motors and stepper motors. The linear motor is the type that most people are familiar with. It is used in electric appliances, locomotives, and most other motor applications. The linear motor is not often interfaced with digital circuitry, but when it is, most of the same considerations apply as with relays.

Linear motors generally draw a large amount of current. As a result, two, three, or more stages of amplification are sometimes necessary. Also, the back-EMF increases with the motor size, so that high-voltage transistors must be used as the final interface.

Stepper motors are more often used with digital systems. A stepper motor moves a discrete amount in response to each digital input. For example, a stepper motor might move one degree per digital pulse. Thus, 360 pulses would be required for a complete revolution of the shaft. A common use of stepper motors is in robots and robotic arms. When the pulse train controlling the motor is fast enough, the motion appears to be continuous.

The tremendous advantage of the stepper motor, besides its digital control, is that its exact position can be known at any time. It can also be instantly reversed.

The actual interface circuit for the stepper motor can be complex and is often specific to the particular stepper motor in use. The manufacturers usually have application notes, specifications, and other documents to aid in the design of a proper circuit.

SUMMARY

1 When logic elements of different families are interconnected, interface problems can arise and circuit modifications must sometimes be made.

2 When interfacing TTL to CMOS, a pullup resistor must be added to the output of the TTL gate since the high-level output of TTL cannot be guaranteed to be higher than the minimum high level necessary for CMOS. A typical value is 2.2 kΩ.

3 If CMOS is powered from a supply voltage greater than 5 V, a pullup from the TTL output to the supply voltage is necessary (for open-collector TTL).

4 A 40109 or similar level-shifting circuit must be used if the CMOS supply is greater than 5 V and open-collector TTL is not used.

5 When the CMOS supply voltage is less than 5 V, a 4049 or 4050 can be used as the interface since its input voltage can exceed its supply voltage.

6 CMOS does not have sufficient current-sinking capability to provide proper low-level output voltage when interfaced to TTL. A 4049 or 4050 can be used as the interface between CMOS and TTL; however, the fanout to standard TTL is only one.

7 CMOS gates can be wired in parallel to provide greater drive currents.

8 Pullup or pulldown resistors, whichever is appropriate, must be used when interfacing switches to logic.

9 The mechanical contacts of switches bounce when closed. If the bouncing will cause problems in the logic circuit, a debounce circuit should be employed.

10 Excessive contact resistance can cause improper operation of TTL circuitry

11 A multiplexer allows several signals to be placed on a smaller number of wires.

12 A demultiplexer decodes the multiplexed signal back to is original form or some other convenient form.

13 It is best to connect LEDs so they turn on with a low logic signal since logic circuits sink more current than they can source.

14 If more current is necessary to drive the LED than the logic circuitry can safely supply, a transistor current amplifier should be used.

15 Multiplexing LED displays allows for fewer drive devices. With multiplexing, the brightness decreases as the duty cycle decreases.

16 Liquid crystal displays have a very low power consumption. They can be seen much easier in daylight than LEDs but cannot be seen at night unless lit from another source.

17 Integrated LCD drivers simplify the interfacing between digital circuits and LCDs.

18 Vacuum fluorescent displays operate much as a vacuum tube triode does. The disadvantage of VF displays is that most need a voltage higher than 5 V for operation.

19 Integrated VF drivers are also available and, as with LCD drivers, simplify interfacing.

20 Some other display devices are CRTs, incandescent displays, and electrochemical displays. A great deal of LSI interface circuitry exists for CRT interfacing.

21 Relays, motors, and solenoids share two characteristics that complicate interfacing: they are electromechanical, and they contain inductive coils.

KEY TERMS

contact bounce	interface	solid state relay
CRT	LCD	stepper motor
current limit resistor	LED	switch bounce
demultiplexer	pulldown resistor	VF
electromagnetic interference (EMI)	seven-segment display	

QUESTIONS AND PROBLEMS

1 Draw the proper circuit for a 7404 TTL INVERTER interfaced to one input of a 4081 CMOS AND gate.

2 A CMOS 4001 NOR gate has two inputs. One input is the output of another 4001, and the other is the output of a 7404 TTL INVERTER. Draw the properly interfaced schematic diagram.

3 Assume that the CMOS gates in problem 2 are powered from a 10 V supply and that the TTL has the standard 5 V supply. Draw the proper interface circuitry for this configuration.

4 The output from a 7400 TTL NAND gate is to be the input for a 4011 CMOS NAND gate. However, the CMOS is powered from a 3 V battery. Design the necessary interfacing to accomplish this.

5 The input to an NMOS logic circuit comes from the output of a CMOS 4011. Draw the schematic so that it is properly interfaced.

6 The output of a 4001 CMOS NOR gate is the input for a 7404 INVERTER. Design the proper interface and draw the schematic.

7 The output of a 4049 CMOS inverting buffer is the input for a 7404 INVERTER. Design the proper interface and draw the schematic.

8 How would you interface a CMOS output if it is to be the input to four different TTL gates?

9 Could you connect the output of a 4049 to *both* inputs of a 7400 (using the 7400 as an INVERTER)?

10 A 4011 CMOS NAND is powered with 10 V. The output of the 4011 is to be inverted and input to a 7400 NAND gate. Design the interface and draw the schematic.

11 Design a TTL circuit that has two active-low toggle switches as inputs and an output that is the NAND function of these switch inputs.

12 Design the circuit of problem 11 using CMOS.

13 A mylar keyswitch, which has a resistance of 400 Ω and is operated as an active-low switch, is to be interfaced to a TTL 7404 INVERTER. Draw the schematic and remember to allow for the switch resistance. Show your calculations for the pullup resistor.

14 Design a contact debounce circuit for an active-high switch.

15 Design the 16-digit keypad to 4-line multiplexer circuit that was mentioned in the text. (You will need a TTL data book to find the pinout of the 74150.)

16 Design an LSTTL circuit that has three active-low switches as inputs. The circuit will illuminate an LED if any of the switches are closed. (You may need to refer to a TTL data book for this problem.)

17 Use the 7447 IC to design a 4-digit, seven-segment LED display circuit. (Assume the inputs to the 7447 are BCD.)

18 Use the 4055 and CMOS logic to design a circuit that displays, on a single-digit seven-segment LCD, the number of active-high switches that are closed. The input to your circuit should be five active-high switches.

19 What type of display device would most likely be found with a personal computer?

20 Of what value is the diode in the relay interface circuit of Fig. 8-29?

21 Would 15 V CMOS or standard TTL be more prone to relay electromagnetic interference? Why?

22 Why is a stepper motor most likely to be designed into a robotic arm?

ANALOG CONVERSIONS

It would be extremely convenient if we were digital creatures: we could plug ourselves directly into computers or other digital systems. Unfortunately, we are analog beings who must be interfaced to digital systems. This chapter looks at some of the basic circuits used to convert analog information to digital information and digital information to analog information. The ability to make the conversions electronically allows us as humans to interface to digital systems.

BINARY WEIGHTED DIGITAL-TO-ANALOG CONVERTERS

The *binary weighted ladder digital-to-analog converter* (DAC) is one of the easiest digital-to-analog converters to understand. It is composed of a series of resistors that double in value, just as binary numbers do from bit position to bit position. Fig. 9-1 shows a 4-bit binary weighted ladder DAC. It is important to note that the output load resistor R_L must be at least ten times the value of the largest resistor in the series. If it isn't, the output voltage will not be accurate.

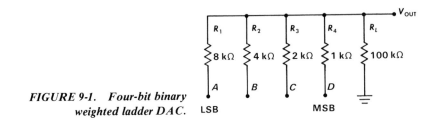

FIGURE 9-1. Four-bit binary weighted ladder DAC.

DETERMINING THE VALUE OF EACH VOLTAGE STEP

Fig. 9-2 illustrates the DAC in Fig. 9-1 with the input voltages of 5 V for a logic 1 and 0 V for a logic 0. The minimum input value, 0001, has been applied so that we may determine the *voltage step,* which is equal to the output voltage of the smallest input. Notice that the least significant bit input position (LSB) is the input with the largest value of resistance. This allows the least current flow for the smallest input value.

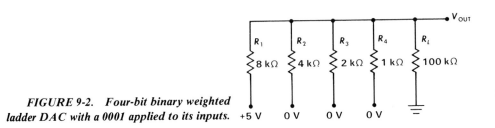

FIGURE 9-2. Four-bit binary weighted ladder DAC with a 0001 applied to its inputs.

Fig. 9-3a shows the DAC in Figs. 9-1 and 9-2 redrawn to determine the output voltage. This is redrawn again in Fig. 9-3b which shows the value of the parallel combination of R_1, R_2, R_3, and R_4. We now have a simple series circuit that can be used to determine the output voltage for the smallest input number, 0001. This value is also the step voltage of this DAC.

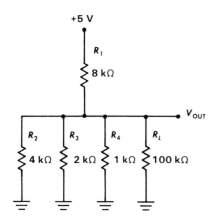

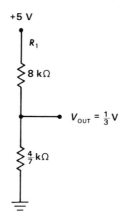

a. The binary weighted ladder of Figure 9-2 redrawn.

b. Figure 9-2 redrawn so that the final output voltage can be determined.

FIGURE 9-3. DAC showing output voltage.

The following equation provides a simpler method of determining the step voltage:

$$V_{STEP} = \frac{V_1 - V_0}{2^n - 1}$$

V_1 is the logic 1 voltage; V_0 is the logic 0 voltage; and n is the number of bits.

If this equation is applied to the converter in Fig. 9-3, we can determine the voltage step:

$$V_{STEP} = \frac{5 - 0}{2^4 - 1} = \frac{5}{15} = \frac{1}{3} V$$

Table 9-1 is the voltage table for the DAC in Fig. 9-3.

Fig. 9-4 illustrates another DAC with differential input voltages of +5 V for a logic 1 and -5 V for a logic 0. It might appear difficult to determine the voltage step and voltage table for this circuit, but that is not the case. The voltage step for this circuit is:

$$V_{STEP} = \frac{5 - (-5)}{2^4 - 1} = \frac{10}{15} = \frac{2}{3} V$$

The voltage step is twice that of the previous example because the difference between the logic 1 and logic 0 voltages has doubled. The voltage table for this

Analog Conversions 259

TABLE 9-1 THE VOLTAGE TABLE FOR THE DAC OF FIG. 9-3.

D	C	B	A	V_{OUT}
0	0	0	0	0.00
0	0	0	1	0.33
0	0	1	0	0.67
0	0	1	1	1.00
0	1	0	0	1.33
0	1	0	1	1.67
0	1	1	0	2.00
0	1	1	1	2.33
1	0	0	0	2.67
1	0	0	1	3.00
1	0	1	0	3.33
1	0	1	1	3.67
1	1	0	0	4.00
1	1	0	1	4.33
1	1	1	0	4.67
1	1	1	1	5.00

circuit starts with −5 V for a 0000 input condition, followed by −4.33 V for a 0001 input condition, etc. We determine each subsequent output voltage by adding the voltage step to the prior output voltage, as illustrated in Table 9-2.

Fig. 9-5 shows a 6-bit binary weighted DAC. Notice that the only difference between this DAC and the DAC in Fig. 9-4 is two additional inputs. The values of the input resistors are two and four times the highest resistance value of the DAC in Fig. 9-4.

Accuracy, Resolution, and Conversion Time of the DAC

The *accuracy* of a DAC is expressed as the percentage difference between the expected output voltage and the actual output voltage. This is usually given by the manufacturer. A typical DAC will generally have an accuracy of ±0.1 percent.

The *resolution* of a DAC is expressed as either the number of input bits the converter contains or the reciprocal of the number of input combinations, expressed as a percentage. For example, an eight-bit binary weighted DAC would have either eight bits of resolution or a resolution of 0.39 percent (1/256).

Conversion time for the DAC is the time required to convert a change at the inputs to the correct analog voltage at the output. Conversion times for most

TABLE 9-2 THE VOLTAGE TABLE FOR THE DAC OF FIG. 9-4.

D	C	B	A	V_{OUT}
0	0	0	0	−5.00
0	0	0	1	−4.33
0	0	1	0	−3.67
0	0	1	1	−3.00
0	1	0	0	−2.33
0	1	0	1	−1.67
0	1	1	0	−1.00
0	1	1	1	−0.33
1	0	0	0	+0.33
1	0	0	1	+1.00
1	0	1	0	+1.67
1	0	1	1	+2.33
1	1	0	0	+3.00
1	1	0	1	+3.67
1	1	1	0	+4.33
1	1	1	1	+5.00

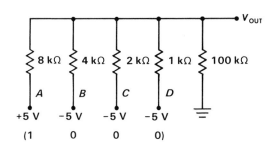

FIGURE 9-4. Binary weighted ladder DAC shown with both positive and negative input voltages.

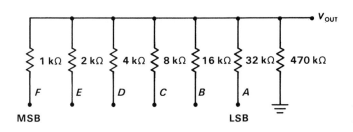

FIGURE 9-5. Expanded binary weighted DAC showing six input connections.

DACs are very short, with some of the newest integrated versions converting from digital to analog in less than 50 ns.

R-2R DIGITAL-TO-ANALOG CONVERTERS

The main problems with the binary weighted ladder are that resistors with values that are binary powers of two are difficult to obtain and these resistor values become very large. For this reason, the binary weighted ladder is found only in applications that require very simple DACs. The R-2R ladder solves these problems by using only two values of resistance: R and 2R.

DETERMINING THE VALUE OF EACH VOLTAGE STEP

The value of the voltage step is determined for the R-2R DAC exactly in the same manner as for the binary weighted DAC. Fig. 9-6 depicts a four-bit R-2R DAC with a binary 0001 applied to its inputs. If the output voltage for this input condition can be determined, then the output voltage can be determined for any input combination.

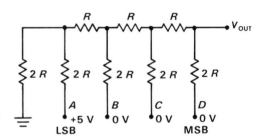

FIGURE 9-6. Four-bit R-2R DAC with a binary 0001 applied to its input connections.

Fig. 9-7 shows the equivalent circuit to that of Fig. 9-6 with the output voltage for this input condition. Because of the nature of the binary number system, the output voltage of Fig. 9-7 can now be used to complete the voltage table for this R-2R ladder (see Table 9-3).

An equation for determining the voltage step for the R-2R ladder follows:

$$V_{STEP} = \frac{V_1 - V_0}{2^4}$$

If this equation is applied to the circuit in Fig. 9-6, the same voltage step is obtained:

$$V_{STEP} = \frac{5 - 0}{2^4} = \frac{5}{16} = 0.3125 \text{ V}$$

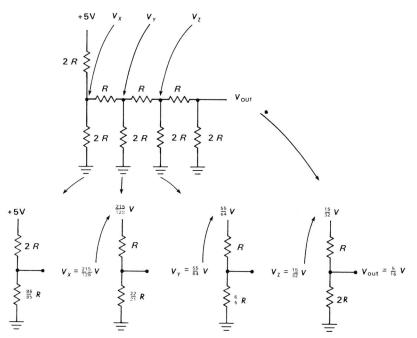

FIGURE 9-7. Equivalent of the circuit in Fig. 9-6, showing how the final output is determined in stages. V_X is determined first, followed by V_Y, V_Z, and V_{OUT}.

TABLE 9-3 THE VOLTAGE TABLE FOR THE *R-2 R* DAC OF FIG. 9-6.

D	C	B	A	V_{OUT}
0	0	0	0	−0.0000
0	0	0	1	−0.3125
0	0	1	0	−0.625
0	0	1	1	−0.9375
0	1	0	0	−1.25
0	1	0	1	−1.5625
0	1	1	0	−1.875
0	1	1	1	−2.1875
1	0	0	0	+2.5
1	0	0	1	+2.8125
1	0	1	0	+3.125
1	0	1	1	+3.4375
1	1	0	0	+3.75
1	1	0	1	+4.0625
1	1	1	0	+4.375
1	1	1	1	+4.6875

Connecting a Circuit to the DAC

Fig. 9-8 shows how transmission gates, or analog switches, can be used to apply either +5 V or ground to an input. The transmission gate is an excellent switch with an on-resistance of less than 100 Ω and an off-resistance of more than 100 MΩ.

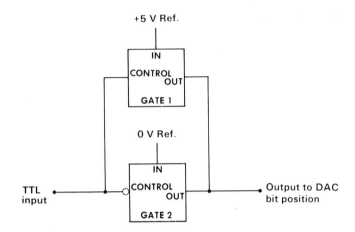

a. Circuit.

FIGURE 9-8.
Transmission gates used to steer either 5 or 0 V to a DAC bit position.

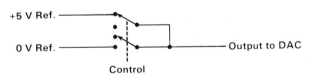

b. Mechanical equivalent of circuit.

When a logic 1 is applied at a bit position, transmission gate 1 is turned on, connecting 5 V to the input of the DAC, and transmission gate 2 is turned off. If a logic 0 is applied at a bit position, transmission gate 1 is turned off and transmission gate 2 is turned on, connecting 0 V to the input. The 5 V source must be precisely regulated to achieve an accurate DAC output. For this reason the 5 V supply should be a precision power supply with an output within 0.1 percent of 5 V.

INTEGRATED DACs

In many applications today it is not practical to build a DAC from resistors and transmission gates. Accurate DACs are, however, available for just a few dollars from almost all the integrated circuit manufacturers.

Fig. 9-9 depicts the DAC0830 8-bit digital-to-analog converter. This device functions from a single power supply with supply voltages ranging from +5 V to +15 V. Because this device can operate from a single +5 V power supply, it is an ideal companion to TTL or CMOS logic circuitry. Table 9-4 details the pin functions of this DAC.

Top View

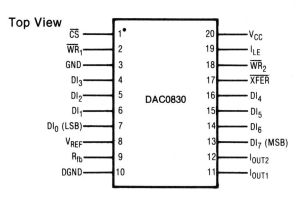

FIGURE 9-9. Pinout of the DAC0830 8-bit digital-to-analog converter. (Courtesy of National Semiconductor Corporation. Authorization has been given by National Semiconductor Corporation, Santa Clara, California, world leaders in Digital Acquisition Circuitry.)

Sample DAC0830 Circuit

Fig. 9-11 shows the DAC0830 connected so that it generates a 0 to +10 V output signal. Notice that the input reference is adjusted to −10 V, the opposite polarity of the output voltage. This reference determines the full-scale output voltage at the output of the op-amp.

The power supply connected to the op-amp must be greater than the maximum expected output voltage. This is due to some losses that occur in the output circuitry of the op-amp. In this case +15 V and −15 V were chosen for the op-amp power supply inputs.

Let's suppose that a binary 1111 1111 is connected to the binary inputs of this converter. Due to the way that the converter is connected, the output of the converter will become +10 V in 1 μs—the settling time for this converter. The output changes because the $\overline{WR}_1$, $\overline{WR}_2$, and $\overline{CS}$ pins were grounded and the ILE pin is connected to +5 V. These three inputs could have been used to delay the change in the output voltage if they were connected to an external timing circuitry. These signals are commonly applied to the DAC0830 by a microprocessor (microprocessors will be covered in Chapter 12).

TABLE 9-4 PIN FUNCTIONS OF THE DAC0830.

Pin Name	Pin Number	Function
$\overline{CS}$	1	Internally combined with the ILE pin to enable the $\overline{WR}_1$ input.
ILE	19	Enables the $\overline{WR}_1$ input. (If ILE = 1 and $\overline{CS}$ = 0, then $\overline{WR}_1$ is active.)
$\overline{WR}_1$	2	Strobes the data inputs into an internal 8-bit holding register (see Fig. 9-10).
$\overline{WR}_2$	18	Enters the data from the internal latch into the 8-bit DAC if $\overline{XFER}$ is a logic 0.
$\overline{XFER}$	17	Enables the $\overline{WR}_2$ input to the DAC.
$DI_0 - DI_7$	4–7, 13–16	The eight digital inputs to the DAC. DI_0 is the LSB input and DI_7 is the MSB.
$I_{OUT(1)}$	11	Equals a value of output current proportional to the input binary code.
$I_{OUT(2)}$	12	Equals the current generated by the input reference voltage minus the current available at $I_{OUT(1)}$.
R_{fb}	9	Feedback resistor provided for the external op-amp that generates the analog output voltage.
V_{REF}	8	Reference voltage for the R-$2R$ ladder.
V_{CC}	20	Power supply input that can be any value between +5 V and +15 V.
AGND	3	Analog ground.
DGND	10	Digital ground

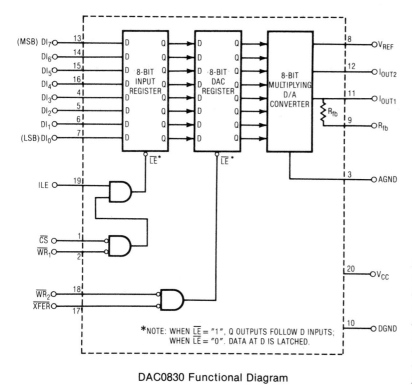

FIGURE 9-10. Internal
structure of the DAC0830.
(Courtesy of National Semi-
conductor Corporation.
Authorization has been given
by National Semiconductor
Corporation, Santa Clara,
California, world leaders in
Digital Acquisition Circuitry.)

DAC0830 Functional Diagram

*NOTE: WHEN $\overline{LE}$ = "1", Q OUTPUTS FOLLOW D INPUTS;
WHEN $\overline{LE}$ = "0". DATA AT D IS LATCHED.

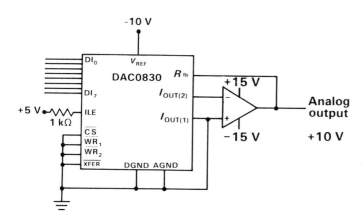

FIGURE 9-11. DAC0830
connected to generate an
output voltage within the
range of 0 to +10 V.

SCALING ADDER DIGITAL-TO-ANALOG CONVERTERS

The *scaling adder* type of digital-to-analog converter is found in applications that do not require a high degree of accuracy or many input bit positions. Fig. 9-12 depicts a simple 4-bit scaling adder DAC. This circuit uses an op-amp to sum the inputs, which are weighted −0.1, −0.2, −0.4, and −0.8. This input weighting allows a full-scale output voltage of −1.5 V if the input voltages are 0 V for a logic 0 and +1 V for a logic 1. With standard TTL levels of 0 V and +5 V, the full-scale output voltage is −7.5.

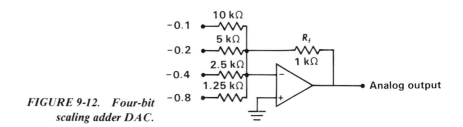

FIGURE 9-12. Four-bit scaling adder DAC.

OPERATION OF THE SCALING ADDER

The voltage gain equation for each input to the scaling adder is the same equation as that used with a simple inverting op-amp (see Fig. 9-13):

$$V_{gain} = -\frac{R_f}{R_i}$$

R_f is the value of the feedback resistor, and R_i is the value of the input resistor. In the circuit of Fig. 9-12, there are four input resistors and one feedback resistor. The gain equation for this circuit is a simple extension of the gain equation for the op-amp:

$$V_{gain} = -\left(\frac{R_f}{R_1} + \frac{R_f}{R_2} + \frac{R_f}{R_3} + \frac{R_f}{R_4} \right)$$

Each input produces an output voltage that is summed with those of the other inputs, as demonstrated by the equation for this circuit. Because R_1 equals 10 kΩ and R_f equals 1 kΩ, the gain due to this input is 0.1. The sum of the gains for all inputs is equal to the total gain for the circuit, −1.5:

$$V_{gain} = -\left(\frac{1}{10} + \frac{1}{5} + \frac{1}{2.5} + \frac{1}{1.25} \right) = -(0.1 + 0.2 + 0.4 + 0.8) = -1.5$$

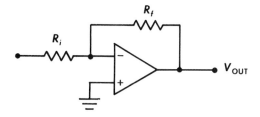

FIGURE 9-13. *Inverting configuration of an op-amp, showing both R_f and R_i.*

INSTANTANEOUS ANALOG-TO-DIGITAL CONVERTERS

The *instantaneous analog-to-digital converter* (ADC) is the most expensive and by far the quickest ADC available today. This type of converter is at times called either a *flash converter* or a *simultaneous converter*.

THE BASIC INSTANTANEOUS ADC

Fig. 9-14 pictures a simple instantaneous ADC constructed from eight comparators and a series voltage divider containing nine resistors. Because the resistors are all of equal value and the power supply voltage is +9 V, the input is resolved in 1 V steps.

If the voltage at the analog input terminal exceeds +1 V, then the output of the first comparator O_1 becomes a 1; if it exceeds +2 V, then both comparator output connections O_1 and O_2 become logic 1s. This occurs because, if a comparator has a larger voltage on its noninverting input terminal than on its inverting input terminal, the output becomes a logic 1. If the inverting input terminal has the larger voltage, the output becomes a logic 0. The truth table for this ADC and its eight binary outputs is illustrated in Table 9-5.

TABLE 9-5 THE TRUTH TABLE FOR THE INSTANTANEOUS ADC ILLUSTRATED IN FIG. 9-14.

V_{IN}	O_1	O_2	O_3	O_4	O_5	O_6	O_7	O_8
0–0.999	0	0	0	0	0	0	0	0
1–1.999	1	0	0	0	0	0	0	0
2–2.999	1	1	0	0	0	0	0	0
3–3.999	1	1	1	0	0	0	0	0
4–4.999	1	1	1	1	0	0	0	0
5–5.999	1	1	1	1	1	0	0	0
6–6.999	1	1	1	1	1	1	0	0
7–7.999	1	1	1	1	1	1	1	0
8–	1	1	1	1	1	1	1	1

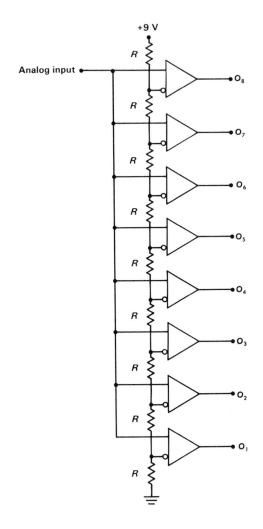

FIGURE 9-14. *8-bit flash converter with a resolution of 1 V.*

A Sample Flash Converter Using a Bar Display

As the truth table shows, the output of this ADC is not a binary value proportional to the input voltage; instead, the number of high output bits indicates the input voltage. Four high outputs indicate +4 V; five high outputs indicate +5 V, etc. This type of output does have an application, as, for example, in a bar meter on a modern stereo or a digital VU meter. Fig. 9-15 depicts a 1–10 V meter constructed from a flash converter and ten LEDs. We could scale this for just about any voltage range by changing the reference voltage. A reference of +1 V would allow the display to indicate voltages from +0.1 to +1 V. To

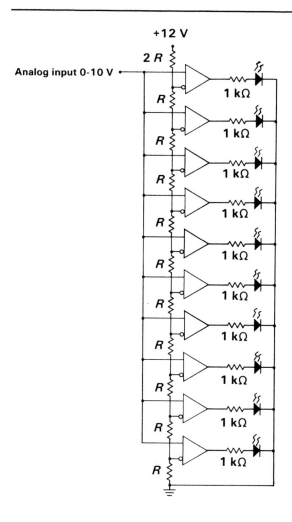

FIGURE 9-15. *A 1-10 V bar graph meter using a 10-bit flash converter.*

achieve a nonlinear range of voltages, as in a digital VU-meter, the values of the resistors in the voltage divider can be changed to develop a logarithmic scale of voltages.

Converting the Outputs to a Binary Number

Fig. 9-16 shows a 7-bit flash ADC connected to an 8-to-3 line priority encoder. The priority encoder takes the seven input lines and converts them to an output that is a binary number from 000 through 111. In this converter, 000 through 111 indicates a voltage of from 0 V through +7 V. Fig. 9-17 illustrates the pinout and truth table for the 8-to-3 line priority encoder. Since the least significant bit of the ADC has been connected to the most significant input of the priority

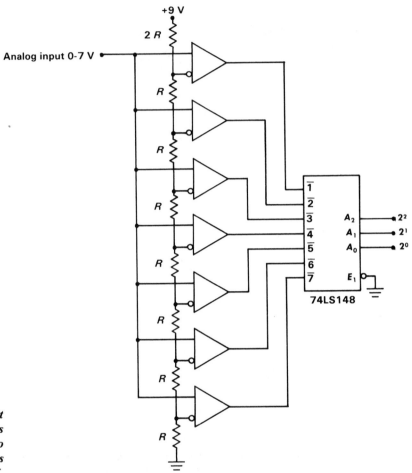

FIGURE 9-16. Seven-bit flash converter that generates a binary output equivalent to the value of the input times 1 V.

encoder, the output is 000 if all the outputs of the comparators are zero, 001 if all are zero except the least significant bit, etc. (This occurs because the outputs of the 8-to-3 line priority encoder are inverted.)

RAMP/SLOPE ANALOG-TO-DIGITAL CONVERTERS

Another, much less expensive, form of the ADC is the *ramp/slope converter*. As its name implies, this converter uses a ramp or sloping waveform to convert from analog to digital.

SN54148, SN54LS148 . . . J OR W PACKAGE
SN74148, SN74LS148 . . . J OR N PACKAGE
(TOP VIEW)

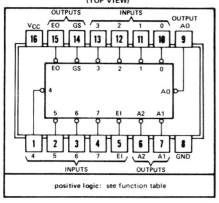

'148, 'LS148
FUNCTION TABLE

INPUTS									OUTPUTS				
EI	0	1	2	3	4	5	6	7	A2	A1	A0	GS	EO
H	X	X	X	X	X	X	X	X	H	H	H	H	H
L	H	H	H	H	H	H	H	H	H	H	H	H	L
L	X	X	X	X	X	X	X	L	L	L	L	L	H
L	X	X	X	X	X	X	L	H	L	L	H	L	H
L	X	X	X	X	X	L	H	H	L	H	L	L	H
L	X	X	X	X	L	H	H	H	L	H	H	L	H
L	X	X	X	L	H	H	H	H	H	L	L	L	H
L	X	X	L	H	H	H	H	H	H	L	H	L	H
L	X	L	H	H	H	H	H	H	H	H	L	L	H
L	L	H	H	H	H	H	H	H	H	H	H	L	H

FIGURE 9-17. Pinout and function table of the 74148 8-to-3 line priority encoder. (Courtesy of Texas Instruments Incorporated.)

THE RAMP GENERATOR

There are two simple and commonly used techniques for generating a ramp for this type of ADC. The first technique is the *charge pump,* which is the less expensive, and the second form is the *synchronous ramp generator,* which uses a DAC.

The Charge Pump Ramp Generator

Fig. 9-18 shows the basic charge pump circuit, which can be broken down into two parts: a constant current source that charges a capacitor and a comparator and switch that discharge the capacitor.

Transistor Q_1 has been biased with a fixed current through its base, which causes this transistor to conduct with a fixed amount of current at its collector. This fixed collector current is the constant current source used to charge the capacitor. If current is allowed to flow into the plates of a capacitor at a

Analog Conversions 273

constant rate, the voltage between the plates will build in a linear fashion. This linear rise, with respect to time, is a *ramp* of voltage.

The comparator and transistor Q_2 discharge the capacitor when it reaches a predefined threshold voltage. When the capacitor charges to the threshold voltage, determined by the reference voltage applied to the comparator, it turns Q_2 on. Because Q_2 is connected directly across the capacitor, the capacitor abruptly discharges to 0 V. The waveform obtained across the capacitor is illustrated in Fig. 9-19.

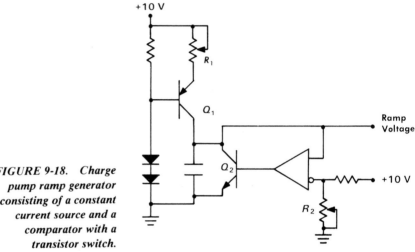

FIGURE 9-18. Charge pump ramp generator consisting of a constant current source and a comparator with a transistor switch.

FIGURE 9-19.   Voltage waveform obtained across the capacitor in the circuit of Fig. 9-18.

The ramp generator is adjusted by potentiometers R_1 and R_2. R_1 varies the length or slope of the ramp, and R_2 adjusts the threshold point or the maximum amplitude of the ramp.

The DAC Ramp Generator

A binary counter and a DAC can also generate a ramp. This method is illustrated in Fig. 9-20 with an 8-bit DAC and an 8-bit binary up counter. As the binary counter counts up to its maximum count from 0000 0000, the output of the DAC will slowly build up and produce a voltage ramp. The ramp generated by this circuit is illustrated in Fig. 9-21.

To adjust this type of ramp generator, we vary the input CLOCK signal to

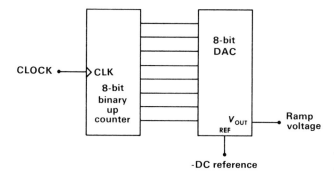

FIGURE 9-20.   *Eight-bit binary up counter and DAC used to generate a ramp voltage waveform.*

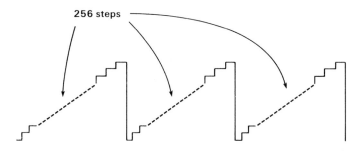

FIGURE 9-21.   *Ramp voltage waveform obtained at the output of the DAC in Fig. 9-20.*

the counter, which changes the length or slope of the ramp. The maximum amplitude of the ramp is modified by the value of the DC reference voltage applied to the DAC.

The Comparator

If a ramp waveform is compared with an unknown analog voltage, the output of the comparator will indicate the *crossover point* of the two signals. These input signals and the comparator are illustrated in Fig. 9-22, along with the

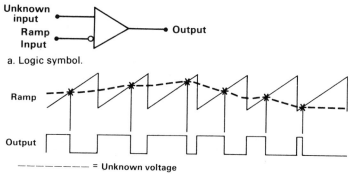

FIGURE 9-22.   *The comparator. Notice how the width of the output pulse changes with a change in the analog input level.*

output waveform. Notice that the output of the comparator remains a logic 1 until the ramp input exceeds the value of the analog input. This circuit will change the analog input into a pulse width modulated (PWM) output. In other words, the output pulse width will be proportional to the amplitude of the analog input voltage.

Gating a Counter and a Latch with the Output of the Comparator

What happens if the output of the comparator is connected to an AND gate that allows some CLOCK pulses to be passed through to a counter? Suppose that ten CLOCK pulses occur during the interval of the ramp applied to the comparator. This means that, if the analog input voltage is +5 V, five CLOCK pulses will be gated through to the counter. If the input voltage is +3 V, then three CLOCK pulses will be gated through to the counter. This relationship is illustrated in Fig. 9-23.

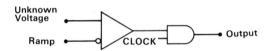

a. Logic circuit.

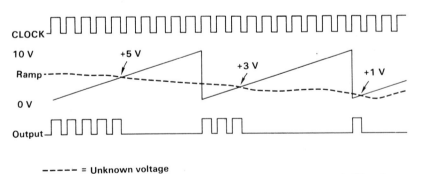

b. Waveforms.

FIGURE 9-23. AND gate attached to the comparator's output, causing CLOCK pulses to be gated through to the output by the unknown analog input voltage.

In most cases, there are at least 1000 CLOCK pulses in the interval of a ramp, so 1–1000 CLOCK pulses would be accumulated for each ramp waveform. Fig. 9-24 depicts a complete ramp/slope ADC that includes a DAC for generating a ramp synchronized with the CLOCK and a counter for accumulating a count and also for holding the output of the ADC. The 555 timer, functioning as an astable multivibrator, must be present or the counter will continuously count out the value of the analog input voltage. The 555 timer also clears the counter so that it can again begin converting from analog to digital. Fig. 9-25 contains the typical timing diagrams for this converter.

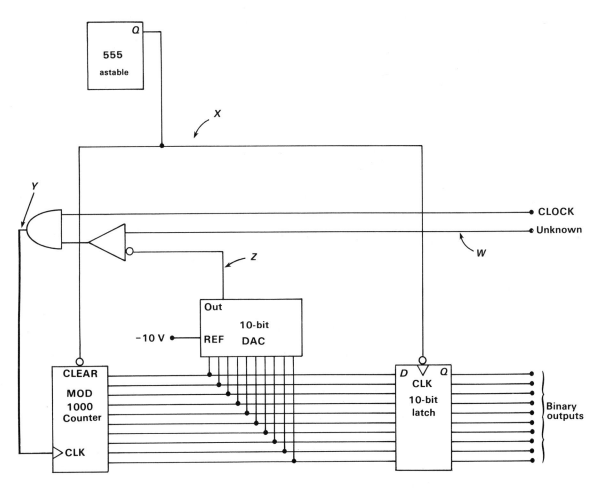

FIGURE 9-24. Complete ramp/slope ADC.

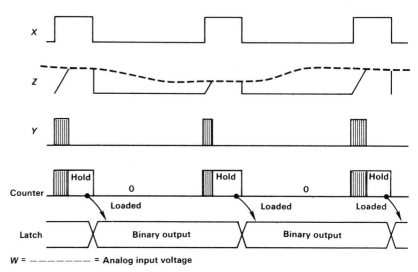

FIGURE 9-25. Timing diagram for the ramp/slope ADC illustrated in Fig. 9-24.

The Resolution and Inherent Error of the ADC

This type of ADC has a resolution equal to the maximum input voltage divided by the maximum count. In the example just presented, the maximum input voltage was +10 V and the counter could accumulate a count of 1000. Thus the resolution for this converter is 0.01 V. That is, it can resolve an input voltage to within 0.01 V.

All ADC's have a built-in error called a *quantizing error*. This error is caused by the comparator's inability to resolve between certain voltage levels and is equal to $\pm 1/2$ bit.

The Ramp/Slope ADC Conversion Time

Because this type of converter uses a ramp generated by a DAC and a counter, it is not nearly as fast as the flash converter. The flash converter can convert a number in a few nanoseconds, but the ramp/slope ADC takes considerably longer—many microseconds. This is because there is a maximum limit to the clocking frequency applied to the counter. With a TTL counter, this limit is about 30 MHz. If the counter counts to 1000, then the maximum conversion time is the period of the CLOCK times 1000. In this case, that amounts to 33.3 μs. The 555 timer or a similar device, as shown in Fig. 9-24, could be adjusted so that a sample is taken once every 33.3 μs maximum. Samples usually occur at a rate no greater than 100 times per second in the type of converter found in most digital multimeters (DMMs).

TRACKING ANALOG-TO-DIGITAL CONVERTERS

The *tracking ADC* (sometimes called *continuous ADC*) is almost identical to the ramp/slope ADC. The difference is that the tracking ADC can follow the input waveform more easily because it uses an up/down counter instead of a standard up counter. With the ramp/slope ADC, each periodic sample requires that the counter start at zero and count up to the unknown input value. The tracking ADC seldom returns to zero to count out a sample. Instead, it continuously counts up with an increase in input voltage and down with a decrease—it *tracks* along with the unknown input voltage.

THE PURPOSE OF THE COMPARATOR

The comparator in this type of ADC operates in a slightly different manner. Remember that, with the ramp/slope ADC, the output of the comparator gates CLOCK pulses through to the counter. In the tracking ADC, the comparator's output controls the direction of the up/down counter.

Fig. 9-26 shows the tracking ADC and a sample analog input and counting sequence. When first turned on, the counter is cleared to zero and begins

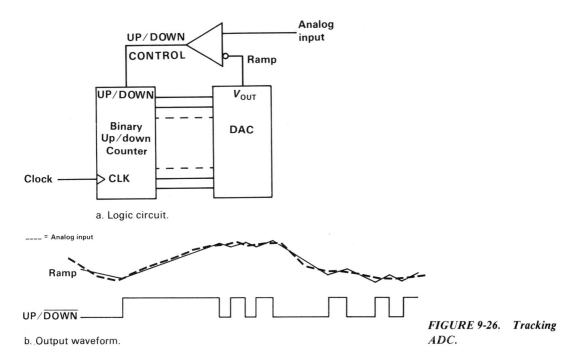

a. Logic circuit.

b. Output waveform.

FIGURE 9-26. Tracking ADC.

counting up to the analog input voltage. If the analog input voltage remains constant, the counter, whose output is connected to a DAC, reaches the value of the unknown input voltage. It then counts one higher, changes direction, and counts down below the input value. At this point, it again changes direction and counts up. The converter "hunts" or oscillates around the unknown input value. If the input abruptly changes in value, the counter follows the change.

CONVERSION SPEED AND ACCURACY

The conversion speed of this type of converter, compared to the ramp/slope converter, is fairly high for a changing input voltage. It is able to follow rapidly changing analog input voltages.

As the converter hunts or oscillates around a fixed input voltage, it allows an error of one to two counts, which in many cases is not objectionable. This error is often indicated in the accuracy specification of the converter as ±1 count.

SUCCESSIVE APPROXIMATION ANALOG-TO-DIGITAL CONVERTERS

Today the *successive approximation register ADC* is the most common type because it is used in almost all commercially available ADC integrated circuits. The successive approximation register ADC, or SAR/ADC, receives its name from the method that it uses to convert from analog to digital.

THEORY OF OPERATION

Suppose that we have a 4-bit SAR/ADC with a resolution of 1 V. The input to this converter could then range from 0 to +15 V. If the most significant bit of the 4-bit input to the DAC is set and the remaining bits are cleared, we can test to see if the unknown input voltage is less than or greater than +8 V (the output voltage of the DAC). This is our first approximation. Refer to Fig. 9-27 for the circuit used for this test and the input bit pattern to the DAC. Since the unknown input voltage is less than +8 V (it is +6 V), the output of the comparator becomes a logic 0. This 0 is steered into the most significant bit of the input to the DAC.

The next bit position is then set and applied to the DAC along with the 0s in the MSB and the other two bit positions (see Fig. 9-27b). The 0100 applied to the input of the DAC generates +4 V on its output. The comparator will generate a logic 1 on its output because the +6 V on the analog input is greater than the +4 V from the DAC. The logic 1 on the output of the comparator is steered back into the next-to-most significant bit for a result of 0100. So far, two approximations have been taken and the outcome is still not certain.

Fig. 9-27c shows the next bit pattern applied to the DAC, 0110, and the output from the comparator. The previous two bits (01XX) have been combined with the next bit position to be tested (XX1X) so that the next approximation can be taken. The number 0110 happens to be the exact value of the unknown analog input. The logic 1 output from the comparator is steered to the next-to-least significant bit position for an outcome of 0110 (assuming that the output of the comparator is a logic 1 for equal input voltages).

One more approximation must be taken because the input voltage could be +7 V. This time a binary 0111 is applied to the DAC; in other words, the LSB of the number is now set (see Fig. 9-27d). This time the output of the comparator is a logic 0, which means that a 0 is gated back to the LSB position. The result is now 0110, which is the correct value for the input voltage of +6 V.

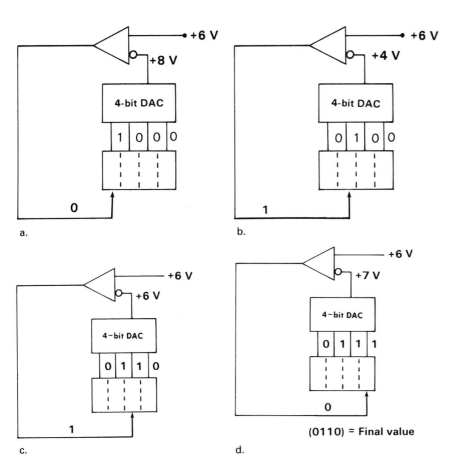

FIGURE 9-27. Basic operation of a 4-bit SAR/ADC showing the four steps required for a conversion.

Conversion Speed

The 4-bit SAR/ADC takes exactly four clocking periods to convert from analog to digital. If the converter is an 8-bit SAR/ADC, it will take eight clocking periods. This type of converter is, therefore, faster than the tracking and ramp/slope ADCs. The only faster ADC is the flash converter, which is still too expensive to find widespread use as an ADC with over eight bits of resolution.

4-BIT SAR/ADC

Fig. 9-28 illustrates a 4-bit SAR/ADC along with the SAR shift register and CLOCK circuitry required to operate it. The START ($\overline{STRT}$) pulse begins

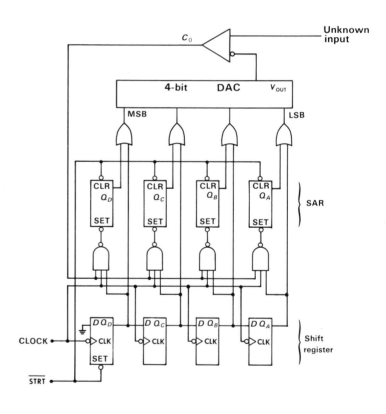

FIGURE 9-28. Complete 4-bit SAR/ADC.

conversion by forcing a logic 1 into the MSB position of the 4-bit shift register, and it also clears the SAR. (See Fig. 9-29 for the timing of the circuit shown in Fig. 9-28.) The output of the shift register is connected through the SAR to the DAC, which generates the first sample voltage for the comparator.

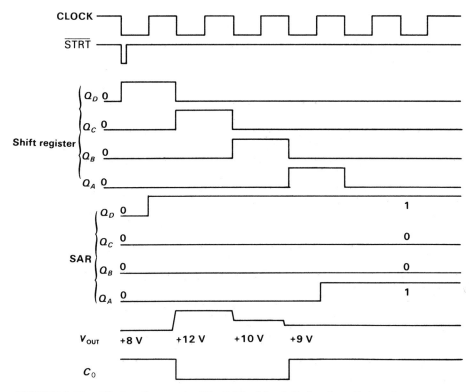

FIGURE 9-29. Timing diagram for Fig. 9-28 with a 9 V signal applied to the analog input.

The output of the comparator is gated back into the SAR on the next CLOCK pulse. This CLOCK pulse also shifts the logic 1 from the MSB to the next bit position. The next sample is taken, and this process repeats until the fourth CLOCK pulse stops the conversion process. At this time the SAR contains the final binary value of the analog input voltage.

AN INTEGRATED SAR/ADC

Fig 9-30 illustrates the pinout of the ADC0805 8-bit analog-to-digital converter. This converter is very easy to use and fairly quick, converting from analog to digital in just 100 μs. Table 9-6 lists the pin functions for this ADC.

TABLE 9-6 PIN FUNCTIONS OF THE ADC0805.

Pin Name	Pin Number	Function
DB_0-DB_7	11–18	The digital representation of the analog input.
DGND	10	Digital ground input.
$V_{REF}/2$	9	No connection.
AGND	8	Analog ground input.
$V_{IN}(-)$	7	Analog differential input.
$V_{IN}(+)$	6	Analog differential input.
CLK IN	4	External CLOCK input or used by this ADC for its own CLOCK generation.
CLK R	19	This connection is used by the CLOCK feedback resistor.
V_{CC}	20	The +5 V power supply connection.
$\overline{CS}$	1	The chip selection input enables both the $\overline{RD}$ and $\overline{WR}$ pin connections.
$\overline{RD}$	2	The $\overline{RD}$ input enables the data output pins DB_0-DB_7.
$\overline{WR}$	3	The $\overline{WR}$ input pin is used to start the analog-to-digital conversion process on its negative edge.
$\overline{INTR}$	5	The $\overline{INTR}$ output indicates the end of a conversion on its negative edge.

The Analog Inputs $V_{IN}(-)$ and $V_{IN}(+)$

These inputs are *differential inputs,* just as the inputs to an op-amp are differential inputs. Differential inputs are inputs whose voltages are summed to produce an internal voltage to the amplifier. For example, if +1 V is applied to $V_{IN}(-)$ and +2 V is applied to $V_{IN}(+)$, then -1 V is applied to the internal amplifier. In most cases the $V_{IN}(-)$ input terminal is grounded and the unknown analog input voltage is connected to the $V_{IN}(+)$ terminal. This connection is illustrated in Fig. 9-31a. An alternate connection, shown in Fig. 9-31b, is used for a 0 code level

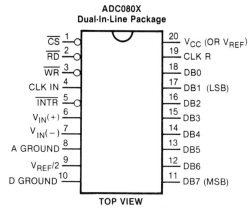

ADC080X
Dual-In-Line Package

Pin		Pin	
$\overline{CS}$	1	20	V_{CC} (OR V_{REF})
$\overline{RD}$	2	19	CLK R
$\overline{WR}$	3	18	DB0
CLK IN	4	17	DB1 (LSB)
$\overline{INTR}$	5	16	DB2
$V_{IN}(+)$	6	15	DB3
$V_{IN}(-)$	7	14	DB4
A GROUND	8	13	DB5
$V_{REF}/2$	9	12	DB6
D GROUND	10	11	DB7 (MSB)

TOP VIEW

FIGURE 9-30. Pinout of the ADC0805 analog-to-digital converter. (Courtesy of National Semiconductor Corporation. Authorization has been given by National Semiconductor Corporation, Santa Clara, California, world leaders in Digital Acquisition Circuitry.)

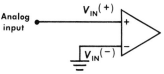

a. The most common connection.

b. An alternative connection that allows the input to be offset from 0 V.

FIGURE 9-31. Connection for the ADC0805 analog-to-digital converter.

other than 0 V. This connection allows the user to set the 0 code level at any analog voltage level. The permissible range of analog inputs for this device is 0 to +5 V when it is operated from a single +5 V power supply.

CLOCK Connections

The conversion CLOCK for the ADC0805 can be provided by an external timing source, or it can be internally generated. An external source is connected to the CLK IN connection, as illustrated in Fig. 9-32a. The allowable frequency range of the external CLOCK is 100–1460 kHz.

If an external clocking source is unavailable, this ADC can operate with an *RC* timing circuit connected as illustrated in Fig. 9-32b.

The Control Inputs, $\overline{RD}$, $\overline{WR}$, and $\overline{CS}$

This chip has been designed to be used with a microprocessor, which is beyond the scope of this chapter; however, it can be used without a microprocessor. For use without a microprocessor, the $\overline{CS}$ and $\overline{RD}$ pins are connected to ground. This selects the ADC and enables the data output connections. The $\overline{WR}$ pin is the only pin that is controlled.

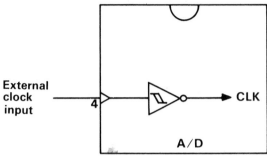

a. Using an external clock.

FIGURE 9-32. Conversion CLOCK for the ADC0805. (Courtesy of National Semiconductor Corporation. Authorization has been given by National Semiconductor Corporation, Santa Clara, California, world leaders in Digital Acquisition Circuitry.)

$$f_{CLK} \cong \frac{1}{1.1\,RC}$$

b. Using an RC constant.

The $\overline{WR}$ pin starts a conversion that begins on the 1-to-0 transition of this pin connection. Fig. 9-33 shows the timing required to operate this ADC from the $\overline{WR}$ pin. It takes some time for the ADC to complete a conversion. During this time, erroneous data may appear at the output connection. For this reason the INTR pin has been provided to indicate that the converter is busy. A logic 1 on the INTR pin indicates that the converter is busy, and a logic 0 indicates that the converter is not busy. The $\overline{WR}$ signal starts the conversion, and the negative edge of the INTR signal indicates that the conversion is complete.

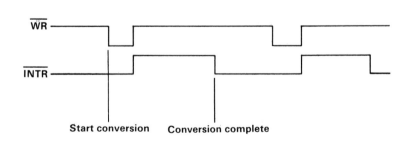

FIGURE 9-33. Timing diagram for the basic operation of the ADC0805 analog-to-digital converter.

A Sample Circuit Using the ADC0805

Fig. 9-34 pictures an ADC0805 connected to a position-indicating potentiometer. The $\overline{WR}$ input is connected to a 555 timer so that 100 samples per second occur. The output of the ADC will indicate the relative position of the pot to within 2°.

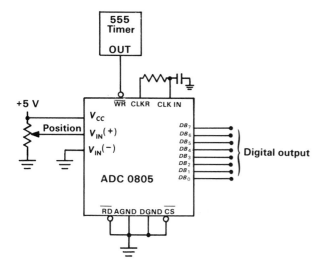

FIGURE 9-34. ADC0805 analog-to-digital converter connected to a position-indicating potentiometer.

SUMMARY

1 There are basically two types of digital-to-analog converters (DACs): the binary weighted and the R-2R ladder.

2 The voltage step for the binary weighted ladder is equal to the difference between the logic 1 and logic 0 inputs, divided by two raised to the number of inputs minus one. A 4-bit DAC with 0 V and +5 V as logic inputs would have a voltage step of 5/15 or 1/3 V.

3 The voltage step for the R-2R ladder is equal to the difference between the logic 1 and 0 inputs, divided by 2 raised to the number of inputs. A 4-bit DAC with 0 V and +5 V as logic inputs would have a voltage step of 5/16 V.

4 The accuracy of a DAC is a measure of the percentage of difference between the expected output and the actual output.

5 The resolution of a DAC or ADC is usually represented by the number of input or output bits. An 8-bit DAC or ADC would have a resolution of eight bits.

6 The conversion time for a DAC or ADC is the amount of time that the device takes to convert from digital to analog or vice versa.

7 The instantaneous ADC or, as it is sometimes called, the flash converter, is by far the quickest ADC available. Its disadvantage is that it is still very expensive and is available only in integrated form with eight bits of resolution.

8 The flash converter is used in many forms of bar displays because of the way that the output appears.

9 The outputs of a flash converter can be converted into binary through connection to a priority encoder.

10 A less expensive form of ADC is the ramp/slope converter, which uses a ramp waveform that is compared with the unknown analog input in order to generate a gate pulse. The gate pulse is used to gate the CLOCK pulse through to a counter.

11 A ramp for the ramp/slope ADC can be generated by either a charge pump or a counter and a DAC.

12 All ADCs have an inherent error caused by the inability of the comparator to resolve the difference between certain voltage levels. This usually causes an error of $\pm 1/2$ bit.

13 The tracking ADC, or continuous ADC, will follow the analog input voltage. This type of converter is ideal for monitoring AC signals.

14 The main problem with the tracking ADC is that, once it has approached a steady value, it hunts around the input voltage level.

15 The SAR/ADC, or successive approximation ADC, is slower only than the flash converter and is commonly found in most integrated ADC circuits.

16 The SAR/ADC can convert from analog to digital in a period of time equal to the number of bits times the CLOCK period.

QUESTIONS AND PROBLEMS

1 Diagram an 8-bit binary weighted DAC. Start with a 1 kΩ resistor for the MSB position.

2 Given a 5-bit binary weighted DAC, determine the voltage step if the logic 1 voltage is +6 V and the logic 0 voltage is 0 V.

3 Given a 6-bit binary weighted DAC, determine the voltage step if the logic 1 voltage is +5 V and the logic 0 voltage is −3 V.

4 Draw a complete voltage table for problem 2.

5 Draw a complete voltage table for problem 3.

6 Express the resolution of a 6-bit DAC as a percentage.

7 Diagram a 7-bit R-$2R$ DAC using 1 and 2 kΩ resistors. (Make sure that you indicate the LSB and MSB inputs.)

8 Given a 6-bit R-$2R$ DAC, determine the voltage step if the logic 1 level is +9 V and the logic 0 level is 0 V.

9 Given a 5-bit R-$2R$ DAC, determine the voltage step if the logic 1 level is +12 V and the logic 0 level is −5 V.

10 Draw a complete voltage table for problem 8.

11 Draw a complete voltage table for problem 9.

12 What is the conversion or settling time for the DAC0830?

13 What is the purpose of R_{fb} in the DAC0830?

14 If V_{REF} is equal to −2.5 V, then what is the full-scale output of the DAC0830?

15 Determine the final output voltage for the scaling adder illustrated in Fig. 9-35.

16 Determine the weight of each input for the scaling adder shown in Fig. 9-36.

17 Given the flash converter and the analog input voltage indicated in Fig. 9-37, determine the logic level of each output connection.

18 How can the outputs of a flash converter be converted to a binary number?

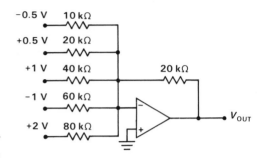

FIGURE 9-35. *Scaling adder for problem 15.*

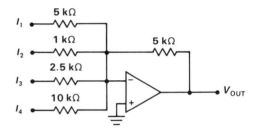

FIGURE 9-36. *Scaling adder for problem 16.*

19 Detail how a ramp voltage waveform can be generated by a charge pump.

20 Explain how R_1 in Fig. 9-18 can vary the length of the ramp waveform developed across the capacitor.

21 Explain how the length of the output pulse width of the comparator in Fig. 9-22 relates to the analog input voltage.

22 How can the slope of the ramp waveform produced by a DAC be changed?

23 What is "inherent error" in an ADC?

24 Contrast the conversion speeds of the flash converter and the ramp/slope ADC.

25 What is the main difference between a ramp/slope ADC and a tracking ADC?

26 Why is the tracking ADC suited to converting AC analog voltages to digital?

27 Explain why a tracking ADC hunts.

28 Briefly describe the operation of a 2-bit SAR/ADC.

29 If a SAR/ADC has a CLOCK frequency of 10 MHz and eight bits of resolution, how long will it take to convert from analog to digital?

30 What is the purpose of the $\overline{WR}$ pin on the ADC0805?

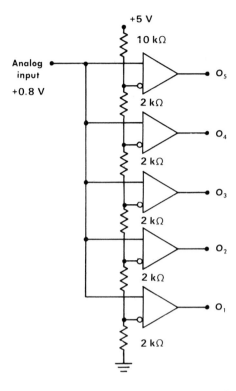

FIGURE 9-37.   Flash converter for problem 17.

31 What does the INTR output pin indicate on the ADC0805?

32 What happens if the $\overline{CS}$ connection on the ADC0805 is connected to +5 V?

33 Explain the purpose of the 555 timer in the circuit of Fig. 9-34.

10

SEMICONDUCTOR MEMORY

Semiconductor memory forms the heart of most computer systems, from the largest main-frame computer to the smallest dedicated-control microcomputer. The use of semiconductor memory is not limited to computer systems, however, as illustrated in Chapter 2. There are other types of memory that use magnetic media for storage; these will be discussed in Chapter 11.

BIPOLAR RAM

The term *Random Access Memory* (RAM) is actually a misnomer, since virtually all semiconductor memory is accessible in a random fashion. The term RAM is commonly used to describe memory that can be both read from and written to, usually at very high speeds. In general, RAM retains data only as long as power is applied (i.e., it is *volatile memory*). The most common uses for RAM include temporary storage of data; storage for high-speed execution of

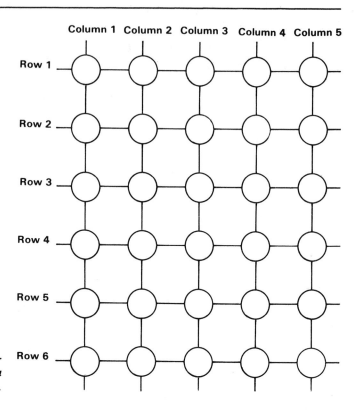

FIGURE 10-1. *Matrix method of addressing. Each circle represents a RAM cell with a unique row—column address.*

programs; buffers for printers, CRT displays, and modems; and "scratch pad" memory, in which data are manipulated.

The information stored in RAM, or any other type of memory, is called *data*. Most memory systems store thousands of bits of data. Therefore, one of our first concerns is how to find a particular bit, or group of bits, from all these thousands.

In the design of memory cells, the location of each bit is specified by a set of inputs called the *address*. Fig. 10-1 illustrates the function of addressing via a matrix. Any data bit can be accessed by placing a 1 in the proper row and column. Notice that this is a unique address, as are all memory addresses. The address does not have to be for a single bit, however. Many microprocessors operate on groups of eight bits (called *bytes*). In the memory system designed for such a microprocessor, the address accesses a byte.

The basic storage element in a RAM is the flip-flop. Fig. 10-2 is the logic diagram of a flip-flop and the circuitry for addressing as well as reading and writing. This circuit represents a simple 1-bit RAM cell. To read or write to the memory we must first enter the correct address (a 1 in the appropriate inputs). To read the data stored in the RAM, only the address must be inserted; the data

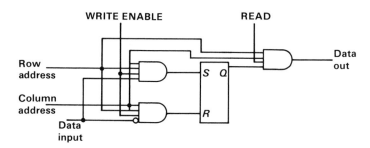

FIGURE 10-2. Single-bit flip-flop RAM.

are then presented at the DATA OUT line. To write to the RAM, we provide the address and then enter the data to be written at the DATA IN line; in addition, the WRITE ENABLE input must be high. This prevents the writing of data when a READ is desired. This is the basic operation of all RAM devices.

THE BIPOLAR RAM CELL

The basic TTL RAM cell is shown in Fig. 10-3. (On an actual RAM IC, additional logic circuitry is necessary.) As with other TTL logic circuits, the TTL RAM cell uses multiple-emitter transistors. When the bases of the transistors are connected as shown, a latch is formed that can store a single bit of information as long as V_{CC} is present. Two of the emitters are used for addressing, and the third is for READ/WRITE operation. Normally, when the cell is not being addressed, the row and column address lines will be low, as will the READ/WRITE lines.

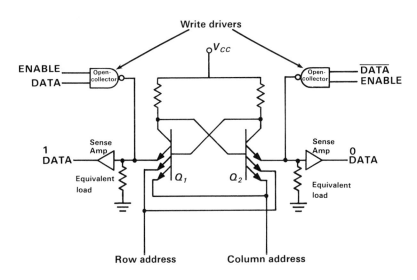

FIGURE 10-3. Basic TTL RAM cell.

If we wish to read from the cell, we must bring both the row and column address lines high and keep both READ/WRITE lines low. Notice that when the address lines are low, the current from the transistor that is in the on state flows out of them. When the address lines are brought high, the emitter current from this transistor is diverted to the READ/WRITE emitter. This emitter is connected to a READ sense amplifier (one of the additional circuit elements on the IC chip). If the 0-storage transistor was on, the READ sense amplifier circuit will place a 0 on the data line for that cell. If the 1-storage transistor was on, the READ sense amplifier circuit will place a 1 on the data line for that cell.

Writing to the cell is similar in operation. Again, the address lines are brought from the low to the high state. This time, however, the READ/WRITE line for the bit opposite of the value to be written (1 or 0) is also brought high. This causes the transistor with all emitters high to turn off, which, due to the cross-connection of the bases, causes the other transistor to turn on. If a 1 is stored and the 0 WRITE line is brought high, a 1 will be written into the cell as the transistors latch in the opposite direction. If, of course, a 1 is stored and the 1 WRITE line is brought high, no change will occur in the cell. Notice that as long as V_{CC} is applied to the RAM cell, the data will remain stored. This type of RAM is called *static* RAM since the data remain static until the user writes over them.

Fig. 10-4 shows the connection diagram and truth table for the National Semiconductor DM7599 Tri-state® RAM. This is a bipolar RAM with its 64-bit capacity arranged in 16 groups of 4 bits (16×4). This is a very small memory and is used in applications that require only a few pieces of information to be stored.

The DM7599 has four address inputs (since 2^4 is 16), four data inputs (D_1–D_4) for WRITE data, a MEMORY ENABLE, and a WRITE ENABLE. The outputs are the four READ bits, S_1 through S_4. To write data to the RAM, both ENABLE lines are brought low and the data to be written to the RAM are

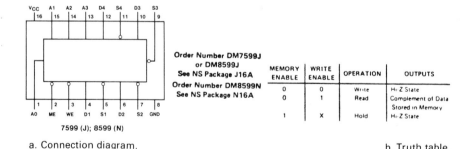

MEMORY ENABLE	WRITE ENABLE	OPERATION	OUTPUTS
0	0	Write	Hi Z State
0	1	Read	Complement of Data Stored in Memory
1	X	Hold	Hi Z State

a. Connection diagram.   b. Truth table.

FIGURE 10-4.   DM7599 bipolar RAM. (Courtesy of National Semiconductor Corporation. Authorization has been given by National Semiconductor Corporation, Santa Clara, California, world leaders in Digital Acquisition Circuitry.)

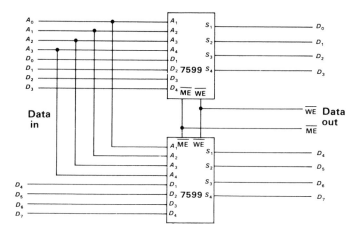

FIGURE 10-5.   Two 7599s connected for
8-bit-wide memory configuration.

entered at the D_1-D_4 lines. The output lines are placed in a high-impedance state during the WRITE operation. The memory can be read if the MEMORY ENABLE is brought low and the WRITE ENABLE is brought high, whereupon the data are presented to the outputs. For both reading and writing, of course, the correct address must be supplied at the address lines. When the MEMORY ENABLE line is brought high, the memory assumes a "hold" mode and all outputs go to the high-impedance state. The necessity of the MEMORY ENABLE function will be explained shortly.

If a memory eight bits wide is necessary, two DM7599s can be connected as shown in Fig. 10-5. Since each address must specify a unique group of eight bits instead of four, the address lines are connected in parallel. The data lines are separated, since each RAM contains half of the group of eight bits for each address.

If greater memory capacity is necessary, these memories can be cascaded to form larger memory systems. For example, Fig. 10-6 illustrates a 32×8 (32 groups of eight bits) memory system using the DM7599. Two pairs of ICs are connected as in Fig. 10-5 to allow eight bits of storage, but the outputs from each of the 16×8 systems are connected in parallel because we now use 32 addresses instead of 16. However, since all DM7599s have only four address lines, a fifth must be found somehow ($2^5=32$). This is where the MEMORY ENABLE line becomes important, as it in effect becomes the fifth address line. Notice that if the A_5 line is a 0, the lower 16-byte memory is enabled since its MEMORY ENABLE is now low. If the A_5 line is a 1, the opposite occurs and the upper 16-byte memory is enabled.

The outputs can be connected because of the Tri-state® ability of the RAMs. Tri-state® is a third state of a digital circuit. Instead of having only the 1 and 0 states, a Tri-state® circuit has a high-impedance state. In this state it can be

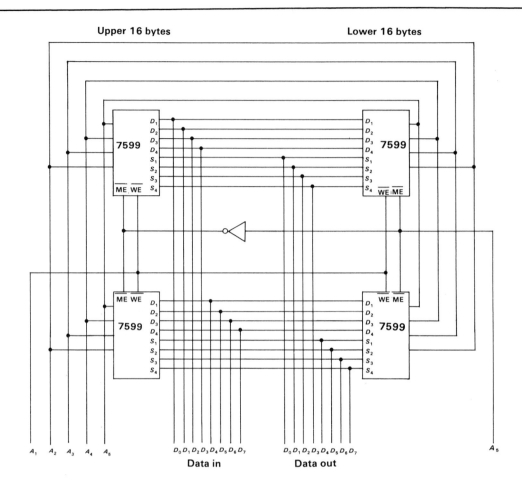

FIGURE 10-6. *A 32 × 8 memory system with 7599s.*

connected to other outputs because it appears as an open circuit and does not influence the other outputs. If these RAMs did not have Tri-state® outputs, we could not cascade them as we have done here. In addition to cascading, we can also design additional logic circuitry to allow even more address lines for even larger memories.

TIMING DIAGRAMS

As we saw in Chapter 7, timing diagrams can aid in the understanding of circuit operation. This is especially true for memory circuits, which have many time-critical inputs and outputs. Fig. 10-7 shows the industry standard symbols for

Waveforms

Waveform Symbol	Input	Output
———	MUST BE VALID	WILL BE VALID
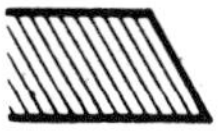 (cross-hatch)	CHANGE FROM H TO L	WILL CHANGE FROM H TO L
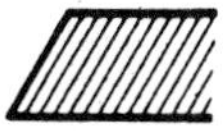 (cross-hatch)	CHANGE FROM L TO H	WILL CHANGE FROM L TO H
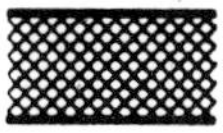 (mesh)	DON'T CARE: ANY CHANGE PERMITTED	CHANGING: STATE UNKNOWN
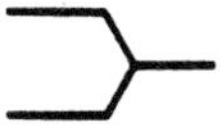 (high-Z)	N/A	HIGH IMPEDANCE

FIGURE 10-7. Standard timing diagram symbols. (Courtesy of National Semiconductor Corporation. Authorization has been given by National Semiconductor Corporation, Santa Clara, California, world leaders in Digital Acquisition Circuitry.)

memory timing diagrams. Notice that the waveforms are drawn with sloped rising and falling edges. This is because real circuits have a finite rise and fall time, rather than the instantaneous transition implied by some timing diagrams.

According to industry standards, the subscripts following the letter t (denoting "time") in a timing diagram represent a "from-to" sequence, possibly followed by one or two letters used descriptively. For example, t_{ASR} is the time for Address Set-up for the READ function. The manufacturer's data books will not only show the timing diagram but will also define the symbolic abbreviations and give numerical values.

ACCESS TIME

One of the most important parameters for RAM, as for any kind of memory, is the *access time*. In memories, a series of gate delays takes place between the time when the address is first presented and the time when valid data are available at the data bus. This time lag is the access time. For a discrete digital or microprocessor system to operate properly, the access time of the memories in the system must be less than the length of the memory READ cycle of the remainder of the system. If not, the remainder of the system must wait in an idle mode while the memory finishes delivering data to the outputs. (The access times for several types of memory are listed in Table 10-2, included in the summary to this chapter.)

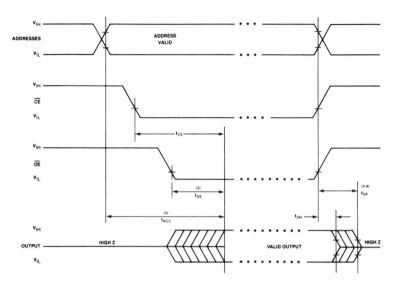

FIGURE 10-8.   Access time (t_A).
(Courtesy of Intel Corporation.)

Fig. 10-8 depicts a typical timing waveform from a data book. The access time is shown as t_A and is the time from the input of a valid address to the output of valid data. Notice that valid data from the previous address was being output at the time of the current address. This exemplifies the importance of a timing diagram and the access time.

MOS STATIC RAM

In addition to bipolar memories, MOS devices can also be configured into memory devices. In fact, the majority of memory devices are MOS. The high input impedance and low power consumption of MOS circuitry, in addition to increasing speed availability, have helped MOS RAMs to dominate the market. MOS memories are available in the static form like the bipolar, but are also available in a dynamic form, which we will study later.

An NMOS static RAM cell is depicted in Fig. 10-9. As with all MOS logic devices, transistors function as loads (instead of the resistors in bipolar RAMs). The operation of the MOS RAM is similar to that of the bipolar RAM. Normally the address lines are low, which causes transistors Q_5-Q_8 to be off, consequently isolating the cell from the DATA 1 and DATA 0 output lines.

In order to read the RAM cell, we must bring the address lines high, causing

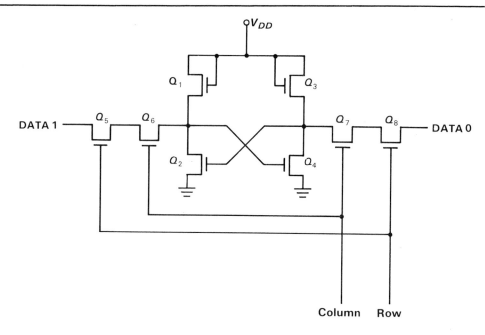

FIGURE 10-9.   NMOS static RAM cell.

Q_5–Q_8 to turn on. If, for example, Q_4 is on and Q_2 is off, addressing the cell will cause a 1 to be asserted at the DATA 1 output and a 0 at the DATA 0 output, indicating that a 1 was stored in the cell. The opposite would occur if Q_4 were off and Q_2 were on: a 1 would appear at the DATA 0 output, indicating a 0 was stored.

In writing to the cell, we take advantage of the fact that MOSFETs can be used in the "reverse" mode (with respect to drain and source); the Q_5–Q_8 transistors are constructed to allow this usage. If Q_4 is turned on and the DATA 0 line is high (1), the high DATA 0 line will cause Q_2 to turn on when the cell is addressed, thereby turning off Q_4 and reversing the state of the RAM cell. As with the bipolar RAM, the READ/WRITE logic is accomplished by additional logic circuitry on the chip.

Also like the bipolar RAM, as long as V_{DD} is supplied to the MOS RAM, the data will remain in the cell: it is a static RAM device. When the power is removed, the data will be lost: thus the MOS RAM is volatile as well.

You may have observed in our discussion of the bipolar and the MOS RAM that there is really no way of determining what state the RAM cell is in. This is the case with all RAMs. Therefore, when power is first applied to a RAM, all cells will contain random and useless data that must be erased or written over before the RAM is used.

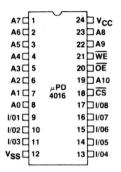

Pin Names	
A0 - A10	Address Inputs
WE	Write Enable
CS	Chip Select
OE	Output Enable
I/01 - I/08	Data Input/Output
Vcc	Power (+ 5V)
Vss	Ground

a. Connection diagram.

FIGURE 10-10.
The μPD4016.
(Courtesy of NEC
Electronics, Inc.)

Truth Table					
CS	OE	WE	MODE	I/O	POWER
H	X	X	Not Selected	High-Z	Standby
L	L	H	Read	Dout	Active
L	H	L	Write	Din	Active
L	L	L	Write	Din	Active

b. Truth table.

MOS STATIC RAM APPLICATION

Fig. 10-10 is a connection diagram and truth table for the NEC μPD4016 2k×8 static RAM. This is a popular memory for which there are many second sources and cross references. This is also a moderately large memory, with a single chip capable of storing 2k bytes of data (1k=1024). In many microcomputer applications, a single 4016 is all that is necessary for a system RAM.

The 4016 is arranged in the same fashion as most recently developed memories. Its data bus is bidirectional; that is, one data bus is used for both reading and writing. This scheme is more compatible with microprocessor applications. The control pins on the 4016 are WRITE ENABLE, CHIP SELECT, and OUTPUT ENABLE. The WRITE ENABLE (WE) operates in the manner discussed earlier. The OUTPUT ENABLE ($\overline{OE}$) is used during the WRITE operation to direct the data bus for use as an input. CHIP SELECT ($\overline{CS}$) is used when memories are cascaded for address logic. All these control inputs are active-low, as signified by the bar over the symbols.

Fig. 10-11 is a complete 6k-byte memory system using 4016s. This can be used in a microprocessor or other application. The address operation becomes clear upon examination of the address logic. A truth table will show how this

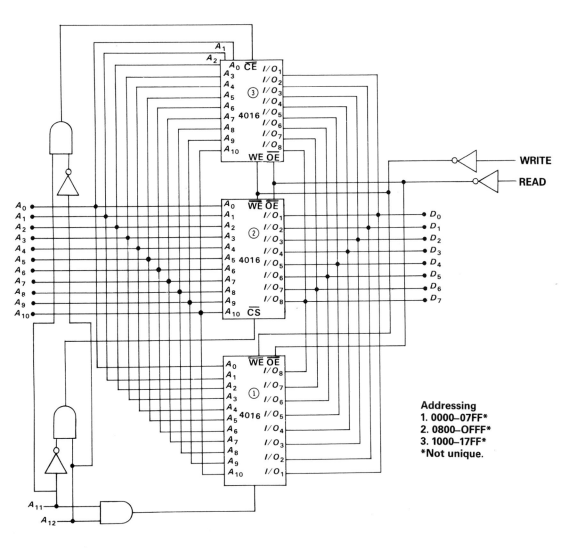

FIGURE 10-11. *A 6k × 8 memory system using 4016 2k × 8 static RAMs.*

Addressing
1. 0000–07FF*
2. 0800–0FFF*
3. 1000–17FF*
*Not unique.

logic allows correct addressing to bring the CHIP SELECT low, thereby activating the memory. Note also that in this system the addresses are *not* unique. Since only 6k bytes of memory are used, duplicate addresses are allowed and become *don't cares*. For example, the addresses for the first 2k×8 memory are 0000–07FF (hex), but its data can also be accessed with (or reside at) 2000–27FF. In a small system, address logic can be simplified in this way, as long as another memory does not have identical addresses. In larger systems, the logic design must allow greater specification of address; if all addresses are to be used, each chip must have a unique address.

DYNAMIC AND PSEUDOSTATIC RAM

Static RAM is capable of storing data as long as power is applied. The price for this convenience is additional transistors, resulting in lower density and, consequently, a higher price per bit. The other, possibly less costly, option is *dynamic* RAM. Dynamic RAM uses fewer transistors but can only hold data for a short period. The actual cost per bit can be higher with dynamic RAM because of additional circuitry, however. Pseudostatic RAM is dynamic RAM with the additional circuitry on the chip, so that it appears to be static.

DYNAMIC RAM

All the RAM devices studied so far have been static RAMs. In the MOS RAM example, several transistors are necessary for each cell. The storage cell in a dynamic RAM requires only one transistor.

Fig. 10-12 depicts a dynamic RAM cell consisting of a single NMOS transistor and a capacitor. The capacitor in a dynamic RAM is the actual storage device. When the cell is addressed, the data stored in the capacitor are presented to the data line for a READ operation. During a WRITE operation, the capacitor is either charged or discharged when addressed to store a 1 or 0, respectively.

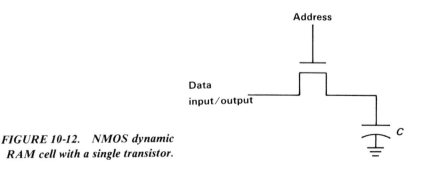

FIGURE 10-12. NMOS dynamic RAM cell with a single transistor.

However, there is a problem with the dynamic RAM: Even though the impedance from drain to source of the MOSFET is high, the charge on the capacitor will eventually leak off, thereby losing the data. To correct this, dynamic RAMs must have the data periodically rewritten. This is called *refreshing* the RAM. The process of refreshing must take place about every 2–3 ms depending on the chip used and consists of reading the stored bit and then rewriting it to the RAM. This can require a fair amount of logic circuitry and thus can make the cost of a dynamic system higher than that of a static system. Also, some speed can be lost when a READ is requested during a REFRESH cycle.

The single-transistor dynamic RAM in Fig. 10-12 has a problem that requires analysis. If a small capacitor is used (which is desirable), a READ operation could destroy the data by discharging the capacitor into the data line. However, if a large capacitor is used, the amount of time necessary to charge the capacitor in order to write a 1 is too long. One solution to this problem is used in most dynamic RAMs: Each time a bit is read, it is immediately written back to the cell to prevent destruction.

Another solution is the 3-cell configuration shown in Fig. 10-13. This design features separate input and output transistors with a buffer in between. To write to the cell, bring the WRITE line high, and the data will be transferred to the capacitor from the input line through transistor Q_1. To read the cell, bring the READ line high and the data will be transferred to the output line through Q_2–Q_3. The extremely high gate impedance of Q_2 keeps the capacitor from losing its data during a READ cycle. The data read from this circuit will be inverted, however, because of Q_2. Although this circuitry is sufficient for the actual memory operation of the dynamic RAM, additional circuitry is necessary on the chip for addressing and refreshing. The extremely high impedance of MOSFETs, such as that exhibited by Q_2, is necessary to keep the capacitor from discharging. For this reason, only MOS dynamic RAMs are practical.

The space saving realized with dynamic RAMs make single chips with large

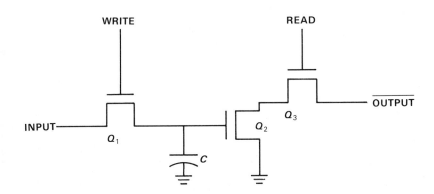

FIGURE 10-13. Three-transistor dynamic RAM.

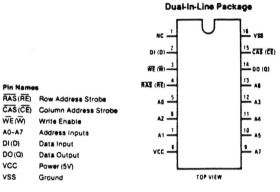

FIGURE 10-14. *Pin assignments for 4164 dynamic RAM. (Courtesy of National Semiconductor Corporation. Authorization has been given by National Semiconductor Corporation, Santa Clara, California, world leaders in Digital Acquisition Circuitry.)*

Pin Names	
$\overline{RAS}$ ($\overline{RE}$)	Row Address Strobe
$\overline{CAS}$ ($\overline{CE}$)	Column Address Strobe
$\overline{WE}$ ($\overline{W}$)	Write Enable
A0–A7	Address Inputs
DI (D)	Data Input
DO (Q)	Data Output
VCC	Power (5V)
VSS	Ground

densities possible. The 4164 dynamic RAM depicted in Fig. 10-14 has a storage capacity of 64k×1 bit, significantly larger than the capacities of the static RAMs studied. This RAM is popular in personal computer systems. The 4164 is a single-transistor dynamic RAM.

The 4164 has some pins that are different from those covered so far. The 4164 uses a conventional WRITE ENABLE control, but it has separate data input and output pins. Also, only eight address lines are supplied for 64k addresses. This is where the Row Address STROBE ($\overline{RAS}$) and Column Address STROBE ($\overline{CAS}$) are used. The eight address lines allow the addressing of 256 locations. Because the address lines are multiplexed to allow the row and column addresses to be fed to the chip at different times (coincident with a Row or Column Address STROBE pulse), 256×256 or 65,536 (64k) addresses are possible with only eight address pins.

The 4164 can be refreshed in one of three ways. The obvious way is to perform a complete memory cycle during the maximum 2 ms REFRESH cycle time. With this method, each data bit is read and rewritten to the RAM. The timing diagram in Fig. 10-15 shows the proper sequence for this method of refreshing.

A second way of refreshing the 4164 is the $\overline{RAS}$-only REFRESH. During this REFRESH, the output remains in the high-impedance state since $\overline{CAS}$ is kept high. As each of the row addresses is pulsed, the data are refreshed, but power dissipation is kept to minimum since no output current flows. One such REFRESH pulse is depicted in the timing diagram of Fig. 10-16.

The third method of refreshing is called the hidden or buried REFRESH and is shown in Fig. 10-17. If valid data are necessary at the output, neither the memory cycle nor the $\overline{RAS}$-only REFRESH can be used. In the hidden REFRESH, $\overline{CAS}$ is held low immediately after a memory READ cycle while each address is strobed to the rows. This keeps the last valid data on the output while refreshing the entire memory.

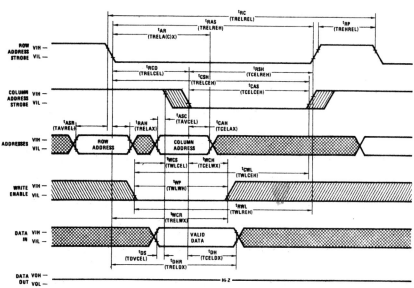

*Symbols in parentheses are proposed industry standard.

FIGURE 10-15. READ-WRITE REFRESH cycle timing diagrams. (Courtesy of National Semiconductor Corporation. Authorization has been given by National Semiconductor Corporation, Santa Clara, California, world leaders in Digital Acquisition Circuitry.)

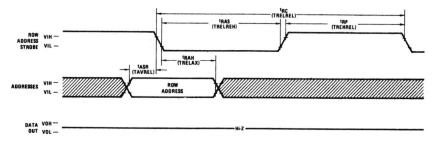

FIGURE 10-16. Timing diagram for $\overline{RAS}$-only REFRESH. (Courtesy of National Semiconductor Corporation. Authorization has been given by National Semiconductor Corporation, Santa Clara, California, world leaders in Digital Acquisition Circuitry.)

Clearly, the REFRESH circuitry can be somewhat complex. Fortunately, some microprocessors, such as the Z80® microprocessor, to be covered in Chapter 12, have on-board REFRESH logic, which makes dynamic RAMs easier to work with.

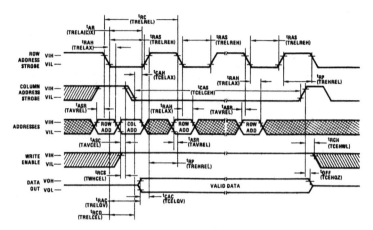

FIGURE 10-17. *Timing diagram for hidden REFRESH cycle. (Courtesy of National Semiconductor Corporation. Authorization has been given by National Semiconductor Corporation, Santa Clara, California, world leaders in Digital Acquisition Circuitry.)*

PSEUDOSTATIC RAM

The pseudostatic RAM uses the advantages of high density, low cost, and low power consumption of the dynamic RAM, but appears from the chip level to be static. The secret is that the REFRESH logic circuitry has been incorporated into the chip itself. This raises the cost per bit, and pseudostatic RAMs are therefore not very popular.

One problem with the pseudostatic RAM is that it cannot be accessed during a REFRESH. In most dynamic circuits this is not a major problem, since the rest of the circuit "knows" when a REFRESH is taking place. With a pseudostatic RAM, this occurs in one of two ways. One method is the synchronous mode, in which the external circuitry sends a set of REFRESH pulses to the RAM during every REFRESH cycle. The external circuit thus "knows" when refreshing is taking place. The asynchronous mode is another solution and involves quizzing the RAM by the external circuitry (usually through an output called READY). If the RAM cannot be accessed, the external circuitry either completes another task and tries again later or waits on a valid READY signal from the RAM.

READ ONLY MEMORIES: ROM, PROM, AND EPROM

The usage of Read Only Memory (ROM) in combinational circuits was covered briefly in Chapter 3. The greatest use for ROM, however, is in microcomputer circuits. In these circuits the data stored in ROM generally compose the operating program for the computer. Since this operating program is not to be changed, ROM is used. ROM is also used to store important data for computer circuits and sequential circuits.

ROM

Read Only Memory is programmed, or written to, once, and the data can never be changed—only read. The manufacturer programs the ROM by actually incorporating the data into the semiconductor IC via a photographic mask. User-programmable ROMS, which will be covered shortly, are also available, so these factory-programmed ROMs are sometimes referred to as "masked ROMs."

Since the actual IC fabrication process differs for each ROM mask, ROMs generally find application in high-volume circuit production. Most manufacturers will not accept an order for less than 1000 ROMs, and many orders are for hundreds of thousands or more. Since the data cannot be changed in a ROM, it is very important that the programming is precisely correct.

Like RAMs, most ROMs are constructed with MOSFET devices. A typical ROM array of 16 cells is shown in Fig. 10-18. The mask production process connects specific gates to the address lines. If a gate is connected to an address line, a 0 will be programmed into that cell. If the gate is left unconnected, that transistor will be permanently off and the cell will represent a 1 due to the active

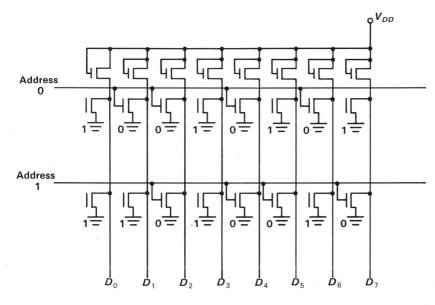

FIGURE 10-18. NMOS ROM. *The information at address 0 is 10010101 and at address 1 is 11010010.*

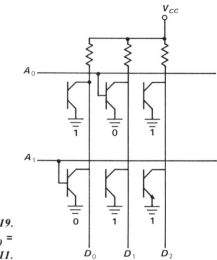

FIGURE 10-19.
Bipolar ROM. A_0 =
101, A_1 = 011.

pullup transistor. Notice that the 16 cells are arranged to output eight bits at each of two addresses. Memories that are eight bits wide are called *Byte-wide®* *memories.* *

When power is removed from a ROM, the only consequence is that the data cannot be read. As soon as power is reapplied, the identical data can be read from the ROM. Thus, the ROM is a *nonvolatile* memory.

Bipolar ROMs are also available for high-speed memory applications. The operation is the same as for the MOS ROM, except that the bases of the transistors are used for the mask information and the pullups are resistors, as shown in Fig. 10-19.

A popular MOS ROM is the 2316, which is arranged in the 2k×8-bit fashion. The connection diagram for the 2316 is depicted in Fig. 10-20. Since no writing is possible, there is no WRITE ENABLE pin. The 2316 has a bidirectional data bus, like the 4016, and contains three CHIP SELECT lines to minimize the amount of external logic necessary to address the ROM. There are 11 address lines, which are sufficient to address all 2k bytes of data (2^{11}=2048). The process of cascading ROMs to form larger memory systems is identical to that for RAMs.

PROM

A PROM is a Programmable Read Only Memory. Unlike masked ROMs, PROMs are user-programmable: that is, they can be programmed outside the factory, but only once. This feature allows the user to utilize ROM when the

*Byte-wide® is a registered trademark of MOSTEK, Inc.

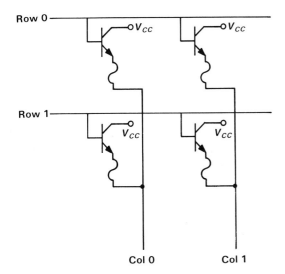

Row 0

Row 1

Col 0   Col 1

FIGURE 10-21. PROM matrix prior to programming.

required volume is insufficient to warrant production of a mask-programmed ROM. PROMs are also very handy in prototyping, in which only a few circuits are to be built for testing. In the event of error, a new PROM can be programmed in a few minutes.

Obviously, the method of programming the PROM is quite different from the creation of a semiconductor mask. In a PROM, the emitter connections of the memory transistors have fuses in series with them. The programming is done by selectively blowing these fuses, thereby isolating the cell.

Fig. 10-21 depicts a typical PROM matrix prior to programming. This PROM is supplied with all 1s. To program a 0, the proper row-column address is supplied, and then sufficient current is allowed to flow out the column address line to blow the fuse. The data are sensed from the column address lines. The fuses used in PROMs generally consist of either a metal link (such as titanium-tungsten or nickel-chromium) or polycrystalline silicon.

The actual programming of a PROM is accomplished on a special programmer that controls the current and the pulsing of the cell to blow the fuse. Although programmers are not difficult to design and construct, most users employ commercially purchased programmers.

PROMs are generally bipolar in construction and have a small memory capacity. MOS PROMs are available, but MOS devices are more frequently designed into EPROMs.

EPROM

The EPROM was invented by Intel Corporation and, next to the microprocessor, is one of the greatest contributions to the digital age. Like the PROM, the

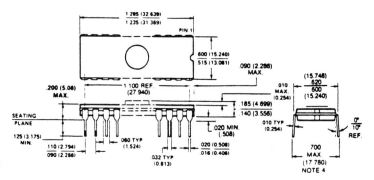

FIGURE 10-22. EPROM package.
(Courtesy of Intel Corporation.)

EPROM allows data to be written into Read Only Memory, but it also allows the data to be erased and reprogrammed if necessary. For this reason, it is called Erasable Programmable Read Only Memory. There are two major types of EPROM: Electrically Erasable (EEPROM) and Ultraviolet Erasable (UVE-PROM). The EPROM is easily distinguished from other ICs by the quartz-glass window on the top of the chip, as shown in Fig. 10-22. Ultraviolet light penetrates the chip through this window to erase the EPROM.

The reasonable price of the EPROM and its ease of modification have led to its inclusion in medium-volume production of several thousand units. There are even EPROM versions of single-chip microcomputers that incorporate the microprocessor and EPROM memory as well as RAM on a single IC chip.

EPROM Topology

The EPROM is made with a special type of MOS transistor called a *floating gate* transistor. As shown in Fig. 10-23, the gate is not physically connected to any other part of the transistor: it is left floating. When a high voltage is placed across the drain and source, electrons are deposited on the floating gate. Since there is no connection to the gate, the electrons remain trapped there, giving the gate a negative charge. This negative charge acts exactly like a low level on a

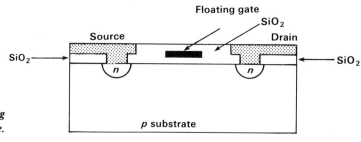

**FIGURE 10-23. EPROM floating
gate structure.**

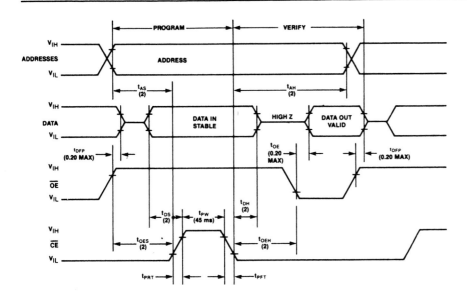

NOTE:
1. ALL TIMES SHOWN IN PARENTHESIS ARE MINIMUM TIMES AND ARE μ SEC UNLESS OTHERWISE NOTED.
2. t_{OE} AND t_{DFP} ARE CHARACTERISTICS OF THE DEVICE BUT MUST BE ACCOMMODATED BY THE PROGRAMMER.

FIGURE 10-24.   Timing waveforms for EPROM programming. (Courtesy of Intel Corporation.)

standard NMOS transistor: it turns the transistor off. With no high voltage between drain and source, no electrons are deposited, and the transistor is on. This is the method of programming an EPROM, and this high programming voltage, supplied by a programmer similar to that used for the PROM, is called V_{PP}. Fig. 10-24 is the timing diagram used for programming a typical EPROM.

The EPROM can be erased through an electro-optical-chemical process. Exposing the floating gate to fairly intense ultraviolet light causes the electrons to migrate from the gate, returning the gate to its original uncharged state. Total erasure of an EPROM usually takes less than an hour. The EPROM can then be reprogrammed for any set of data desired.

If the EPROM is to be used as permanent or semipermanent memory, an opaque label is usually placed over the window to avoid inadvertent erasure by exposure to UV or sunlight. When this label is in place, data retention is virtually permanent. The EPROM can be "worn out" by too many erasures, but since an EPROM is generally good for hundreds of erasures, this is seldom a problem.

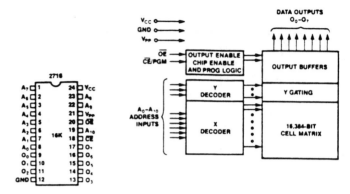

Pin Names	
A_0-A_{10}	ADDRESSES
$\overline{CE}$/PGM	CHIP ENABLE/PROGRAM
$\overline{OE}$	OUTPUT ENABLE
O_0-O_7	OUTPUTS

Mode \ Pins	$\overline{CE}$/PGM (18)	$\overline{OE}$ (20)	V_{PP} (21)	V_{CC} (24)	Outputs (9–11, 13–17)
Read	V_{IL}	V_{IL}	+5	+5	D_{OUT}
Standby	V_{IH}	Don't Care	+5	+5	High Z
Program	Pulsed V_{IL} to V_{IH}	V_{IH}	+25	+5	D_{IN}
Program Verify	V_{IL}	V_{IL}	+25	+5	D_{OUT}
Program Inhibit	V_{IL}	V_{IH}	+25	+5	High Z

FIGURE 10-25. The 2716 EPROM sheet. (Courtesy of Intel Corporation.)

27XX Family of EPROMs

The most popular family of EPROMs is the 27XX (or 27CXX for CMOS versions). The last two digits of the product number denote the bit size in k-bits. For example, the 2716 (see Fig. 10-25) has 16k bits of data capacity arranged in Byte-wide® fashion, as are all members of the 27XX family (i.e., 2k×8 bits, the same as for the 4016 RAM and 2316 ROM). Present family members include the models 2708, 2716, 2732, 2764, 27128, and 27256. The 2708 requires a multiple power supply and is not used in new designs.

The 2716 is supplied with all 1s programmed in the memory cells. Upon erasure, it returns to this state. The act of programming changes a 1 to a 0 by depositing electrons on the floating gate of a particular memory cell. The connection diagram and truth table for the control inputs are shown in Fig. 10-25. A comparison of the pin diagram of the 2716 with that of the 2316 (see Fig. 10-20) will reveal that they are identical except for the control inputs. If a system designed with the 2716 is put into high-volume production, the 2316 can be plugged right in at a lesser cost, as long as the control inputs are used carefully.

To program the 2716, and other family members, the V_{PP} voltage is brought to 25 V DC and the $\overline{OE}$ pin is brought high. Then the address and data information are presented to the chip, and the $\overline{CE}/PGM$ input is strobed with a 50 ms pulse. This programs the eight memory cells located at that address with the data presented on the data lines. A high on the $\overline{CE}/PGM$ input when 25 V are not present on the V_{PP} line causes the outputs to assume the high-impedance state.

NONVOLATILE SEMICONDUCTOR MEMORY

The problem with the nonvolatile memories discussed so far is that none can be erased rapidly enough for use as a READ/WRITE memory. This section deals with memory devices that can be written to and read from like RAMs but have the nonvolatility of ROMs.

Basically, there are two types of memory with these properties: Electrically Erasable Programmable Read Only Memory (EEPROM or E²PROM) and the Nonvolatile RAM (NOVRAM®).* The EEPROM is generally used in large memory applications and is available in pinouts identical or similar to those of the 27XX family of EPROMS. In fact, many experts believe that the EEPROM will replace the EPROM in most applications by the end of the decade. The NOVRAM® is a very fast, usually small, memory device. The NOVRAM® actually has two memory devices on one chip: a fast RAM and an EEPROM.

THE EEPROM

The EEPROM can be classified as "read mostly memory." The EEPROM is generally programmed outside the circuit and then inserted in it for operation. Data in the EEPROM can be altered as necessary by the circuit, and all data

*NOVRAM® is a trademark of Xicor, Inc.

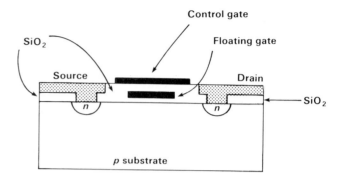

FIGURE 10-26. *EEPROM floating gate and control gate structure.*

contained in the EEPROM are nonvolatile. Alternatively, an EEPROM can be plugged into the host circuit and programmed by the host.

Although some EEPROMs use an MOS technology called Metal-Nitride-Oxide-Semiconductor (MNOS), the majority are fabricated with *n*-channel floating-gate technology like the EPROM (the MNOS parts are called EAROM for Electrically Alterable ROM). The main difference between the EPROM and the EEPROM structure is the control gate, illustrated in Fig. 10-26, which is used to write or erase data from the floating gate.

The 2716-compatible EEPROM is the 2816. Many 2816s (such as Xicor's) allow READ and WRITE functions from a single 5 V supply. Others, such as the Intel version, require high voltage (21 V) for all WRITE functions. All the available products require high voltage for the BYTE ERASE and CHIP ERASE functions, and all are compatible with the Intel high-voltage WRITE scheme. There is a total of six modes of operation for the 2816, as shown in Table 10-1. The CHIP ERASE function has the same effect as UV erasure of an

TABLE 10-1 OPERATING MODES FOR THE 2816.

$\overline{CE}$	$\overline{OE}$	$\overline{WE}$	Mode	I/O	Comment
1	X*	X	Standby	High Z	Lowest power consumption
0	0	1	READ	Data	Data output to bus
0	1	0	BYTE WRITE	Data	Data input on bus
0	1	1	R/W INHIBIT	High Z	Deactivates outputs during other memory addresses
0	1	HV	BYTE ERASE	High	Inputs held high
0	1	HV	BYTE WRITE	Input	Follows BYTE ERASE
0	HV	HV	CHIP ERASE	High	Erases all addresses to high

*Note: X denotes *don't care*.

EPROM. For the Xicor, the BYTE ERASE performs the same function as the BYTE WRITE function if all data are high; for the Intel, the BYTE ERASE must be performed before the WRITE function can be executed.

The writing or erasing of the EEPROM occurs as follows (refer to Fig. 10-26). Whether the on-chip logic handles the erasure or two separate steps are necessary, the floating gate must be erased before it can be written to. The erasure is accomplished when a high voltage is placed on the control gate, which causes electrons to migrate from the floating gate. After the floating gate has been erased, it is programmed in the same manner as the EPROM. As with the EPROM, the data residing on the floating gate are permanent since the gate is insulated from all other parts of the transistor. Byte writing takes about 10 ms to accomplish, so, unlike the RAM, the EEPROM cannot be used in a high-speed circuit.

Reading from the EEPROM is exactly the same as for the EPROM, or any other memory device covered in this text. Since a READ operation has absolutely no effect on the charge held on the floating gate, unlimited READ cycles are possible. Like the EPROM, the EEPROM can eventually wear out through repeated erasure. However, EEPROMs are good for thousands of ERASE cycles.

In addition to Byte-wide® writing and erasing, the entire chip can be erased with one command. The CHIP ERASE function returns all cells to the 1 state and requires approximately 10 ms.

NOVRAM®

The NOVRAM®, as mentioned earlier, actually consists of two memory devices on one chip. The NOVRAM® contains an NMOS static RAM and also an EEPROM. The NOVRAM® is sometimes called a "shadow RAM" because the EEPROM is a shadow of the RAM addresses.

A block diagram of the Xicor X2212 256×4-bit NOVRAM® is shown in Fig. 10-27, and its pinout configuration is shown in Fig. 10-28. During normal operation, only the RAM section is used. Since this is a standard NMOS static RAM circuit, full-speed operation at 300 ns access time is possible. When desired, a STORE command is sent to the 2212, causing all data stored in the RAM to be stored at the same addresses in the EEPROM. This requires 10 ms. Thus the data are now nonvolatile. When it is desirable that the data stored in EEPROM be reinserted in the RAM, an ARRAY RECALL can be issued to the 2212 and all data in the EEPROM will be recalled to the proper address in the RAM. The ARRAY RECALL takes only 1.2 μs. The data stored in EEPROM remain so that multiple recalls will produce the same result until another STORE is executed.

The main application for the NOVRAM® is in systems that must use data at very high speeds but cannot afford to lose the data upon a power outage. A

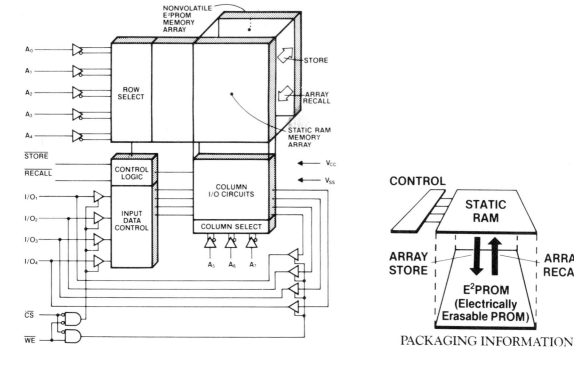

a. Functional diagram.

b. Memory organization.

FIGURE 10-27. Xicor X2212 NOVRAM® block diagram. (Courtesy of Xicor.)

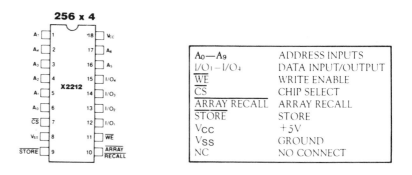

A_0—A_9	ADDRESS INPUTS
I/O_1–I/O_4	DATA INPUT/OUTPUT
$\overline{WE}$	WRITE ENABLE
$\overline{CS}$	CHIP SELECT
$\overline{ARRAY\ RECALL}$	ARRAY RECALL
$\overline{STORE}$	STORE
V_{CC}	+5V
V_{SS}	GROUND
NC	NO CONNECT

FIGURE 10-28. Xicor X2212 pin assignments. (Courtesy of Xicor.)

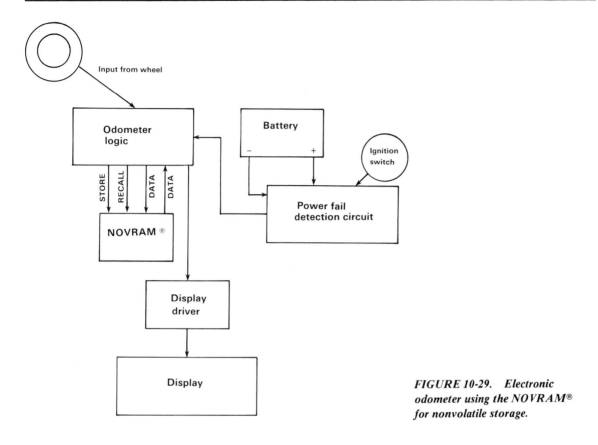

FIGURE 10-29. Electronic odometer using the NOVRAM® for nonvolatile storage.

good example of such a system is an electronic odometer in an automobile. Mechanical odometers never forget the data stored on them because they are advanced by gears. However, in an electronic odometer, the data are stored in semiconductor memory. Since the data must be updated continuously, an EPROM or EEPROM is not totally suitable. Of course, when the ignition is off or the battery removed from the car, the data must be retained. When one of these power-down events occurs, a NOVRAM ® could write its RAM data into EEPROM and wait for power to be restored before returning the mileage data to the RAM. This simple system is illustrated in block diagram form in Fig. 10-29. Other applications suited to the NOVRAM® include utility meters, vending machines, office equipment, and appliances.

BATTERY-BACKED-UP RAM

One other type of memory system that can be classified as nonvolatile, in the context used in this section, is a RAM that has a battery to retain its data in the

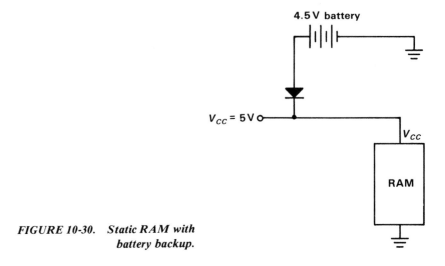

FIGURE 10-30. *Static RAM with battery backup.*

event of a power failure. Fig. 10-30 illustrates a system with this feature. When the power supply voltage drops below the threshold (one diode drop, or $V_{CC} - 0.7$ V), the battery cuts into the circuit and powers the RAM. Many RAMs are available with data retention down to a supply of 2 V or less, and these are ideal for this type of application, especially in CMOS. Note that only static RAMs can be used in battery-backed-up systems.

OTHER TYPES OF SEMICONDUCTOR MEMORY

In addition to the memories studied so far, there are two other types of semiconductor memory as well. Although these types of memory are not extremely popular at this time, some applications for them do exist, and changes in technology could make them cost-effective in the future. These are shift register memory and the charge-coupled device.

SHIFT REGISTER MEMORY

Due to the availability and low cost of RAMs, shift register memories are not frequently used. The basic circuit of a shift register memory is depicted in Fig. 10-31. The WRITE logic transfers information into the shift register in serial format. The READ logic retrieves the data in the same serial format. This is referred to as a First-In, First-Out (FIFO) memory. (Last-In, First-Out (LIFO) memory will be discussed in Chapter 12.)

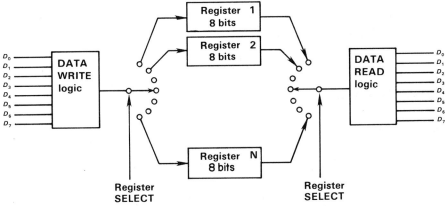

FIGURE 10-31. Shift register memory.

Shift register memories are useful in communication systems and in any system in which data are transferred between two circuits. The shift register allows each circuit to retrieve data at its convenience. A common communication device, the Universal Asynchronous Receiver-Transmitter (UART), is a good example of a system that uses a shift register memory.

CHARGE-COUPLED DEVICES

The charge-coupled device (CCD) was introduced in 1969. However, applications for the CCD have not been abundant. Only recently has a major application of the CCD been devised, and in this area the CCD is not used exactly as memory. This application is imaging.

The CCD operates on the principle of transferring charge "packets." The CCD is an MOS device with three closely spaced gate electrodes. The "packet" of charge is passed from one gate to another in a "dynamic memory" fashion. The presence of charge represents a 1, and the absence of charge is a 0. CCDs are generally configured in a shift register format. The difficulty in using CCDs is that, in order to move the charge packet, successively higher voltages must be impressed on the gates. Therefore, three supply voltages are necessary.

Some uses for the CCD are digital filters and time delays. In the digital filter, signals from several CCD cells combine to perform the filtering function desired. The time-delay application is rather obvious: cascading the CCD cells causes the data to travel for a longer period of time, resulting in a time delay from input to output.

The main application for CCDs, as mentioned above, is imaging. In imaging, radiation rather than gate voltages causes the charge packet to form. Thus the intensity of the charge packet is proportional to the radiation impinging on that particular cell. When the charge packets are shifted out, an electrical video signal results; thus CCDs are often used in solid state television cameras. These devices are especially useful with low light and infrared light; consequently, they find a great deal of application in military systems.

SUMMARY

1 Information stored in a memory is called data. The location of a particular piece of data is specified by its address. Data are stored as single bits or groups of bits; a group of eight bits is called a byte.

2 RAM (Random Access Memory) is general-purpose READ/WRITE memory. RAM is a volatile memory: the data are lost when power is removed.

3 The basic RAM storage element is the flip-flop, which is a 1-bit memory.

4 In a static RAM, data will remain stored as long as power is applied to the RAM.

5 Bipolar RAM is faster than MOS RAM but consumes more power.

6 Memory can be expanded with the CHIP SELECT pins and external address decoding logic.

7 The access time is the propagation delay in a memory between the time the address is applied to the chip and the valid data are available on the data bus. The approximate access times for several memory devices are compared in Table 10-2.

8 Dynamic RAM can store data for only a short time. Therefore, a dynamic RAM must have its data refreshed periodically.

TABLE 10-2 COMPARATIVE ACCESS TIMES FOR VARIOUS MEMORY DEVICES.

Memory Type	Access Time (Typical)
Bipolar RAM	30–50 ns
NMOS static RAM	100–200 ns
NMOS dynamic RAM	50–100 ns
CMOS static RAM	100–200 ns
Bipolar ROM/PROM	30–70 ns
MOS ROM/PROM/EPROM	250–500 ns
EEPROM	300–500 ns
NOVRAM®	200–300 ns

9 The storage element in a dynamic RAM is a capacitor.

10 The simplest dynamic RAM storage cell has only a single transistor. This allows dynamic RAM to be packaged more densely, cost less per bit, and consume less power than comparable static RAMs. The REFRESH circuitry, however, sometimes eliminates any cost savings.

11 A pseudostatic RAM is a dynamic RAM with REFRESH circuitry on the chip.

12 Once programmed, ROMs cannot have their data changed. The data in a ROM are written in the manufacturing process with a photographic-type mask.

13 PROMs are ROMs that can be programmed by the user. Once programmed, the data are permanent and cannot be altered.

14 EPROMs are user-programmable ROMs that can be erased by exposure to UV light. EPROMs use a special MOS transistor called a floating gate MOSFET.

15 Some EPROMs are available in the same pinout arrangement as ROMs to facilitate prototyping.

16 EEPROMs are similar to EPROMs except that they can be electrically erased and written over. Since they can be written over in the circuit, EEPROMs are sometimes referred to as read mostly memory.

17 A NOVRAM® contains a RAM and an EEPROM. The RAM portion is used during normal operation, and data can be stored in the EEPROM with a command signal. The most common application of the NOVRAM® is in systems where the data must not be lost during power interruptions.

18 Some RAMs have a battery backup circuit to act as a nonvolatile memory.

19 Shift register memory is often used in communication systems.

20 CCDs are most often found in imaging devices for solid state television cameras.

QUESTIONS AND PROBLEMS

1 What is RAM generally used for?

2 What does the term *volatile memory* mean?

3 What are data?

4 What is an address?

5 Define the term *byte*.

KEY TERMS

access time	EEPROM	PROM
address	EPROM	RAM
byte	FIFO	refresh
Byte-wide®	floating gate	ROM
charge packet	LIFO	static RAM
data	nonvolatile	UART
dynamic RAM	NOVRAM®	volatile

6 Using the DM7599 (see Fig. 10-4), design a 32×12-bit RAM memory system.

7 Could you interface an NMOS EPROM to a system with a memory cycle time of 100 ns? (Refer to Table 10-2.)

8 What is the access time of a memory?

9 Using the 4016 memory, design a 4k×16 memory system. Make the addresses unique, and design the logic so that the addresses of the memories are 3000–3FFF (hex).

10 Repeat problem 9, but locate the memories at 4000–4FFF(hex). The addresses do not have to be unique.

11 Why does a dynamic RAM have to be refreshed?

12 How many transistors are in a minimal dynamic RAM cell?

13 How is a ROM programmed?

14 What are some reasons for using a ROM over a PROM or EPROM?

15 How are PROMs programmed?

16 How are EPROMs programmed?

17 How are EPROMs erased?

18 Design a 6k×8 Read Only Memory system using the 2716. (Refer to Fig. 10-19.)

19 How is an EEPROM erased?

20 Give some examples of when an EEPROM would be preferable to a ROM or EPROM.

21 When would you most likely write data to the EEPROM portion of a NOVRAM®?

22 Would there be any problem with using a NOVRAM® whose data were written to the EEPROM every 100 ms if the system life was to be ten years?

23 Why would you want to use CMOS RAMs instead of bipolar RAMs in a battery-backed-up RAM system?

24 Would a shift register memory (such as a UART) be used in serial or parallel communications systems?

25 In block diagram form, design a system that has a parallel-loaded shift register memory with a serial output and that feeds a signal back to the parallel input system when it is ready for the next data word. (Assume that the parallel system uses bytes.)

11

MAGNETIC MEMORY

Magnetic memory is very useful for storing vast quantities of digital data. It comes in the form of magnetic cores, bubbles, tape, flexible disks, hard disks, and, although not magnetic, laser disks. The main advantage of modern magnetic and laser disk storage is their low cost per bit and permanence. It is permanent or nonvolatile because the information is stored either in the form of magnetic energy or optically, and these types of storage are not vulnerable to power supply aberrations. This chapter explains the operation and details of each of these types of magnetic memory and also optical disk memory.

MAGNETIC CORE MEMORY

Magnetic core memory has been around since the mid-1950s. It was one of the first forms of fast magnetic memory to be used in large main-frame computer systems. Today it is still found in a few applications, and it will remain in use for at least five or more years in older products.

OPERATION OF A MAGNETIC CORE

FIGURE 11-1.
Magnetic core.

Fig. 11-1 illustrates a small magnetic core shaped like a doughnut. The shape of this magnet allows it to retain a fairly low magnetic field for a very long time. If you remember the horseshoe magnets in grade school science class, you will recall that the teacher was very strict about keeping a piece of metal across the end of the magnet called a *keeper*. This is because a magnet will lose its strength if the keeper remains off for a long period of time. The horseshoe magnet with its keeper in place is the basic shape of a magnetic core. The magnetic core has its own built-in keeper, which allows it to retain a magnetic field for a very long time.

A Magnetic Core is a 2-state Device

Fig. 11-2 depicts two magnetic cores that each have a piece of wire threaded through their centers. Notice that the current flowing through the wire generates a magnetic field that will magnetize the core. If the current and its magnetic field magnetize the core in a counter-clockwise direction, a logic 0 is stored (see Fig. 11-2a). If the core is magnetized in a clockwise direction, a logic 1 is stored (see Fig. 11-2b). Because a core is constructed of a magnetizable material, the magnetic field will remain stored in the core even when the current is removed. Thus this is a type of nonvolatile storage.

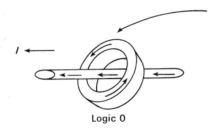

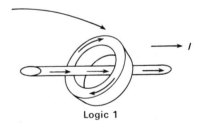

a. Magnetized in a counterclockwise direction.

b. Magnetized in a clockwise direction.

FIGURE 11-2.   Magnetic core.

Characteristics of a Magnetic Core

One method of illustrating the operation of the magnetic core is to plot the current required to magnetize it against the amount of magnetic energy or *flux density* stored in the core. This plot is called a *hysteresis curve* and is illustrated in Fig. 11-3. Notice from this point that there are two different directions of magnetic flux that can be stored in the core. These appear at the point on the graph where the current is zero.

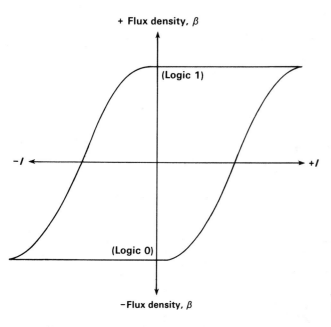

+ Flux density, β

(Logic 1)

$-I$ ◄————————►$+I$

(Logic 0)

$-$Flux density, β

FIGURE 11-3. *Hysteresis curve that illustrates the operation of a magnetic core.*

Suppose that the core is resting, or quiescent, at the logic 1 level. If the current through the core is increased in a positive direction, the level of flux density will change only slightly. When the current is returned to the zero level, the core will remain magnetized in the same direction. This is its *current stable* state. If the current is now increased in the negative direction, the core "flips" states and becomes magnetized in the opposite direction (logic 0). This is where it will remain if the current is reduced to zero.

Magnetic cores vary widely in size and speed and even in their application. They are normally used in a digital system as nonvolatile memory, but a communications designer might use them as inductors placed around conductors in a system to reduce parasitics. Magnetic cores for memory and all other purposes range approximately from 3 to 500 mils in diameter and operate or switch in 0.2–2000 μs.

SELECTION OF CORES IN A MATRIX

Fig. 11-4 depicts 16 magnetic cores arranged into a 4×4 core matrix. Although this is much too small to be practical, it does illustrate how a single core can be addressed or selected.

Notice that wires X_0 and Y_1 have some current flowing through them. Each wire carries *one-half* of the current required to magnetize a core. Therefore, only one of the cores in this matrix will receive enough current flow to become

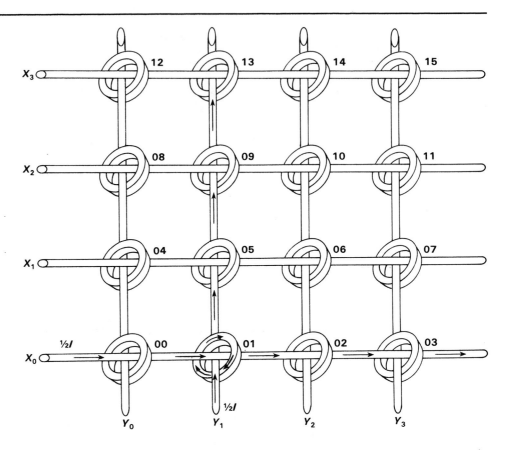

FIGURE 11-4. *A 4 × 4 core memory matrix.*

magnetized. In this example, core 01 is magnetized in the direction indicated, and cores 00, 02, 03, 05, 09, and 13 will remain unchanged even though some current flows through them.

If current is allowed to flow through lines X_1 and Y_3 then core 07 will be activated. It is important to note that only one X line and one Y line carry current at a given time.

Writing a 1 or 0 into the Selected Core

In order to write a 1 or a 0 into a selected core, we must make some provision to allow the currents to flow in different directions. The current must flow in one direction to store a logic 0 and in the opposite direction to store a logic 1. We can switch the direction of current by connecting a *bidirectional current driver* to each selection line.

One method of directing the flow of current through a selection line is to use a *steering network* on each end of each selection line. This circuit is illustrated in Fig. 11-5. Fig. 11-5a shows how a logic 1 is stored in a core, and Fig. 11-5b shows a logic 0 being stored. This illustration shows only one selection line. In practice, at least two lines would be threaded through each of the cores.

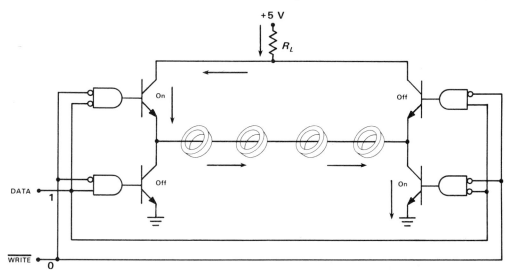

a. Storing a logic 1 in a magnetic core.

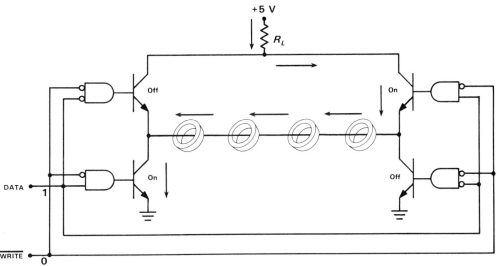

b. Storing a logic 0 in a magnetic core.

FIGURE 11-5. Current-steering network.

Address Inputs to the Core Matrix

Fig. 11-6 shows how a 4-bit binary address is converted to eight selection lines for the 16 magnetic cores in the 4×4 core matrix. The address has been broken down into two halves: the two most significant address bits are used to generate the four active-low X signals, the the two least significant address bits generate the four active-low Y signals. The decoding is accomplished by a 74LS139 dual 2-to-4 line decoder.

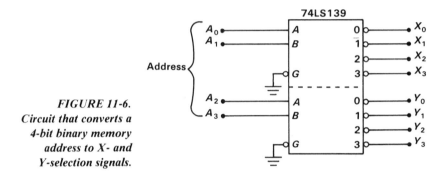

FIGURE 11-6.
Circuit that converts a 4-bit binary memory address to X- and Y-selection signals.

READING INFORMATION FROM A MAGNETIC CORE

In order to read information from a core memory, we must thread another wire, the *sense winding*, through each core (see Fig. 11-7). In addition to this extra wire, we also require a sense amplifier to sense the state of a core in the matrix.

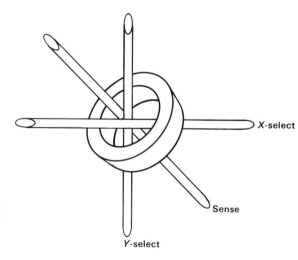

FIGURE 11-7.
X-select, Y-select, and sense wires threaded through a magnetic core.

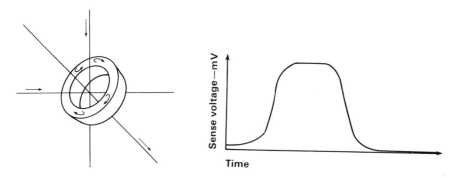

a. With changed polarity—output voltage on the sense line is plotted.

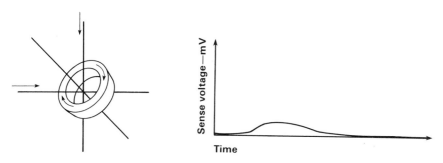

b. With no changed polarity—output voltage on the sense line is plotted.

FIGURE 11-8. Magnetic core.

Whenever the polarity of a core's magnetic field is reversed, the output of the sense winding typically rises to about 25 mV. The change in the magnetic field induces a current flow into the sense winding and a subsequent output voltage from the sense amplifier as illustrated in Fig. 11-8a. If the core remains magnetized with the same polarity, the output voltage of the sense winding at the input to the sense amplifier will be a 0.5–2 mV. This occurs because a very weak magnetic field cuts across the sense winding and generates a very weak output (see Fig. 11-8b).

The Sense Amplifier

The sense winding and the sense amplifier are illustrated in Fig. 11-9. The input to the sense amplifier uses the common mode configuration because noise

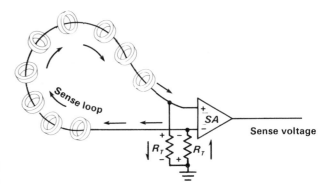

FIGURE 11-9. *Sense loop and current-sensing amplifier.*

across a sense winding is normally a common mode voltage which is blocked by this amplifier configuration. (A signal applied in common mode is applied to each input terminal.) A signal on the sense winding causes a current flow in the sense loop, generating a positive-going voltage on the positive input to the amplifier and a negative-going voltage on the negative input. Since the amplifier will amplify the positive-going input without inversion and the negative-going input with inversion, it in effect amplifies twice the input voltage. Noise, on the other hand, generates an in-phase voltage on each input terminal, and these voltages will cancel at the output of the amplifier.

Destructively Reading Data From a Core

When a magnetic core is read, the contents of the core are destroyed. For reading, the selected core is put at the logic 0 state. If the output of the sense amplifier rises in response to a change of polarity, then the core obviously contained a logic 1. If the output of the sense amplifier remains about the same, then the core contained a logic 0. In either case, the core is changed to a logic 0, and its contents have been destructively read out. Therefore, a READ for a core memory always consists of two WRITEs. The first WRITE stores a 0 in the core and the second restores the previous logic level to the core.

MULTIPLE-BIT CORE MEMORY

In most cases core memory is arranged in 8- or 9-bit word widths. (If nine bits are used, the ninth bit normally contains parity information.) Each bit position in a word is stored in a different *core bit plane*. Fig. 11-10 illustrates a 16×4 (16 words with 4 bits per word) core memory and illustrates the four core bit planes required for this memory organization. All the core bit planes in this type of memory use the same set of current drivers and selection logic. The only things

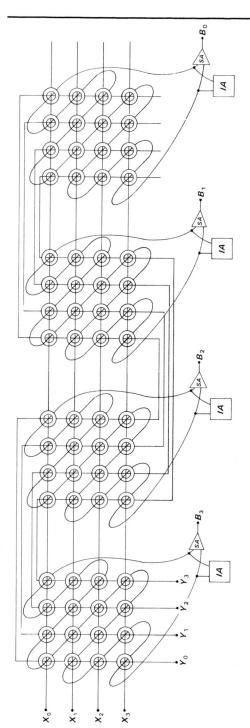

FIGURE 11-10. A 16 × 4 core memory system illustrating the four core bit planes, the sense amplifiers, and the inhibit amplifiers.

that are not common to all of the core bit planes are the sense windings, sense amplifiers, and inhibit amplifiers.

An *inhibit amplifier* is an amplifier used during the READ operation to inhibit the rewriting of logic 1s to bit planes that should remain logic 0s. It does this by causing a current flow in the sense winding of inhibited bit planes. The inhibited core bit planes receive a canceling magnetic field from the sense line so that the selected core remains a logic 0.

Let's suppose that a 0101 binary is stored in one of the 16 memory locations. In order to read this from the memory, all bits in the selected location are cleared to logic 0s (0000). The least significant bit and the next-to-most significant bit are then returned to the logic 1 level. However, because the current drivers are common to all the core bit planes in the memory, some method must be used to prevent all the bits from being changed to logic 1s. In this example, the inhibit amplifiers for the most significant bit position and the next-to-least significant bit position (the 0 bit positions) are enabled. Fig. 11-11 shows the effect of the inhibit amplifier and the current it generates through the sense windings of the four selected cores.

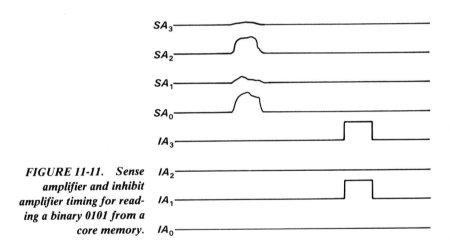

FIGURE 11-11. Sense amplifier and inhibit amplifier timing for reading a binary 0101 from a core memory.

MAGNETIC BUBBLE MEMORY

Magnetic bubbles are spots of magnetic energy maintained by a strong, permanent magnetic field. The term *bubble* comes from the fact that, under an electron microscope, these spots look like bubbles.

THE BUBBLE MEMORY DEVICE

Fig. 11-12 illustrates the basic structure of a magnetic bubble memory device. The device consists of a nonmagnetic yttrium-iron garnet substrate with an epitaxial layer of magnetic yttrium-iron garnet. The bubbles are maintained on the surface of the magnetic garnet layer with a strong, permanent magnet; they can be moved from point to point on the surface with an external rotating magnetic field. There are also other structures on the magnetic garnet surface that control the motion of the bubbles. These structures, constructed from a *thin-film permalloy*, are sometimes called *propagating elements* because they move magnetic bubbles across the surface of the device.

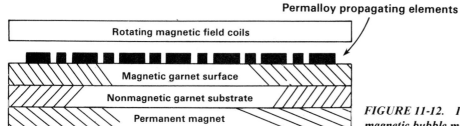

FIGURE 11-12. Internal structure of a magnetic bubble memory device.

The Permalloy Structure of the Propagating Elements

The purpose of the permalloy structure in a bubble memory device is to control the direction of the flow of the bubbles. Some common shapes for these structures are the T-bar pattern and the chevron pattern (see Fig. 11-13).

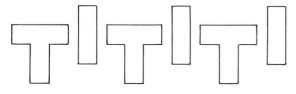

a. T-bar structure.

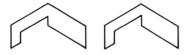

b. Chevron structure.

FIGURE 11-13. Two typical patterns used as propagating elements in a magnetic bubble memory.

Fig. 11-14 shows how a rotating magnetic field can move the magnetic bubbles across the surface of the magnetic garnet. In this figure the bubble has a north magnetic polarity, which interacts with the polarity of the magnetic field introduced to the T-bar structure and with the polarity of the external rotating magnetic field.

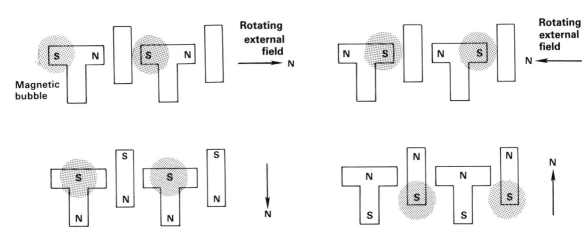

FIGURE 11-14. *Effect of the rotating magnetic field on the position of the bubbles in a magnetic bubble memory.*

CREATING AND DESTROYING MAGNETIC BUBBLES

The magnetic bubble represents a logic 1; lack of a bubble represents a logic 0. Most manufacturers use a *generator* to create bubbles and a *replicator/annihilator* to destroy or reproduce bubbles (see Fig. 11-15).

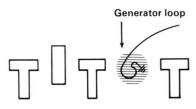

a. Large current generates a bubble in the generator loop.

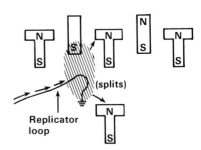

b. Small current stretches the bubble at the replicator loop and produces two bubbles with two separate pathways.

FIGURE 11-15. *Magnetic bubbles.*

The generator is a loop of wire placed near the surface of the magnetic garnet. When a current pulse passes through the generator loop, a magnetic bubble is created on the surface of the garnet. The replicator/annihilator accepts a bubble as it moves across the surface of the device, and passes it off the magnetic garnet, effectively annihilating the bubble. To replicate a bubble, a low-level current pulse is applied to the replicator/annihilator, which elongates the original bubble until it splits into two bubbles of equal size. One of the two bubbles travels off the device, and the other remains on the surface of the magnetic garnet.

THE INTERNAL ORGANIZATION OF A BUBBLE MEMORY DEVICE

Fig. 11-16 shows how the magnetic bubble memory is organized into sections called *loops*. A loop is a path on the surface of the device that continually recirculates information in the same way as a ring counter. Most bubble memory devices are organized into a major loop and many minor loops. The *major loop* transfers information from the minor loops to the user, and the *minor loops* store the information.

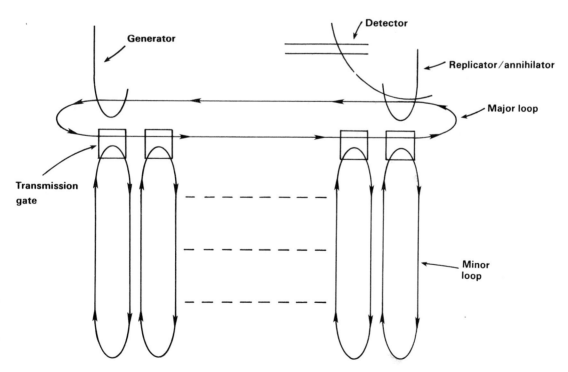

FIGURE 11-16.   Internal constuction of a magnetic bubble memory device.

The output of each minor loop has a device similar to a *replicator/annihilator* called a *transmission gate*. The transmission gate allows data from a minor loop to be replicated onto the major loop during a READ operation. When data are to be written, it takes information from the major loop and steers it to the minor loops. The data are written into the minor loops in parallel; that is, if a magnetic bubble device contains 256 minor loops, then each time data are transferred to the minor loops, 256 bits are transferred at one time (one bit to each minor loop). This simplifies memory control because there is no requirement for a memory address input. Because there are no address connections, extreme care must be taken when the device is turned on or off. It must only be turned off (powered down) if the position of the information in the minor loops is known. This allows the information stored in the minor loops to be used at a later time.

Reading the Bubble Memory Device

For the information in a bubble device to be read, bubbles must be replicated. In the process of replication, one bubble goes to the major loop and one passes a *detector*, a magneto-resistive element that changes in resistance when a bubble passes near it. The output of the detector is connected to a bridge circuit, which translates its change in resistance into a change in voltage. Fig. 11-17 shows the bridge circuit and sense amplifier used to detect the presence of a magnetic bubble. Reading also involves positioning the bit to be read over the detector, a process that may take as long as 10 ms in some magnetic bubble memory devices.

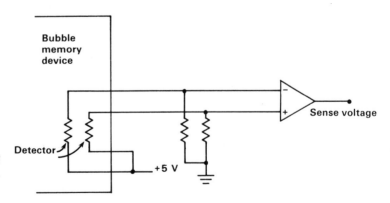

FIGURE 11-17. Detector bridge, which is constructed with two external resistors and the internal magneto-resistive detector.

WHERE ARE BUBBLE MEMORIES USED?

Bubble memory devices are finding application in portable computer terminals, development systems, and military systems. They may replace magnetic disk systems if the price ever comes down. In 1984, it cost over 1000 dollars for a

1M-byte magnetic bubble memory system consisting of eight 1M-bit devices and the interface circuitry. Because of its slow speed, magnetic bubble memory is unlikely to replace semiconductor memory as the main memory for small computer systems, but it may replace disk systems in many applications.

MAGNETIC TAPE MEMORY

Magnetic tape memory is commonly used to back up or duplicate information stored in the main memory of a computer system or on another magnetic tape or disk. Backing up data is important because mistakes and power supply transients can destroy digital data.

STORING DIGITAL DATA ON A MAGNETIC SURFACE

There are many methods of storing digital information on the surface of a magnetic tape. Two of the commonly used methods are *frequency-shift keying* (FSK) and *nonreturn-to-zero* (NRZ). The FSK method converts digital data to audio tones for storage on a standard audiotape, and NRZ stores digital data directly on the tape.

FSK Data Storage

FSK data storage is used in low-cost digital systems such as inexpensive home computers. The FSK system allows an ordinary audio cassette recorder to be connected directly to the computer to store programs and data.

Fig. 11-18 shows how a digital signal can be converted to audio for use in a low-cost computer system. Each bit of digital data is converted to one of two different tones through a digital multiplexer. The reason this technique is called frequency-shift keying is that the output of the multiplexer is shifted from one frequency to another by the data. The output of the multiplexer passes through a low-pass filter, which removes all the high-frequency components. The signal is then amplified and fed to the cassette recorder's microphone input.

FSK is capable of recording approximately 600 binary bits of information per second without severe dropout problems. *Dropouts* are missing bits of information caused by poor contact between the magnetic tape and head or flaws on the surface of the magnetic tape.

FSK Data Retrieval

Because FSK data are stored as two different tones indicating a logic 1 and a logic 0, the data are rather easy to retrieve. Fig. 11-19 shows a *phase-locked*

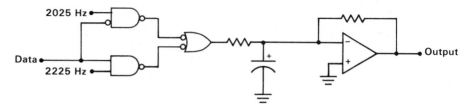

a. Circuit that generates two tones for FSK recording.

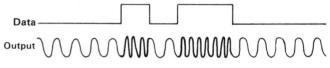

b. Voltage waveforms produced from a digital data stream.

FIGURE 11-18. Conversion of digital signals to audio signals.

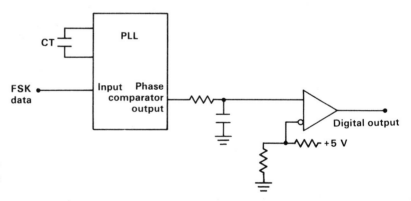

FIGURE 11-19. PLL tone decoder that generates digital data from FSK data.

loop (PLL) that is used to extract the original binary data. The PLL operates at a frequency that lies halfway between the logic 1 and the logic 0 frequencies. When a logic 1 tone is received at the input of the phase comparator, the output of the phase comparator becomes more positive, and when a logic 0 tone is received the output approaches ground. After passing through a low-pass filter, the signal is converted to a digital signal by a voltage comparator.

NRZ Data Storage

NRZ refers to a recording method whereby the flux density on the tape never returns to the zero level. The tape is always completely saturated in one direction or the other to ensure that the signal is much stronger than any noise that might be present in the system or on the tape. To record this type of information, we must convert the digital data to a current that can magnetize the tape. Fig. 11-20 represents a section of magnetic tape containing NRZ data. The arrows in this illustration indicate a magnetic field and its polarity.

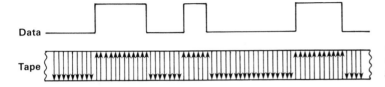

Data

Tape

FIGURE 11-20. *Representation of digital data stored on a magnetic tape in NRZ form.*

Fig. 11-21 shows a circuit used to generate the current for the magnetic tape head in an NRZ recording system. When the WRITE ENABLE input is active and a logic 1 is applied to the RECORD DATA input, current flows through Q_1 and the RECORD head to ground; and when a logic 0 is applied, current flows through Q_2 and the RECORD head to ground. This sequence of events produces opposing magnetic fields at the head for recording NRZ data.

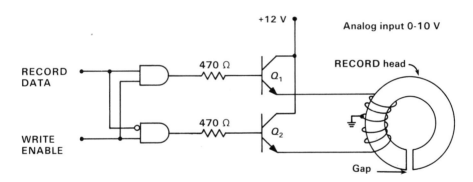

FIGURE 11-21. *Circuit that generates NRZ data from digital data.*

NRZ Data Retrieval

In order to retrieve information from a tape containing NRZ data, we must be able to convert changes in flux level back to digital data. Fig. 11-22 shows the output of a magnetic READ head as a magnetic tape containing NRZ data passes beneath it. The output is not a digital squarewave but a series of positive- and negative-going pulses because the field changes only polarity and not amplitude. A magnetic READ head, which is a transformer, will not transduce the signal properly because it only produces an output voltage when the magnetic field on the tape changes.

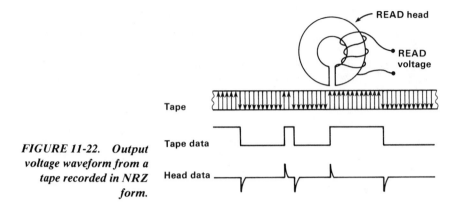

FIGURE 11-22. Output voltage waveform from a tape recorded in NRZ form.

Fig. 11-23 shows a simple circuit used to retrieve this type of data from a magnetic tape and the waveforms generated. An amplifier amplifies and buffers the data from the magnetic READ head and provides two complementary output signals. These complementary signals are clamped to remove the negative pulses. The signals are then applied to the inputs of a NOR gate latch. The latch is set or cleared in response to the pulses, and the output of the latch reconstructs the original digital data.

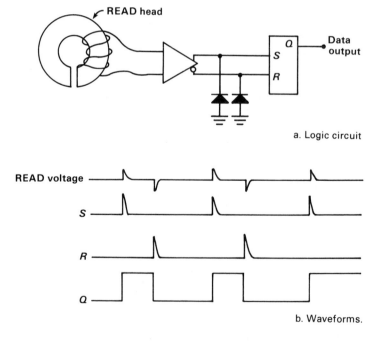

FIGURE 11-23. Circuit that will convert NRZ data from a magnetic tape to a digital data stream.

DATA FORMATS ON MAGNETIC TAPE

Because a tape may contain several minutes' worth of digital data, some technique must be found to locate the desired information in a relatively short period of time. This is normally accomplished by recording on a separate track of the magnetic tape a signal called a *block mark*. Blocks of data and block marks are indicated on the two-track magnetic tape illustrated in Fig. 11-24.

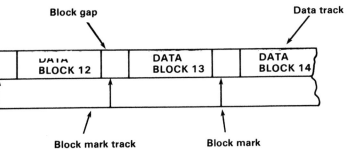

FIGURE 11-24. *Format of data blocks and block marks on a digital tape.*

Most blocks of data are between 1k and 16k bytes in length. To find a block on a tape, we first look in a *directory* located in the first block of the tape, which contains the block number of the desired data. We can then scan the tape at high speed for the correct block of data.

FLOPPY DISK MEMORY

Modern computers require a low-cost means of storing digital data. The magnetic tape is one such means, but its main problem is finding data in a reasonably short period of time. Data location and retrieval by means of a block mark may take 5–10 s or even longer.

The floppy (flexible) disk allows a high density of data storage and a relatively short access time, making it ideal for use in many digital systems. Access times for many floppy disks are much less than 1 s.

A *floppy disk* is a thin, circular piece of plastic coated with a magnetizable film and enclosed in a protective jacket. Fig. 11-25a illustrates the jacket for an 8-inch or a 5¼-inch mini-floppy disk. In addition to these two sizes, a new 3½-inch micro-floppy disk is available. The protective jacket of the floppy disk contains a WRITE *protect notch* that can be covered with a piece of mylar tape to prevent writing. It also contains an *index hole* which is used by the disk drive to locate the beginning of the disk. The disk itself also contains the index hole. As the disk spins, a sensor picks up the hole and signals the drive.

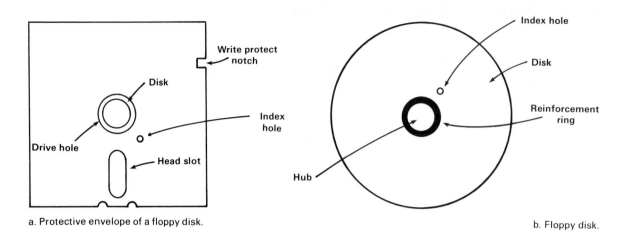

a. Protective envelope of a floppy disk.

b. Floppy disk.

FIGURE 11-25.   Floppy disk and jacket.

Fig. 11-25b illustrates the disk located inside the protective cover. The disk is turned inside the cover at 360 rpm by the disk drive, which engages the disk at its *hub* or center hole. Because the hub is subject to a great deal of stress, most disk manufacturers now reinforce it with a mylar ring.

Disk Data Format

Fig. 11-26 shows how the data are organized on the surface of a floppy disk. The information is recorded in concentric rings called *tracks*. The tracks are subdivided into smaller sections called *sectors*. A typical 8-inch disk contains 77 tracks with 26 sectors per track. Each of the sectors in this format normally contains 128 bytes of data. (A byte usually holds an 8-bit number or an ASCII coded character.) A typical 5¼-inch mini-floppy disk contains 40 tracks, with 10 sectors per track, and 256 bytes per sector. This disk format is called *single-density*. Other formats contain twice or four times the number of bytes per sector and are called *double density* or *quad-density*, respectively. The single-density 8-inch disk holds approximately 256k bytes of information, and the single-density 5¼-inch mini-floppy disk holds about 128k bytes.

The Data Format of a Track

Fig. 11-27a shows how the data in a track of a mini-floppy disk are arranged in fields for various purposes. There are different fields for each sector. The

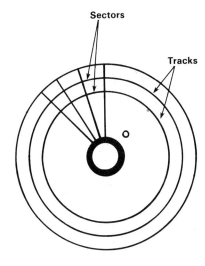

Sectors

Tracks

FIGURE 11-26. Format of the tracks and sectors on a floppy disk.

Final gap	Post index gap	Sector 0 ID field	Post ID gap	Sector 0 data field	Post-data-field gap	Sector 1 ID field	

a. Track on a mini-floppy disk.

ID address mark	Track address	Head address	Sector address	Sector length	CRC (MSB)	CRC (LSB)

b. Sector ID field.

Data address mark	DATA	CRC (MSB)	CRC (LSB)

c. Sector data field.

FIGURE 11-27. Data formats.

sectors are normally numbered 00–09 for a mini-floppy and 00–25 for an 8-inch floppy disk.

The *post-index gap* (see Fig. 11-27a) is a gap in recorded data that is typically 18 bytes long. This gap allows the computer time to detect the index mark and prepares it to read the sector ID field that follows. (The seven bytes in the sector

ID field are identified in Fig. 11-27b.) The *post-ID gap* is typically 17 bytes long and allows the computer time to decode the sector ID field.

The *data field* stores the digital sector information. The format of the data field in each sector appears in Fig. 11-27c. The *post-data-field gap* is typically 27 bytes long and is used to allow the computer time to set up for the next sector ID field. The *final gap* is typically 30 bytes long and allows the computer to locate the index mark signaling the start of a new track.

CRC Checks

The *cyclic redundancy check* (CRC) is an error-checking technique used to check data in both the data field and ID field of a sector. The CRC is generated by a series of left shifts and Exclusive ORs. The circuit required to generate a 16-bit CRC is illustrated in Fig. 11-28. Each time a new bit of information is read from the disk, it is Exclusive-ORed with bit position 5 to produce bit position 6, bit position 11 to produce bit position 12 and bit position 16 to produce bit position 0. (Bit position 16 is the CARRY output from the last shift.) After the outputs of the Exclusive OR gates have settled, the register shifts left. This process generates a unique bit pattern for the data recorded in the sector or ID field.

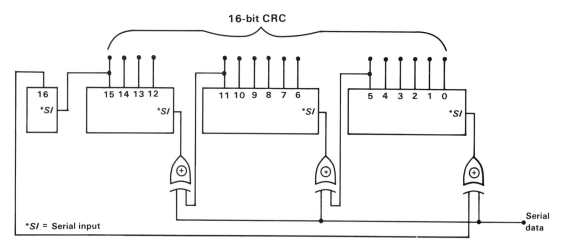

FIGURE 11-28.   *Circuit used to generate the 16-bit CRC for the serial data flowing to or from the disk READ/ RECORD head.*

The CRC is generated each time that new information is written to the disk and also when it is read from the disk. If the CRC calculated as data are read matches the CRC stored on the disk, the data are valid. CRC checks will miss only one error for every 10^{14} bits read from a disk.

Disk Handling

Disks are very easy to destroy if improperly handled. It is important to remember that the READ/RECORD head is located at the bottom of most mini-floppy disk drives. In other words, when you insert the disk into the drive, the bottom side of the disk will be read or written to. This may not be what you expect, so be careful with the bottom slot in the protective cover.

It is also important that the disk be kept as dust-free as humanly possible. Dust and even smoke particles can scratch the surface of the disk, causing permanent data loss. For this reason, it is a poor practice to use the reverse side of a disk in a single-sided disk drive. The top surface of the disk is pressed against the READ/RECORD head by a felt pressure-pad that can pick up a great deal of dust and dirt. If you turn the disk over to use the other side, the dust and dirt on the pressure-pad can scratch the surface and cause data loss.

THE FLOPPY DISK DRIVE

The disk drive mechanism contains electronic circuitry that converts the digital data to current for the READ/RECORD head; converts electrical signals from the READ/RECORD head to digital data; and controls the drive motor, head-loading mechanism, and head-positioning mechanism.

In most modern disk drives the head-positioning mechanism consists of a stepper motor that moves the head toward the center of the disk or out to its edge. The head assembly moves about 0.025 inch per step from track to track. The only signal provided by the drive to indicate positioning is the head home position indicator, which is also called the track 0 indicator. Track 0 is the outermost track on the disk, and it is typically where the beginning of the *disk-operating system* (DOS) begins.

The head-loading mechanism presses the head against the disk. Mini-floppy disk drives do not normally have this mechanism and subsequently suffer an increase in the amount of wear on the disk. Luckily the mini-floppy disk only spins when data are written or read. It typically takes about 50 ms to load the head in an 8-inch disk drive.

The drive motor in a typical disk system is a DC motor that rotates at 300 rpm. Its speed is controlled by a feedback circuit and a tach generator built into the motor itself. The speed of the motor can vary considerably without producing data errors.

READ/RECORD Electronics

Most disk systems use NRZ recording, but the data are formatted or encoded. Unlike magnetic tape, the content of a disk is not pure data: CLOCK information is stored on the disk along with the data. Two techniques are used for disk recording: *frequency modulation* (FM) and *modified frequency modulation* (MFM).

Fig. 11-29 shows how data and clocking information are interwoven in the FM recording technique. The CLOCK bit pattern and the DATA bit pattern form a 16-bit number stored for each byte of data. This means that half of the disk consists of clocking information. The CLOCK and DATA bits are separated by a 2 μs interval so that it takes 4 μs to store one bit of information (the *bit time* of this system). It is therefore possible to store 250,000 bits of information per second on a floppy disk.

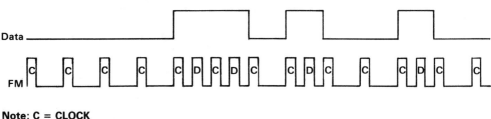

Note: C = CLOCK
D = DATA

FIGURE 11-29.   FM disk data.

Fig. 11-30 shows the format of data for MFM. Unlike the FM signal, the MFM signal doesn't always contain a CLOCK pulse. It contains a CLOCK pulse only if no data have been written in the previous bit time and no data are to be written in the present bit time. Since a bit of information can be recorded once every 2 μs, 500,000 bits of data can be stored in 1 s. This doubles the speed, as well as the density, of this system. Notice that since no CLOCK is present except when two 0s (or empty data positions) appear in a row, the density has

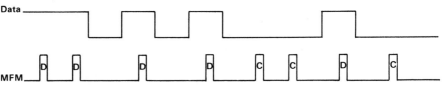

Note: C = CLOCK
D = DATA

FIGURE 11-30.   MFM disk data.

been doubled. MFM is thus often used in double-density systems, and FM is often used in single-density systems.

Disk Drive Maintenance

Disk drives are extremely reliable and require very little maintenance. Once a year the motor speed should be adjusted and the head-positioning mechanism checked, and once a month the head assembly should be cleaned.

Adjusting the motor speed is a very simple task because a strobe disk has been placed on the flywheel of the disk hub drive (see Fig. 11-31). A neon or fluorescent lamp can be used with the strobe disk to time the motor. If the marks on the disk do not appear stationary, the potentiometer provided should be used to adjust the motor speed.

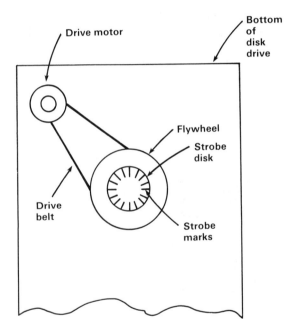

FIGURE 11-31. Underside of a mini-floppy disk drive showing the strobe disk, flywheel, drive belt, and DC motor.

Head-positioning adjustments require a factory alignment disk that is used with an oscilloscope. The data are monitored as they are read from the factory disk. The head-positioning adjustment should not be attempted without this disk.

The head on a disk drive can be cleaned with a cleaning disk or a cotton swab and alcohol. The cleaning disk is the easier method. If a cotton swab is used, the disk drive must be disassembled to expose the head.

HARD DISK MEMORY

Floppy disks are useful for storing short programs and data files, but they become filled with data very quickly. Applications such as word processing and data-based management require vast amounts of storage.

The hard disk has been around longer than the floppy disk, but its cost was prohibitive until recently. A new form of hard disk called a *Winchester drive* has reduced the cost of this type of memory. The READ/WRITE head used for this drive is called a Winchester-type flying head. (The first hard disk drive manufactured by IBM, which could store 30M (30 million) bytes of data per platter, was nicknamed the 30-30; hence the name "Winchester drive.")

THE HARD DISK MEDIUM

A typical hard disk drive contains an 8-inch or 14-inch aluminum platter coated with a lubricated magnetic oxide material. Some hard disk drives contain multiple disk platters stacked one on top of the other. Unlike the floppy disk, the hard disk medium is not removable from the disk drive. The disks are housed in a sealed unit to prevent their contamination by dust, dirt, smoke, oil, etc. Fig. 11-32 shows the 4-disk stack that is found in some of the larger hard disk memory systems.

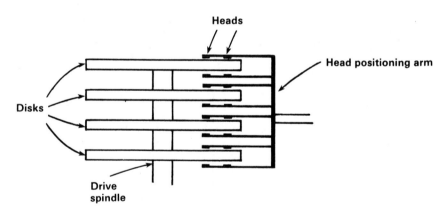

FIGURE 11-32. Four-disk stack showing the heads, drive spindle, and head-positioning arm.

Because the hard disk is not removable from the drive, it can be spun at a much higher speed than the floppy disk. Typically, a hard disk spins at about 3000 rpm, and the floppy disk spins at 300 rpm. This means that the data can be read or written at a much higher rate of speed and with a much higher bit density.

Fig. 11-33 shows the surface of a typical hard disk, illustrating the tracks, sectors, READ/WRITE heads, and fixed CLOCK READ head. One significant change from the floppy disk is that there are often two or more READ/WRITE heads per disk surface. This reduces the amount of time required to move the head assembly from track to track and increases the amount of storage on the surface of the disk.

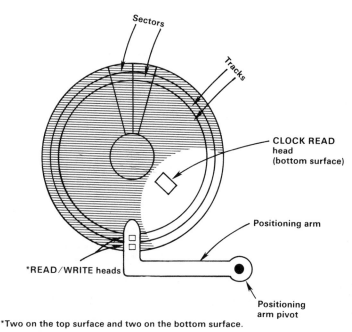

CLOCK READ
head
(bottom surface)

Positioning arm

*READ/WRITE heads

Positioning
arm pivot

*Two on the top surface and two on the bottom surface.

FIGURE 11-33. Hard disk drive show-ing the position of the CLOCK head, READ/WRITE heads, tracks, and sec-tors within a track.

This type of disk system uses a fixed CLOCK track on the bottom surface of the disk. This track generates the WRITE CLOCK and WRITE DATA and decodes the MFM READ data.

The READ/WRITE heads in a hard disk drive are called flying heads. This type of head floats above the disk's surface on a cushion of air generated by the disk's rotation. This increases the life of the magnetic oxide coating because almost no physical contact occurs between the head and the disk surface. The only time that the head ever comes in contact with the disk surface is when the disk stops, speeds up, or slows down. Some drive manufacturers have a special area on the surface designed for head landings called the *head crash area.*

The Hard Disk Data Format

A typical 14-inch hard disk drive has about 200 tracks for recording data and contains about 18,000 bytes of information per track. Because there are multiple heads, the effective number of tracks is raised. In the 4-head drive shown in Fig.

11-33, for example, the number of available tracks is 800. Thus the number of bytes that can be stored on this single 14-inch disk is 800 × 18,000 or 14.4M bytes. The addition of a second disk platter increases the storage capacity to 28.8M bytes.

Fig. 11-34 illustrates the typical data format in a hard disk track. This is very nearly the same as the format of a floppy disk track, except that there may be more sectors. If 128-byte sectors are used, there are at least 60 sectors per track. Remember that quite a large area of the disk's surface is used for storing gaps, ID fields, and other information. When data are to be written or read from the disk, the user must specify the track, head, and sector numbers. In the floppy disk system, the user specifies only the track and sector numbers because there is only one head (unless the drive is double-sided).

FIGURE 11-34. *Track format of a typical hard disk system.*

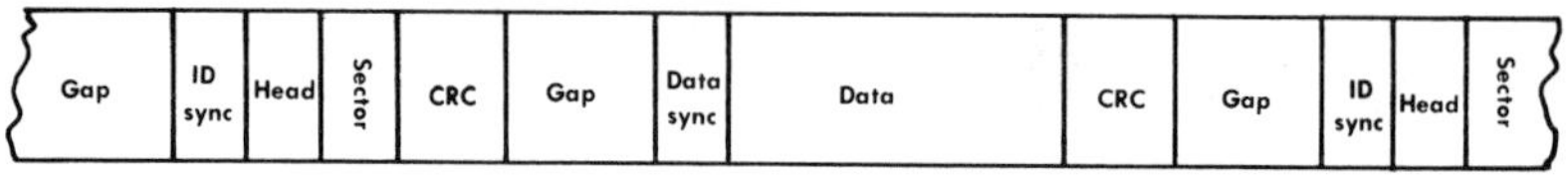

| Gap | ID sync | Head | Sector | CRC | Gap | Data sync | Data | CRC | Gap | ID sync | Head | Sector |

THE HARD DISK DRIVE

The hard disk drive mechanism is a little more delicate than the floppy disk drive because of the Winchester flying heads. While the floppy disk drive can be operated in any position and is also portable, the hard disk drive must be operated in a nearly horizontal or vertical position and is subject to damage from transportation and vibration.

There is no user maintenance of the hard disk drive. The drive mechanism is sealed and requires no periodic maintenance such as head cleaning. The typical *mean time between failures* (MTBF) for a hard disk drive is 8000 hours. If the drive is operated for 10 hours a day, then it will run approximately 800 days before a failure occurs.

OPTICAL DISK MEMORY

Today's computer systems demand extremely large storage capacities at relatively low costs. A new low-cost method for storing digital data is the optical disk. This technology is a spinoff of the laser videodisk technology developed by Pioneer Electronics Corp. of Japan. The current product is a write-once optical disk designed specifically for archiving large volumes of data. In the near future this technology will probably evolve into READ/WRITE optical storage.

OPTICAL DISKS

The optical disk is constructed from light-weight plastic coated with a photo-sensitive emulsion. The emulsion is covered with a thin coating of glass or plastic that protects it from contamination by dust, smoke, and oil.

The optical disk uses a laser beam to store and read data. To store data, a high-intensity beam of light causes the photosensitive emulsion to change its ability to reflect light, which enables the information to be read from the disk at some other time. The only disadvantage of recording data in this method is that the data cannot be erased. Even with this limitation, this system is still usable for storing permanent files. One optical disk can store an incredible 400,000 typewritten pages of information.

For reading information off the disk, a low-intensity laser beam is focused through the plastic or glass coating onto the photosensitive substrate. The change caused by the original high-intensity beam in the amount of light reflected by the photosensitive emulsion can be detected by a photosensor and converted back to digital data.

THE OPTICAL DISK DRIVE

Fig. 11-35 shows the Optimem 1000 from the Shugart Corporation. This drive has a capacity of 1G (one billion) bytes of data per side of the removable disk. Data are recorded at the rate of 5M bits per second, which is about ten times faster than the double-density floppy disk drive. Each side of the disk is formatted with 40,000 tracks, each track contains 25 sectors, and each sector contains 1k bytes of data.

FIGURE 11-35. *Optical disk drive. (Courtesy of Shugart Corporation.)*

SUMMARY

1 Magnetic memory is the most common form of READ/WRITE nonvolatile memory. Examples of this type of memory include the magnetic core, the magnetic bubble, and the magnetic disk.

2 A magnetic device is a bistable device capable of storing one bit.

3 The operation of a magnetic core is described graphically with a hysteresis curve.

4 A magnetic core can be selected if half the current required to magnetize the core is passed through an X-select wire and the other half is passed through a Y-select wire.

5 The current is passed in one direction to write a logic 1 into a magnetic core and in the opposite direction to write a logic 0.

6 The memory address input to a magnetic core memory is converted to X- and Y-select lines through a decoder.

7 The sense winding in a core memory reads information by sensing a small current flow that is induced whenever a core switches states.

8 The sense amplifier converts the small current flow from the sense winding to a digital signal. This amplifier is also used to cancel noise that may appear on the sense winding.

9 A core bit plane is a bit position of a word in a core memory.

10 The inhibit amplifier in a core memory is used to inhibit the writing of a logic 1 in a READ operation.

11 Magnetic bubbles are tiny magnetic spots on the surface of a memory device. These spots can be moved around on the surface of the bubble memory device through the use of a rotating magnetic field.

12 The T-bar or chevron propagation patterns are elements that are used to move bubbles in a bubble memory device.

13 The major loop in a bubble memory is used to transfer information to or from a minor loop. The minor loop is where data are stored.

14 Data are read from a bubble memory through the use of a magneto-resistive detector and written into the device through the use of a bubble generation loop.

15 Magnetic tape memory has its information stored in one of two different ways: FSK and NRZ.

16 Block marks are used on magnetic tape to locate information stored in the blocks.

17 A floppy disk is available in three sizes: 8-inch, 5¼-inch, and 3½-inch.

18 Data are stored on the surface of a floppy disk in tracks, which are subdivided into sectors.

19 Information is recorded on the surface of the floppy disk in either FM or MFM format.

20 A floppy disk stores 128k–256k bytes of data, and a hard disk stores many millions of bytes of data.

21 Hard disk drives use a Winchester flying head. This head floats on a cushion of air created by the rotating surface of the disk.

22 Optical disks can store 1G byte of data or 400,000 typewritten pages.

KEY TERMS

bidirectional current driver

block mark

bubble generator

chevron propagating element

core bit plane

cyclic redundancy check (CRC)

destructive read(out)

directory

DOS

double-density

flexible disk

floppy disk

frequency modulation (FM)

frequency shift keying

hard disk

head crash

hub

hysteresis curve

ID field

index hole

inhibit amplifier

keeper

magnetic bubble

magnetic core

magneto-resistive detector

major loop

Mean Time Between Failures

micro-floppy disk

mini-floppy disk

minor loop

modified frequency modulation (MFM)

nonreturn-to-zero

propagating elements

READ/RECORD head

replicator/annihilator

sector

select lines

sense amplifier

sense windings

single-density

steering network

T-bar propagating element

thin-film permalloy

track

transmission gate

Winchester drive

Winchester flying head

WRITE protect notch

QUESTIONS AND PROBLEMS

1 Explain how a core is magnetized in the counterclockwise direction.

2 Draw and explain the operation of a magnetic core's hysteresis curve.

3 What two lines would be activated to select core 09 of Fig. 11-4?

4 Explain how the circuit in Fig. 11-5a operates.

5 Convert the memory address decoder in Fig. 11-6 to a circuit that will decode a 5-bit memory address.

6 Whenever a core switches states, it generates a small current flow in the sense winding. What voltage would you expect at the input to the sense amplifier?

7 Describe how the sense amplifier cancels noise and amplifies the sense voltage.

8 What is meant by the term *destructive readout*?

9 Why does it take two WRITEs to do a READ with a core memory?

10 The inhibit amplifier is used to prevent what logic level from being written into a magnetic core?

11 What is a core bit plane?

12 If a core memory is organized as a 1k × 8 memory, how many bits are in each word? How many words are there? How many core planes would be required?

13 Magnetic bubbles are maintained with what device?

14 What two types or shapes of propagating elements are used in bubble memories?

15 What devices are used to create and destroy a magnetic bubble?

16 Describe how the rotating magnetic field moves the magnetic bubbles shown in Fig. 11-14.

17 What are the major and minor loops in a magnetic bubble memory?

18 What device is used to sense the presence of a magnetic bubble?

19 How are bubbles transferred between the minor and major loops in a magnetic bubble memory?

20 What two methods are used to place information on the surface of a magnetic tape?

21 Describe the operation of the circuit in Fig. 11-18a.

22 Why doesn't the READ/RECORD head reproduce data stored in NRZ form?

23 Explain how the circuit in Fig. 11-21 is able to change the direction of current flow through the RECORD head.

24 Give a brief description of how the PLL can be used to detect an FSK signal.

25 Explain how the circuit of Fig. 11-23 converts the pulses read from the tape into digital data.

26 What is a block mark?

27 In what sizes are floppy disks available? (List the dimensions and also the names.)

28 What is the purpose of the WRITE protect notch in a floppy disk, and how would a disk be protected against a WRITE?

29 What is the index hole in a floppy disk used to indicate?

30 Define the following terms as they apply to floppy disks: track, sector, CRC, hub, single-density, sector ID field, and post-data-field gap.

31 If the CRC register of Fig. 11-28 contains a 0110 000111 011000 and the bit 16 position is cleared, what is the result if the following data bits are entered: 0,0,1, and 1?

32 Why is it a poor practice to use both sides of a floppy disk in a single-sided disk drive?

33 What does the home position indicator indicate?

34 Describe FM recording as it applies to a disk system.

35 Describe MFM recording as it applies to a disk system.

36 How is the motor speed of a floppy disk drive adjusted?

37 Why is a head-cleaning disk normally employed instead of a cotton swab and alcohol to clean the head of a floppy disk drive?

38 Hard disk drives normally contain how many heads per platter?

39 What type of head is found in a modern hard disk drive?

40 What does the CLOCK head and the signal READ from the disk accomplish in a hard disk system?

41 What information would the user give to a hard disk drive that would not be given to a single-sided floppy disk drive?

12

INTRODUCTION TO MICROPROCESSORS

Microprocessors are now being used in digital design to replace sequential logic and combinational logic circuits. Although you will probably never design a microprocessor, you will be required to use them in the systems that you work with. A microprocessor may be called a programmable logic element because it can be programmed to function as any circuit covered in this text.

This chapter will introduce a few common microprocessors: the Intel 8085A, the Motorola MC6800, and the Zilog Z80®* microprocessor.

*Zilog Z80® microprocessor is a registered trademark of Zilog, Inc., with whom Bobbs-Merrill Co. is not associated.

THE ELEMENTS OF A COMPUTER SYSTEM

An understanding of the computer will aid in understanding the microprocessor because a computer uses the microprocessor or a similar device as its brain or central control unit.

A COMPUTER

A computer is a machine that, in modern times, is constructed from integrated circuits and at least one microprocessor. This machine is called a computer because one of its functions is to solve or compute answers to problems. In fact, each of the first computers was designed to solve a specific problem, and that was all it could do. To solve a new problem, the machine had to be rewired. Today we "rewire" a machine by changing its *program*, a list of instructions stored in the computer's memory in sequential order. The programs stored in modern computers are not limited to problem solving: some control machines, test other electronic equipment, cook a turkey, and do just about anything else that you can imagine.

THE BLOCK DIAGRAM OF A COMPUTER SYSTEM

Fig. 12-1 illustrates the block diagram of a typical digital computer system. The most important part, located at the computer's center, is the *central processing unit* (CPU). The CPU directs the operation of the memory, the input/output (*I/O*) equipment, and the auxiliary memory. The paths connecting the CPU with the other blocks are called *buses*. The buses contain:

1. data that are transferred between blocks in the block diagram,
2. addressing information used to select a particular *I/O* device or location in the memory,
3. control signals for controlling the blocks in the block diagram.

The CPU

The CPU indirectly controls the memory, *I/O,* and auxiliary memory through a program stored in the memory. Direct control is accomplished with a *control bus* that is used to issue READ and WRITE commands to the memory, *I/O,* and auxiliary memory. In addition to control, the CPU is capable of performing some simple arithmetic operations and logic functions with its *arithmetic and logic unit* (ALU). Its most common arithmetic operations are addition and

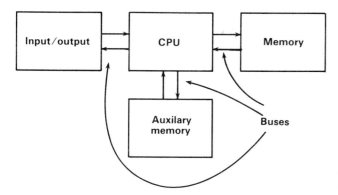

FIGURE 12-1. *Block diagram of a computer system.*

subtraction, and its most common logic functions are AND, OR, NOT, and Exclusive OR.

The CPU is also used in a computer system to make decisions. The decisions that most CPUs make are based on numerical tests of binary data. The tests that most CPUs can perform include checks for the following numerical information:

1. a zero or nonzero condition,
2. a negative or positive number,
3. a CARRY after an addition or a BORROW after a subtraction,
4. even and odd PARITY,
5. an OVERFLOW following an arithmetic operation.

The Memory

The memory in a computer system is used to store the results from a program, store the data used by a program, and store the program itself for the CPU. Most modern computer systems use either semiconductor RAM or, occasionally, magnetic cores for memory. Because these memories are very fast, a program stored in this type of memory is executed by the CPU in a very short period of time. Large computer systems typically execute about 10 million instructions per second.

The Auxiliary Memory

Auxiliary memory is used to store large volumes of data. Auxiliary memory devices (discussed in Chapter 11) include the disk memory and the bubble memory. The floppy disk is used to store data that are temporary in nature; the hard disk, magnetic bubble memory, and optical disk are used to store data that are more permanent. Examples of programs that use permanent data are:

1. word processors,
2. spelling dictionaries,
3. spread sheet programs,
4. data-based management systems,
5. the BASIC language,
6. assemblers,
7. the disk-operating system (DOS).

The Input/Output Equipment

The computer can solve problems at a tremendous rate of speed and access vast quantities of data, but it isn't very useful unless it can communicate information to us. The *I/O* equipment is used to transfer data into or out of the computer system. The most common forms of *I/O* equipment include:

1. printers,
2. keyboards,
3. video display terminals (VDTs),
4. modems,
5. magnetic badge readers,
6. numerical displays.

In fact, any device that generates or accepts an electrical signal can be used as a computer *I/O* device.

THE ELEMENTS OF A MICROPROCESSOR

A microprocessor is the modern, miniaturized version of the CPU. It has not replaced all CPUs, but if predictions are correct, it will do so in the near future. The only difference between the microprocessor and the discrete CPU found in large main-frame computer systems is speed. Microprocessors are capable of performing 1–2 million operations per second, and large main-frame CPUs can perform over 10 million operations per second.

THE BLOCK DIAGRAM OF A TYPICAL MICROPROCESSOR

Fig. 12-2 shows the internal structure of a typical microprocessor. It doesn't appear to be a very complicated machine because it isn't. Usually, the complicated part of the computer system is the program that directs the microprocessor to perform its task.

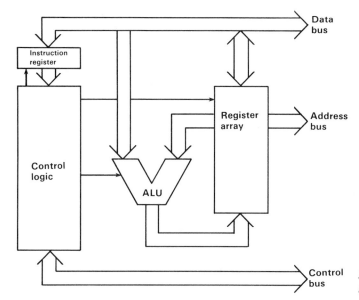

FIGURE 12-2. *Internal structure of a typical microprocessor.*

The ALU

The heart of the microprocessor is its arithmetic and logic unit, which is assigned the task of performing all of the arithmetic and logic. These operations include:

1. addition,
2. subtraction,
3. logical addition (OR),
4. logical multiplication (AND),
5. Exclusive OR,
6. inversion,
7. shifting and rotating numbers.

Many ALUs are capable of both 8- and 16-bit arithmetic, and some of the new microprocessors are capable of 32-bit arithmetic.

Numbers are operated on in one of two different formats: signed and unsigned. The *unsigned number* contains no sign information or sign bit. In other words, if the number is unsigned it is either an 8-bit or a 16-bit positive number. If the unsigned number format is used, then 0000 1010 represents decimal 10, 1000 0001 is decimal 129, and 1100 0000 is decimal 192.

Signed numbers have one of their bits reserved for a sign bit. The most significant bit position of an 8- or 16-bit number is used as the sign bit position. If the number is positive, the sign bit is a 0, and if the number is negative, the sign bit is a 1. The following 8-bit signed numbers are all positive because the left-hand bit position is a 0: 0000 0010 (2 decimal), 0110 0100 (100 decimal), and 0100 0000 (64 decimal). The largest positive 8-bit number (0111 1111) is equal to decimal 127.

Negative numbers have a 1 in their sign bit position, but the remaining bits do not represent magnitudes of two as with positive binary numbers. A negative number is the twos complement of the positive number. For example, if 0000 0001 (+1) is converted to a negative number by twos complementing, the result is 1111 1111 (–1). All microprocessors use the twos complemented form of a positive number to represent a negative number. Other examples of negative numbers are 1111 1100 (–4), 1100 0111 (–57), and 1110 1000 (–24). The largest 8-bit negative number that can be stored is 1000 0000 (–128).

The Register Array

The register array is found in almost all microprocessors and consists of a group of registers used for storing temporary data, addressing memory, and holding the answer from the ALU. Fig. 12-3 illustrates the register array of a microprocessor.

The *accumulator* register is found in all microprocessor register arrays. The purpose of the accumulator is to accumulate results from the ALU. Some microprocessors contain one accumulator, some contain two, and some contain many more; but in all cases the function is the same. If two numbers are added or subtracted, the answer will appear in the accumulator.

The *program counter* is another important register common to all microprocessors. The program counter tracks the instructions or steps of a program. It is so named because it tracks the program by counting up through the memory, which contains the steps of the program in sequential order.

The *condition code register* or flag register is found in all microprocessors and indicates the condition of the microprocessor and its ALU. The conditions that the bits in this register indicate are: sign, carry, parity, overflow, and zero. The condition code bits are used by the microprocessor in the decision-making

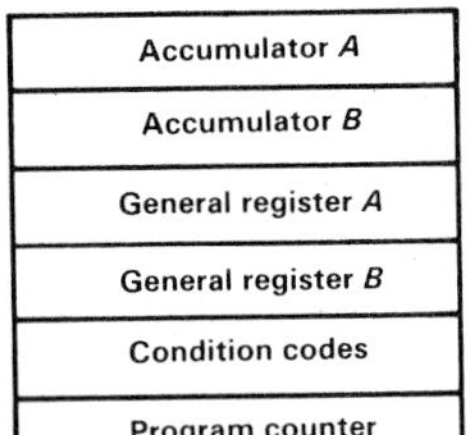

| Accumulator *A* |
| Accumulator *B* |
| General register *A* |
| General register *B* |
| Condition codes |
| Program counter |

FIGURE 12-3. Register array of a typical microprocessor.

functions discussed earlier. These decisions are based on numerical tests, and the code bits or "flags" indicate the outcome of these tests.

The Instruction Register

The *instruction register* holds the current instruction so that the microprocessor can decode it. After an instruction has been retrieved from the location specified by the program counter, it is placed into the instruction register. It is then decoded by the control logic and executed by the microprocessor.

The Control Logic

The *control logic* controls all the components shown in Fig. 12-2 and all events in a microprocessor-based system. When a microprocessor has power applied to it, the control logic stores a specific address in the program counter. This address is then output to the memory via the address bus. The control logic then reads the instruction at that address and transfers it to the instruction register. Next, the control logic determines from the instruction register what instruction is to be executed. After executing this instruction, it outputs the address of the next instruction from the program counter and repeats the above steps.

THE BUS STRUCTURE OF A MICROPROCESSOR

All microprocessors have three buses: an address bus, a data bus, and a control bus. These three buses are used by the microprocessor to communicate to the memory, *I/O*, and auxiliary memory.

The Address Bus

The *address bus* is used to single out an *I/O* device or a location in the memory. A modern microprocessor contains a 16-, 20-, or 24-bit address bus. If the address bus is 16 bits wide, the microprocessor can directly address 64k words of memory; a 20-bit bus provides access to 1M words; and a bus of 24 bits holds the addresses of 16M words.

The Data Bus

The *data bus* is used to transfer information between the microprocessor and the memory, *I/O*, or auxiliary memory. It is bidirectional in most microprocessors, since information must be transferred both into and out of the microprocessor. For example, a READ from the memory will transfer data from the memory into the microprocessor, and a WRITE to memory will transfer data out of the microprocessor into the memory.

The width of the data bus varies from one microprocessor to another. Data buses eight bits wide are usually found in an 8-bit microprocessor, and 16-bit buses are usually found in a 16-bit microprocessor.

The Control Bus

The *control bus* controls the memory, *I/O*, and auxiliary memory through a signal or signals that indicate a READ or a WRITE operation. READ and WRITE are the only two functions that these ancillary components are capable of performing. In addition to these basic control signals, there are controls for more complicated and less-used operations such as the INTERRUPT and the bus arbitration or direct memory access (DMA).

TYPES OF MICROPROCESSORS

The fixed-word-width microprocessor, the single-component microcomputer, and the bit-slice microprocessor are the three basic types of microprocessor available today. This section covers the characteristics, sizes, and applications of these microprocessors.

FIXED-WORD-WIDTH MICROPROCESSORS

The *fixed-word-width microprocessor* is probably the type that is most familiar to everyone. It is available in 4-, 8-, 16-, and 32-bit versions, and it was the first type of microprocessor to be integrated into a single integrated circuit.

Fixed-word-width Microprocessor History

In 1971 the Intel Corporation constructed the Intel 4004, the first microprocessor. This device was a 4-bit microprocessor that could address 4k "nibbles" (4-bit numbers) of memory and execute 50,000 instructions per second. It was used in early video games, calculators, and similar machines.

In 1972 the Intel Corporation began shipping the Intel 8008, which was the first 8-bit microprocessor. This device greatly expanded the features of the 4004 by doubling the word size and increasing the memory space to 16k bytes (8-bit numbers).

In December 1973 Intel began supplying the 8080, which ushered in the microprocessor age. The 8080 operates at ten times the speed of the 4004 and 8008, it addresses 64k bytes of memory, and it is constructed from NMOS logic, making it TTL-compatible. (The earlier 4004 and 8008 were constructed from PMOS logic, which meant that they were not TTL-compatible.)

Since these early days of microprocessor history, almost all the major semiconductor manufacturers have produced microprocessors, with varying degrees of success. These early manufacturers and their products include: Motorola's MC6800, National Semiconductor's IMP-4, Fairchild Semiconductor's F8, and Rockwell-International's PPS-4.

Modern Fixed-word-width Microprocessors

There are currently about 100 different microprocessors in production. Some of the most common types appear in Table 12-1.

THE SINGLE-COMPONENT MICROCOMPUTER

Unlike the fixed-word-width microprocessor, the single-component microcomputer contains memory and *I/O* in addition to a microprocessor. The block diagram for a single-component microcomputer is illustrated in Fig. 12-4.

TABLE 12-1 A COMPARISON OF FIXED-WORD-WIDTH MICROPROCESSORS.

Part number	Manufacturer	Number of bits	Memory size	Speed*
6502	Commodore	8	64k × 8	3 μs
4004	Intel	4	4k × 4	20 μs
8008	Intel	8	16k × 8	20 μs
8080A	Intel	8	64k × 8	2 μs
8085A	Intel	8	64k × 8	1.3 μs
8086	Intel	16	512k × 16	600 ns
8088	Intel	16	1024k × 8	600 ns
iAPX–286	Intel	16	500M × 16	400 ns
iAPX–432	Intel	32	256G × 32	400 ns
6800	Motorola	8	64k × 8	2 μs
6809	Motorola	8	64k × 8	2 μs
68000	Motorola	16	16M × 16	600 ns
68008	Motorola	16	16M × 8	600 ns
TMS1000	Texas Inst.	4	512 × 8	15 μs
TMS9900	Texas Inst.	16	32k × 16	4.7 μs
Z80	Zilog	8	64k × 8	1.6 μs
Z8001	Zilog	16	64k × 16	400 ns

*Note: Speed refers to the minimum execution time of an instruction.

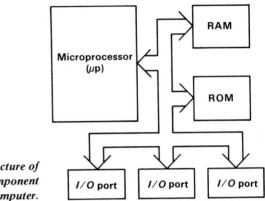

FIGURE 12-4. *Internal structure of*
a typical single-component
microcomputer.

Notice how the entire computer system is combined into one integrated circuit. Table 12-2 compares some of the most common single-component microcomputers. Single-component microcomputers find wide application in such products as:

1. CRT terminals,
2. printers,
3. plotters,
4. emission-control systems in automobiles,
5. electronic games.

TABLE 12-2 A COMPARISON OF SINGLE-COMPONENT MICROCOMPUTERS.

Part number	Manufacturer	I/O Pins	ROM size	RAM size
8048	Intel	24	1k × 8	64 × 8
8096	Intel	40	8k × 8	232 × 8
MK3870/12	MOSTEK	32	1k × 8	128 × 8
MK3870/42	MOSTEK	32	4k × 8	128 × 8
MK68200	MOSTEK	40	4k × 8	256 × 8
6802	Motorola	0	0	128 × 8
6803	Motorola	13	0	128 × 8
6805R2	Motorola	24	2k × 8	64 × 8
14805E2	Motorola	16	0	112 × 8
68705P3	Motorola	20	1804 × 8	112 × 8

BIT-SLICE MICROPROCESSORS

Bit-slice microprocessors are probably the least known of the three types because of their cost and the amount of time required to develop a bit-slice system. Bit-slice microprocessors come in widths of four and eight bits. These 4- or 8-bit bit-slice sections can be cascaded to produce microprocessors with word widths of up to 200 bits.

The main advantages of a bit-slice microprocessor are flexibility and speed. Bit-slice microprocessors are constructed from high-speed bipolar (advanced Schottky) or IIL logic circuitry that allows them to function at two to ten times the speed of a conventional microprocessor.

The disadvantage of a bit-slice microprocessor is that it usually requires over 10,000 different instructions, as opposed to a few hundred for most fixed-word-width microprocessors. Table 12-3 compares a few of the common bit-slice microprocessors.

Bit-slice microprocessors mainly find application in systems that require very high speeds or very complex instructions. Some examples of such systems are data analyzers, logic analyzers, and floating-point arithmetic processors.

TABLE 12-3 A COMPARISON OF SOME BIT-SLICE MICROPROCESSORS.

Part number	Manufacturer	Slice size	Speed	CLOCK
2901	Advanced Micro Devices	4	115 ns	9 MHz
74S481	Texas Instruments	4	115 ns	9 MHz
74AS888	Texas Instruments	8	40 ns	25 MHz

MICROPROCESSOR PROGRAMMING

Because the microprocessor is a programmable logic element, the user must be familiar with basic programming practices. But an exhaustive study of microprocessor programming would take an entire textbook. This section will introduce some of the techniques and terminology used in microprocessor programming.

PROGRAM DEVELOPMENT AIDES

An understanding of microprocessor programming aides, which are often called *microprocessor trainers*, should aid in the understanding of programming itself. Fig. 12-5 illustrates the SKD-85, a typical microprocessor trainer based on the Intel 8085A microprocessor. No matter which microprocessor you become familiar with, the learning process will most likely involve a microprocessor trainer.

The microprocessor trainer contains a keyboard for entering the programs and data in hexadecimal form and also for controlling the trainer. It also contains a 6- or 8-digit LED display that allows the user to view memory data, display results from a program, and look at various other information about the operation of the microprocessor. The typical trainer also includes some type of interfacing component that allows it to be connected to external devices such as CRT terminals, printers, and other digital circuitry.

The microprocessor trainer is one of the best ways to learn programming because it gives the user control over the microprocessor through use of the *machine language* of the microprocessor. *Machine language* consists of the binary (usually coded in hexadecimal) instructions that actually direct or control the operation of the microprocessor.

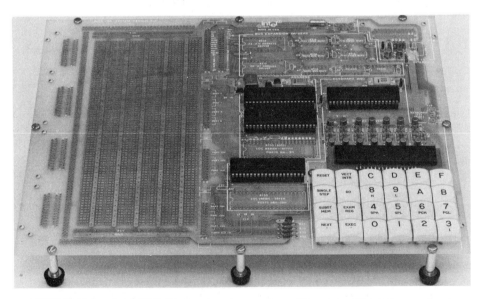

FIGURE 12-5. *Intel SDK-85 microprocessor trainer. (Courtesy of Intel Corporation.)*

PROGRAMMING LANGUAGES

A microprocessor, like any computer system, must be programmed in a specific language. Many languages are available for this purpose, ranging from the machine language, which is the most difficult to understand and use, to *high-level languages,* which are the easiest to understand and use.

Machine language is the most basic form, and most difficult, of programming language. This language requires that commands be entered into a microprocessor as binary (usually coded in hexadecimal) instructions, which directly control the operation of the microprocessor.

Assembly language differs little from machine language as far as ease of programming is concerned. Its main advantage over machine language is that it requires the entry of instructions in a symbolic form rather than in binary or hexadecimal. An example of a *symbolic command,* or *mnemonic code,* is ADD, which may be 0100 0000 in the binary machine language of a particular microprocessor.

Interpretive languages are high-level languages that allow a programmer to write a program in a pseudo-English form. An *interpreter* converts this pseudo-English code into a special internal pseudo-machine language code or *token.* One of the most common forms of interpreters used with microprocessors is BASIC. A typical BASIC command is PRINT 6 + 4. This command causes the computer to calculate the sum of six and four and print the answer on either a printer or a CRT screen. To accomplish this with machine language or assembly language would take many steps.

Compiled languages are used in much the same manner as interpretive languages. BASIC, for example, can be either an interpretive or a compiled language, depending on how the computer uses the pseudo-English code. The interpreter generates a pseudo-machine code (token), whereas the *compiler* generates the actual binary machine code. This may appear to be only a slight difference, but as far as the machine operation is concerned, it is a big difference. Machine language from a compiler will be executed 100–1000 times faster than pseudo-machine code from an interpreter.

WRITING A PROGRAM

Writing a program is not as difficult as it sounds. The first step is simply to think about what you want to do. This may seem self-evident, but programmers who don't think about what they want to do make many mistakes. Once you have thoroughly defined the goals of the program, consider what information must be provided, what data or operations will accomplish the task, and, finally, where in the output the answer will be placed.

Flowcharting Symbols

In writing a program, it is useful to develop a flowchart of the steps required to solve the problem. A flowchart is like a road map: it lets you know where you came from and where you are going. For simple programs, flowcharts may seem unnecessary, but for complicated programs they are essential. Without a flowchart a complicated program will take much longer to write; in fact, it may be impossible to write. Fig. 12-6 shows six of the most common flowchart symbols. It is not an exhaustive list because there are 20–30 flowchart symbols. These symbols are explained below.

1. *Terminal*: This symbol indicates where the program beings, ends, or pauses. It normally contains one of the following words: START, STOP, or PAUSE.

2. *Process*: This symbol indicates any arithmetic or nonarithmetic process used in solving a problem.

3. *Predefined Process*: This symbol is used whenever the same process occurs many times in a program. The symbol contains the name of the process.

4. *Decision*: This symbol is used to ask questions in a program. The possible outcomes might be yes and no, 1 and 0, positive and minus, or true and false.

5. *Input/Output*: This symbol is used to read or write data from an external source or to write data to an external source.

6. *Connector*: Flowcharts can become very lengthy, and rather than have the interconnecting lines criss-cross a page, we use a connector. Connectors usually come in pairs, both containing the same number or letter.

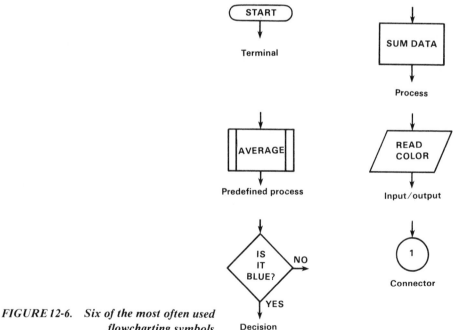

FIGURE 12-6. Six of the most often used flowcharting symbols.

A Sample Problem Using a Flowchart

Suppose that you work for the post office and you are assigned a new employee. Your task is to explain how to sort the mail for the proper zip-code locations in your small city. Suppose also that your trainee isn't too bright. You explain over and over how to sort the mail, but this person still doesn't understand. Finally, out of desperation you decide to use a flowchart, which you learned about in your programming class, to explain how to sort the mail.

The sorting procedure requires that the trainee look at a piece of mail, find the zip code, look at the last two digits, and place the piece of mail in one of three bags. A flowchart for this task, shown in Fig. 12-7, does the trick, and

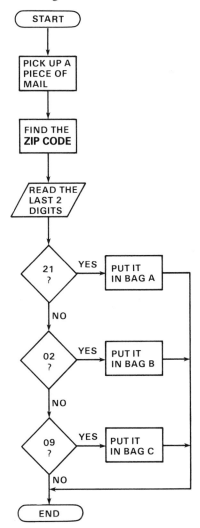

FIGURE 12-7. Flowchart for sorting the zip code of a piece of mail.

your trainee is trained. (It takes quite a while to sort the mail this way, but it can be done.)

This example doesn't apply directly to a computer program, but it should give you an idea of how useful flowcharts can be for even simple tasks. Flowcharts delineate a task in a clear and concise way.

Once a flowchart is written for a computer program, it is coded into a programming language. Flowcharts look the same for any computer language: they need not be rewritten or modified to fit a specific language.

INTRODUCTION TO THE 8085A

The 8085A microprocessor (see Fig. 12-8), manufactured by the Intel Corporation, is a popular 8-bit microprocessor found in applications as diverse as home computer systems, automobile fuel-injection control systems, and automated bank tellers. The pinout of the Intel 8085A appears in Fig. 12-9. This section will introduce this device for those of you who will continue your education on this microprocessor. Later sections in this chapter are provided for students who plan to use the MC6800 from Motorola or the Z80® microprocessor from Zilog.

MEMORY AND I/O ORGANIZATION

The memory and *I/O* are organized separately in most 8085A-based systems. Fig. 12-10 illustrates the directly addressable areas for the memory and *I/O* in an 8085A-based system. Notice that the memory locations are numbered from 0000H through FFFFH and the *I/O* space divisions are numbered from 00H through FFH (H designates a hexadecimal quantity). Each location in the memory and *I/O* space contains an 8-bit number. This means that the 8085A can address 64k bytes of memory and 256 different *I/O* locations or devices.

Fig. 12-11 shows how a memory system would be set up for a dot-matrix printer. The bottom area of the memory contains the program stored in a ROM. The ROM also contains tables for converting ASCII code to dot-matrix code and diagnostic programs for troubleshooting the printer. The RAM is used for the *stack,* where memory bytes are stored in sequential order, and also to store temporary information used by the printer such as tab positions.

The *I/O* space and a list of the external devices are illustrated in Fig. 12-11. Notice that there are only a few *I/O* devices in the dot-matrix printer: the print head, the head-positioning motor, the paper-advance solenoid, and a serial interface to a computer system. The paper-advance solenoid and head-positioning motor are controlled by *I/O* port 41H, the print head character pattern is controlled by port 40H, and the serial interface is controlled by ports 00H and 01H.

FIGURE 12-8. *Intel 8005A microprocessor. (Courtesy of Intel Corporation.)*

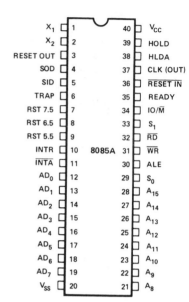

FIGURE 12-9. Pinout of the Intel 8085A microprocessor. (Courtesy of Intel. Reprinted by permission of Intel Corporation, copyright 1983.)

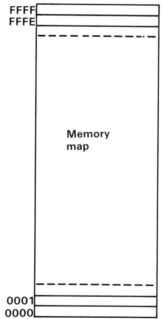

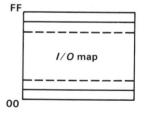

FIGURE 12-10. Memory and I/O maps of the Intel 8085A microprocessor.

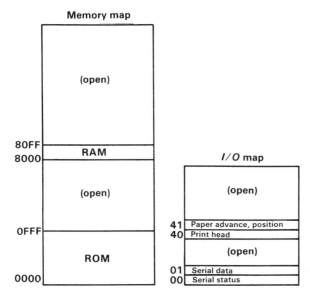

FIGURE 12-11. *Memory and I/O maps of an 8085A-based dot-matrix printer.*

MEMORY AND I/O CONTROL SIGNALS

The memory and *I/O* control signals of the 8085A are READ ($\overline{RD}$), WRITE ($\overline{WR}$), and *I/O* OR MEMORY (IO/$\overline{M}$). With these three signals and an address from the address bus, it is possible to control any *I/O* or any memory device.

If the IO/$\overline{M}$ signal is at a logic 1 level, then an *I/O* device is being accessed by the 8085A; if it is a 0, memory is being accessed. For writing to a memory location, a 0 is placed on $\overline{WR}$, a 1 is placed on $\overline{RD}$, and a 0 is placed on IO/$\overline{M}$. (See Table 12-4 for the complete truth table of these control signals.)

THE PROGRAMMING MODEL OF THE 8085A

Fig. 12-12 shows the internal programming model of the 8085A. It is very important that this structure be understood completely for success in machine or assembly language programming.

The General-purpose Registers

This microprocessor contains six 8-bit general-purpose registers: *B, C, D, E, H,* and *L*. General-purpose registers are used to hold intermediate results in computations or for any other purpose the programmer sees fit. These can function as six general-purpose 8-bit registers, or they can be grouped into pairs

TABLE 12-4 THE TRUTH TABLE FOR THE 8085A CONTROL SIGNALS, $\overline{RD}$, $\overline{WR}$, AND IO/$\overline{M}$.

$\overline{RD}$	$\overline{WR}$	IO/$\overline{M}$	Function
0	0	0	none
0	0	1	none
0	1	0	MEMORY READ
0	1	1	*I/O* READ
1	0	0	MEMORY WRITE
1	0	1	*I/O* WRITE
1	1	0	none
1	1	1	none

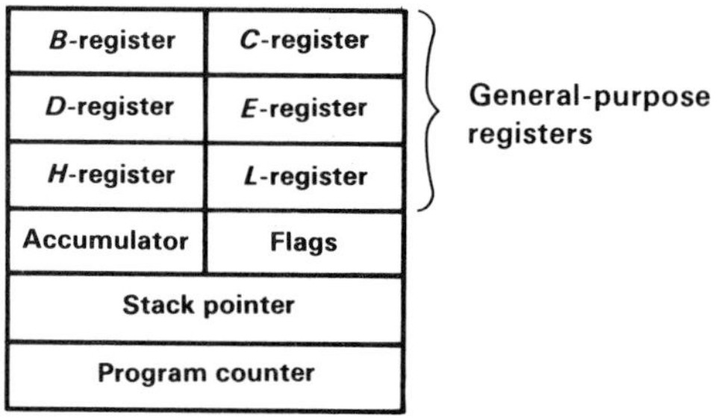

| B-register | C-register | }
|---|---|
| D-register | E-register |
| H-register | L-register |
| Accumulator | Flags |
| Stack pointer | |
| Program counter | |

General-purpose registers

FIGURE 12-12. *Programming model of the Intel 8085A microprocessor.*

of registers (*BC, DE,* and *HL*). These *register pairs* are general-purpose registers that are capable of holding a 16-bit number. A number in a register pair is used as data or to point to (*index*) data in the memory.

The Accumulator Register (A)

The accumulator register is a very important register in all microprocessors. The 8085A contains one accumulator register designed to hold the results from almost all 8-bit arithmetic and logic operations. The only exceptions are for instructions used to increment (add one to) or decrement (subtract one from) the contents of any register or register pair. Whenever a number is added, subtracted, ANDed, ORed, or Exclusive ORed, the answer will end up in the accumulator.

The Flag Register

The flag register was introduced as the condition code register. This register indicates various conditions of the result of an arithmetic or logic operation. It does *not* indicate the value of a particular register or memory location.

Fig. 12-13 shows the flag bit positions for the 8085A. A description of each flag bit follows:

1. *Carry (C)*: The carry flag indicates whether or not a CARRY has occurred from the most significant bit in an addition or if a BORROW has occurred in a subtraction. A 1 in the carry flag bit position indicates a CARRY or BORROW, and a 0 indicates no CARRY or BORROW. (All logic operations clear this bit position, and it is not affected by increment or decrement instructions.)

2. *Parity (P)*: The parity flag indicates the parity of the result of an arithmetic or logic operation. If the *P*-flag bit is a 1, then the parity—a count of the number of 1s, expressed as even or odd—is even. If the *P*-flag bit is a 0, then the result is odd parity.

3. *Auxiliary Carry (AC)*: The *AC*-flag bit indicates whether or not a CARRY occurred from the least significant half-byte (nibble) in an arithmetic or logic operation. This flag bit is cleared by all logic operations except AND, which sets it. (It is not affected by increment or decrement instructions.)

4. *Zero (Z)*: The zero flag indicates whether or not the result of the most recent arithmetic or logic operation is zero. A 1 in the *Z*-flag indicates a zero, and a 0 indicates a nonzero condition.

5. *Sign (S)*: The *S*-flag indicates the sign of the result of the most recent arithmetic or logic operation. A 0 indicates a positive result and a 1 indicates a negative result.

FLAG WORD

D_7	D_6	D_5	D_4	D_3	D_2	D_1	D_0
S	Z	X	AC	X	P	X	C

X: Undefined

FIGURE 12-13. The flag word bits of the Intel 8085A microprocessor. (Courtesy of Intel. Reprinted by permission of the Intel Corporation, copyright 1983.)

The Stack Pointer Register (SP)

The *stack pointer register* is used by the 8085A to keep track of its last-in, first-out (LIFO) stack. The *SP* contains a number that points to, or indexes, the stack. Whenever a program directs the 8085A to place information on its stack, the 8085A decrements the *SP*, stores a byte of data in the memory location indexed by the *SP*, decrements the *SP* again, and places a second byte of data in the memory location indexed by the *SP*. When data are removed from the stack, this sequence is reversed, with the stack pointer incremented instead of decremented. This allows the 8085A in effect to reuse the same area of memory for its LIFO stack.

The Program Counter (PC)

The program counter is used by the microprocessor to find the next step in a program, the address of which is always maintained in the *PC*. When the 8085A retrieves an instruction from the memory, it automatically adds 1 to the *PC* so that the next instruction is retrieved from the next memory location in sequential order.

THE INSTRUCTION SET

There are three basic forms of instructions used by the 8085A: 1-, 2-, and 3-byte instructions. There are 204 one-byte, 18 two-byte, and 24 three-byte instructions. Fig. 12-14 shows the three forms of instructions and the type of information placed in each.

The first byte is always an *op-code* (operation code), which indicates what command is to be executed by the 8085A. In the 1-byte instruction, no other information besides the op-code is required.

In the 2-byte instruction, the second byte almost always consists of an 8-bit number (D_8) that is used when the op-code is executed. For example, ADI 08H is a 2-byte instruction with an op-code of C6H and data of 08H (ADI is a hex value of C6 in machine language). This command is used to add 08H to the accumulator value, and the result is returned to the accumulator.

In the 3-byte command, the second and third bytes represent either data (D_{16}) or a memory address (ADR). For example, in the LDA 1000H command, an op-code of 3AH is followed by the 16-bit memory address 1000H. This command loads the accumulator with the number stored in memory location 1000H. With the 8085A, when an address or data are placed in a 3-byte command, the least significant eight bits (low-order portion) immediately

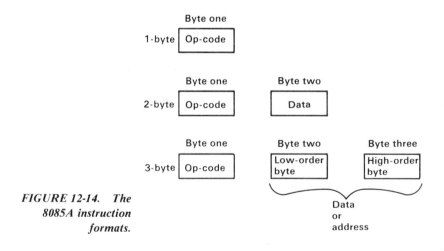

FIGURE 12-14. The 8085A instruction formats.

follow the op-code, and the most significant eight bits (high-order portion) follow the least significant eight bits. The LDA 1000H command is encoded into machine language and stored as 3A00-10.

8085A Instruction Types

The complete set of instructions for the 8085A is provided in Fig. 12-15. An explanation of the instructions appears below.

1. MOVE, LOAD, and STORE: Instructions in this group are used for data transfer between registers and between registers and memory.

2. STACK OPS: The STACK OPERATION instructions use the LIFO stack for storing data and the return addresses of subroutines.

3. JUMP: The JUMP instructions either conditionally or unconditionally load a 16-bit address into the program counter. If the program counter is modified, the instruction flow is modified. This is why this grouping of instructions is called JUMP.

4. CALL: Instructions in this group allow the 8085A to go to a *subroutine*, a reusable group of instructions stored in the memory once and used at several points in a program.

5. RETURN: These instructions are used to return control to the program at the end of a subroutine by retrieving the return address from the LIFO stack. (The return address is placed on the stack by a CALL or an RST.)

6. RESTART: These are 1-byte CALL instructions. The address of the subroutine called by this instruction is AAA000 binary. (RST calls the subroutine beginning at location 001000 binary or 0010H.)

7. INCREMENT and DECREMENT: This group of instructions allows the contents of a register, register pair, or memory location to be increased or decreased by one.

8. ADD: The ADD instructions allow data to be added to the accumulator or to the HL register pair (the DAD instruction is used for the latter purpose).

9. INPUT/OUTPUT: The IN and OUT commands transfer data to and from the accumulator, respectively. The second byte of the command points to the I/O address (00H–FFH).

10. SUBTRACT: The SUBTRACT instructions allow an 8-bit number to be subtracted from the accumulator.

11. SPECIAL: The SPECIAL instructions set or complement the carry flag (STC or CMC), complement the accumulator (CMA), and adjust the answer following a BCD addition (DAA).

Mnemonic	D7	D6	D5	D4	D3	D2	D1	D0	Operations Description
MOVE, LOAD, AND STORE									
MOVr1,r2	0	1	D	D	D	S	S	S	Move register to register
MOV M,r	0	1	1	1	0	S	S	S	Move register to memory
MOV r,M	0	1	D	D	D	1	1	0	Move memory to register
MVI r	0	0	D	D	D	1	1	0	Move immediate register
MVI M	0	0	1	1	0	1	1	0	Move immediate memory
LXI B	0	0	0	0	0	0	0	1	Load immediate register Pair B & C
LXI D	0	0	0	1	0	0	0	1	Load immediate register Pair D & E
LXI H	0	0	1	0	0	0	0	1	Load immediate register Pair H & L
STAX B	0	0	0	0	0	0	1	0	Store A indirect
STAX D	0	0	0	1	0	0	1	0	Store A indirect
LDAX B	0	0	0	0	1	0	1	0	Load A indirect
LDAX D	0	0	0	1	1	0	1	0	Load A indirect
STA	0	0	1	1	0	0	1	0	Store A direct
LDA	0	0	1	1	1	0	1	0	Load A direct
SHLD	0	0	1	0	0	0	1	0	Store H & L direct
LHLD	0	0	1	0	1	0	1	0	Load H & L direct
XCHG	1	1	1	0	1	0	1	1	Exchange D & E, H & L Registers
STACK OPS									
PUSH B	1	1	0	0	0	1	0	1	Push register Pair B & C on stack
PUSH D	1	1	0	1	0	1	0	1	Push register Pair D & E on stack
PUSH H	1	1	1	0	0	1	0	1	Push register Pair H & L on stack
PUSH PSW	1	1	1	1	0	1	0	1	Push A and Flags on stack
POP B	1	1	0	0	0	0	0	1	Pop register Pair B & C off stack
POP D	1	1	0	1	0	0	0	1	Pop register Pair D & E off stack
POP H	1	1	1	0	0	0	0	1	Pop register Pair H & L off stack
POP PSW	1	1	1	1	0	0	0	1	Pop A and Flags off stack
XTHL	1	1	1	0	0	0	1	1	Exchange top of stack, H & L
SPHL	1	1	1	1	1	0	0	1	H & L to stack pointer
LXI SP	0	0	1	1	0	0	0	1	Load immediate stack pointer
INX SP	0	0	1	1	0	0	1	1	Increment stack pointer
DCX SP	0	0	1	1	1	0	1	1	Decrement stack pointer
JUMP									
JMP	1	1	0	0	0	0	1	1	Jump unconditional
JC	1	1	0	1	1	0	1	0	Jump on carry
JNC	1	1	0	1	0	0	1	0	Jump on no carry
JZ	1	1	0	0	1	0	1	0	Jump on zero
JNZ	1	1	0	0	0	0	1	0	Jump on no zero
JP	1	1	1	1	0	0	1	0	Jump on positive
JM	1	1	1	1	1	0	1	0	Jump on minus
JPE	1	1	1	0	1	0	1	0	Jump on parity even

Mnemonic	D7	D6	D5	D4	D3	D2	D1	D0	Operations Description
JPO	1	1	1	0	0	0	1	0	Jump on parity odd
PCHL	1	1	1	0	1	0	0	1	H & L to program counter
CALL									
CALL	1	1	0	0	1	1	0	1	Call unconditional
CC	1	1	0	1	1	1	0	0	Call on carry
CNC	1	1	0	1	0	1	0	0	Call on no carry
CZ	1	1	0	0	1	1	0	0	Call on zero
CNZ	1	1	0	0	0	1	0	0	Call on no zero
CP	1	1	1	1	0	1	0	0	Call on positive
CM	1	1	1	1	1	1	0	0	Call on minus
CPE	1	1	1	0	1	1	0	0	Call on parity even
CPO	1	1	1	0	0	1	0	0	Call on parity odd
RETURN									
RET	1	1	0	0	1	0	0	1	Return
RC	1	1	0	1	1	0	0	0	Return on carry
RNC	1	1	0	1	0	0	0	0	Return on no carry
RZ	1	1	0	0	1	0	0	0	Return on zero
RNZ	1	1	0	0	0	0	0	0	Return on no zero
RP	1	1	1	1	0	0	0	0	Return on positive
RM	1	1	1	1	1	0	0	0	Return on minus
RPE	1	1	1	0	1	0	0	0	Return on parity even
RPO	1	1	1	0	0	0	0	0	Return on parity odd
RESTART									
RST	1	1	A	A	A	1	1	1	Restart
INCREMENT AND DECREMENT									
INR r	0	0	D	D	D	1	0	0	Increment register
DCR r	0	0	D	D	D	1	0	1	Decrement register
INR M	0	0	1	1	0	1	0	0	Increment memory
DCR M	0	0	1	1	0	1	0	1	Decrement memory
INX B	0	0	0	0	0	0	1	1	Increment B & C registers
INX D	0	0	0	1	0	0	1	1	Increment D & E registers
INX H	0	0	1	0	0	0	1	1	Increment H & L registers
DCX B	0	0	0	0	1	0	1	1	Decrement B & C
DCX D	0	0	0	1	1	0	1	1	Decrement D & E
DCX H	0	0	1	0	1	0	1	1	Decrement H & L
ADD									
ADD r	1	0	0	0	0	S	S	S	Add register to A
ADC r	1	0	0	0	1	S	S	S	Add register to A with carry
ADD M	1	0	0	0	0	1	1	0	Add memory to A
ADC M	1	0	0	0	1	1	1	0	Add memory to A with carry
ADI	1	1	0	0	0	1	1	0	Add immediate to A
ACI	1	1	0	0	1	1	1	0	Add immediate to A with carry
DAD B	0	0	0	0	1	0	0	1	Add B & C to H & L
DAD D	0	0	0	1	1	0	0	1	Add D & E to H & L
DAD H	0	0	1	0	1	0	0	1	Add H & L to H & L
DAD SP	0	0	1	1	1	0	0	1	Add stack pointer to H & L
INPUT/OUTPUT									
IN	1	1	0	1	1	0	1	1	Input
OUT	1	1	0	1	0	0	1	1	Output

FIGURE 12-15.   The instruction set for the 8085A Microprocessor (Courtesy of Intel Corporation. Reprinted by permission of Intel Corporation, copyright 1983.)

Mnemonic	Instruction Code D$_7$ D$_6$ D$_5$ D$_4$ D$_3$ D$_2$ D$_1$ D$_0$	Operations Description
SUBTRACT		
SUB r	1 0 0 1 0 S S S	Subtract register from A
SBB r	1 0 0 1 1 S S S	Subtract register from A with borrow
SUB M	1 0 0 1 0 1 1 0	Subtract memory from A
SBB M	1 0 0 1 1 1 1 0	Subtract memory from A with borrow
SUI	1 1 0 1 0 1 1 0	Subtract immediate from A
SBI	1 1 0 1 1 1 1 0	Subtract immediate from A with borrow
SPECIALS		
CMA	0 0 1 0 1 1 1 1	Complement A
STC	0 0 1 1 0 1 1 1	Set carry
CMC	0 0 1 1 1 1 1 1	Complement carry
DAA	0 0 1 0 0 1 1 1	Decimal adjust A
CONTROL		
EI	1 1 1 1 1 0 1 1	Enable Interrupts
DI	1 1 1 1 0 0 1 1	Disable Interrupt
NOP	0 0 0 0 0 0 0 0	No-operation
HLT	0 1 1 1 0 1 1 0	Halt
RIM	0 0 1 0 0 0 0 0	Read Interrupt Mask
SIM	0 0 1 1 0 0 0 0	Set Interrupt Mask

Mnemonic	Instruction Code D$_7$ D$_6$ D$_5$ D$_4$ D$_3$ D$_2$ D$_1$ D$_0$	Operations Description
LOGICAL		
ANA r	1 0 1 0 0 S S S	And register with A
XRA r	1 0 1 0 1 S S S	Exclusive OR register with A
ORA r	1 0 1 1 0 S S S	OR register with A
CMP r	1 0 1 1 1 S S S	Compare register with A
ANA M	1 0 1 0 0 1 1 0	And memory with A
XRA M	1 0 1 0 1 1 1 0	Exclusive OR memory with A
ORA M	1 0 1 1 0 1 1 0	OR memory with A
CMP M	1 0 1 1 1 1 1 0	Compare memory with A
ANI	1 1 1 0 0 1 1 0	And immediate with A
XRI	1 1 1 0 1 1 1 0	Exclusive OR immediate with A
ORI	1 1 1 0 1 1 1 0	OR immediate with A
CPI	1 1 1 1 1 1 1 0	Compare immediate with A
ROTATE		
RLC	0 0 0 0 0 1 1 1	Rotate A left
RRC	0 0 0 0 1 1 1 1	Rotate A right
RAL	0 0 0 1 0 1 1 1	Rotate A left through carry
RAR	0 0 0 1 1 1 1 1	Rotate A right through carry

NOTE

DDS or SSS: B 000, C 001, D 010, E 011, H 100, L 101, Memory 110, A 111.

*All mnemonics copyrighted © Intel Corporation 1976.

12. CONTROL: The CONTROL instructions enable or disable the INTER-RUPTs (EI or DI) and halt the microprocessor (HLT).

13. LOGICAL: These instructions perform the logic operations AND, OR, and Exclusive OR on the data in the accumulator. The COMPARE instruction is a SUBTRACT that affects only the flag bits.

14. ROTATE: These instructions allow the information in the accumulator to be rotated to the right or left (RRC or RLC). Data can also be rotated through the accumulator and carry flag if the RAR or RAL instruction is selected.

Addressing Modes

The 8085A can address data in four different modes. Data can be addressed: in a register, in a memory location, immediately following an op-code, or through an index register. These modes are explained below.

1. *Immediate addressing*: One of the simplest methods of addressing data is the immediate mode. In immediate mode addressing, the data used in an instruction immediately follow the op-code. Referring to Fig. 12-15, you will notice that the symbolic or mnemonic op-code for the SUBTRACT

IMMEDIATE instruction is SUI. If you wanted to subtract 12H from the contents of the accumulator, you would use the SUI 12H instruction.

2. *Direct addressing*: This form of addressing data requires that the op-code be followed by the memory address of the data. There are only a few of these instructions, which are listed in the MOVE, LOAD, and STORE grouping in Fig. 12-15. An example is the STA instruction, which is the mnemonic code for storing the accumulator. If the STA 2000H instruction is executed, the data in the accumulator are copied and transferred to memory location 2000H.

3. *Register addressing*: Register addressing is used in the bulk of the arithmetic, logic, and MOVE instructions. The destination register is DDD and the source register is SSS. The destination and source codes for the registers are listed at the end of the op-codes list in Fig. 12-15 as a note. Suppose it is necessary to copy or move the number in the *B*-register into the *D*-register. To do this, we use the MOV instruction and code DDD with 010 and SSS with 000. The symbolic form of this command is MOV D,B. (Intel symbolic code requires the destination register to be listed first.)

4. *Indexed addressing*: This type of addressing is probably the hardest to understand because the address of the data is not stored with the op-code, but in one of the register pairs, *BC, DE,* or *HL.* If the 8085A executes a MOV M,A instruction, the contents of the accumulator (*A*) are copied into the memory location indicated by the *HL* pair. Whenever an M (memory) is coded into an instruction, the *HL* pair addresses or indexes the memory. The *BC* and *DE* pairs only index memory for the STAX and LDAX instructions.

Example 12-1. Suppose that the data in memory locations 1000H through 1004H are to be added together and the sum is to be stored in memory location 1005H.

Solution. This procedure is listed in symbolic assembly language in the program that follows. The meaning of each command is stated on the right.

```
START: LDA 1000H    Load accumulator with data at 1000H
       LXI H,1001H  Index 1001H
       ADD M        Add contents of 1001H to accumulator
       INX H        Index 1002H
       ADD M        Add contents of 1002H to accumulator
       INX H        Index 1003H
       ADD M        Add contents of 1003H to accumulator
       INX H        Index 1004H
       ADD M        Add contents of 1004H to accumulator
```

INX H	Index 1005H
MOV M,A	Save sum at 1005H
HLT	Halt execution

INTRODUCTION TO THE MC6800

The MC6800 microprocessor (see Fig. 12-16), manufactured by the Motorola Corporation, is a popular 8-bit microprocessor that, like the 8085A, is found in such applications as automobile fuel-injection control systems and automated banktellers. This section is included for those of you who will continue your education on this microprocessor. If you plan to study on the 8085A from Intel or the Z80® microprocessor from Zilog, please refer to either the preceding or the following section.

MEMORY AND I/O ORGANIZATION

The memory and I/O are organized into one area in MC6800-based systems. Fig. 12-17 illustrates the directly addressable areas for the memory and I/O in an MC6800-based system. Notice that the memory and I/O exist in the same space and are addressed from $0000 through $FFFF. The dollar sign ($) is used

PIN ASSIGNMENT

```
 1 ⊏ Vss      RESET ⊐ 40
 2 ⊏ HALT      TSC  ⊐ 39
 3 ⊏ φ₁        N.C. ⊐ 38
 4 ⊏ IRQ        φ₂  ⊐ 37
 5 ⊏ VMA       DBE  ⊐ 36
 6 ⊏ NMI       N.C. ⊐ 35
 7 ⊏ BA        R/W  ⊐ 34
 8 ⊏ Vcc        D₀  ⊐ 33
 9 ⊏ A₀         D₁  ⊐ 32
10 ⊏ A₁         D₂  ⊐ 31
11 ⊏ A₂         D₃  ⊐ 30
12 ⊏ A₃         D₄  ⊐ 29
13 ⊏ A₄         D₅  ⊐ 28
14 ⊏ A₅         D₆  ⊐ 27
15 ⊏ A₆         D₇  ⊐ 26
16 ⊏ A₇        A₁₅  ⊐ 25
17 ⊏ A₈        A₁₄  ⊐ 24
18 ⊏ A₉        A₁₃  ⊐ 23
19 ⊏ A₁₀       A₁₂  ⊐ 22
20 ⊏ A₁₁       Vss  ⊐ 21
```

FIGURE 12-16. Pinout of the Motorola MC6800 microprocessor. (Courtesy of Motorola, Inc.)

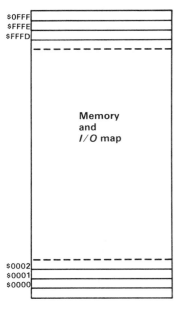

$0FFF
$FFFE
$FFFD

Memory
and
I/O map

$0002
$0001
$0000

FIGURE 12-17. Memory and I/O map of the MC6800 microprocessor.

to designate a hexadecimal quantity in this system. Each location in the memory and I/O space contains an 8-bit number, which means the MC6800 can address 64k bytes of memory and I/O space.

Fig. 12-18 shows how a typical memory system would be set up for a dot-matrix printer. The top area of the memory contains the program stored in a ROM. The ROM also contains tables for converting ASCII code to dot-matrix code and diagnostic programs for troubleshooting the printer. The RAM is used for the stack and also to store temporary information used by the printer such as tab positions.

Notice that there are only a few I/O devices in a dot-matrix printer: the print head, the head-positioning motor, the paper-advance solenoid, and a serial interface to a computer system. I/O location $8003 controls both the paper advance and head position, location $8002 supplies the print head with the print code, and locations $8001 and $8000 control the serial interface.

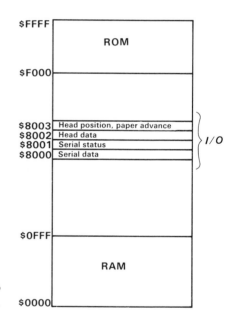

FIGURE 12-18. Memory and I/O map of an MC6800-based dot-matrix printer.

MEMORY AND I/O CONTROL SIGNALS

The memory and I/O control inputs of the MC6800 are READ/WRITE (R/$\overline{W}$), VALID MEMORY ADDRESS (VMA), and CLOCK PHASE 2 (E). With these three signals and an address from the address bus, the MC6800 can control any I/O or any memory device.

The memory or I/O is activated only when both the VMA and E signals are at their logic 1 levels. When both are active, the $R/\overline{W}$ control signal indicates the type of data transfer, a READ or a WRITE.

THE PROGRAMMING MODEL OF THE MC6800

Fig. 12-19 shows the internal programming model of the MC6800. It is very important that this structure be learned completely for successful machine or assembly language programming.

The Accumulator Registers

The accumulator registers, as explained earlier, are very important in microprocessors. The MC6800 contains two accumulator registers (*A* and *B*), which receive the results from all 8-bit arithmetic and logic operations. The only exceptions to this are instructions used to increment (add one to) or decrement (subtract one from) the contents of a memory location or the index register. Whenever a number is added, subtracted, ANDed, ORed, or Exclusive ORed, the answer will end up in the accumulator.

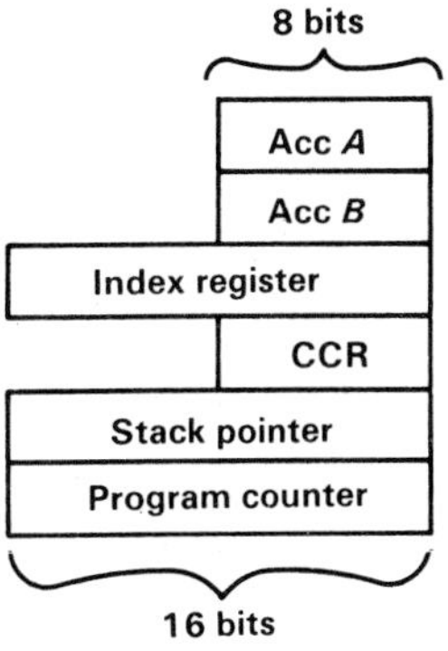

The Index Register (IX)

The index register indexes, or points to, a location in the memory. Indexed instructions use the code IX to locate the information that will be added, subtracted, etc.

FIGURE 12-19.
Programming model of the Motorola MC6800 microprocessor.

The Condition Code Register (CCR)

The condition code register indicates various conditions of the result of an arithmetic, logic, or data transfer operation. The CCR also indicates the status of the INTERRUPT structure of the MC6800.

Fig. 12-20 shows the CCR bit positions. A description of each CCR bit follows:

1. *Carry (C)*: The carry bit indicates whether or not a CARRY from the most significant bit position has resulted from the most recent addition, or whether a BORROW has occurred in the most recent subtraction. A 1 in the carry CCR bit indicates a CARRY or BORROW, and a 0 indicates no CARRY or BORROW.

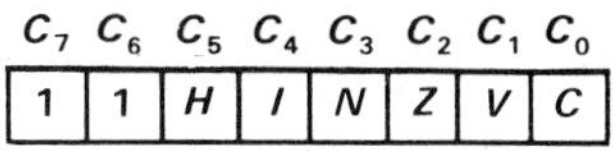

C_7 C_6 C_5 C_4 C_3 C_2 C_1 C_0

1	1	H	I	N	Z	V	C

FIGURE 12-20.   Condition code register (CCR) of the MC6800.

2. *Overflow (V)*: The overflow CCR bit indicates that the result of an arithmetic operation has exceeded the capacity of an accumulator. If the largest positive number (+127) or the largest negative number (−128) is exceeded, the overflow CCR bit will be set.

3. *Zero (Z)*: The zero CCR bit indicates whether or not the result of the most recent arithmetic, logic, or data transfer operation is zero. A 1 in the Z-bit indicates a zero, and a 0 indicates a nonzero condition.

4. *Negative (N)*: The N-bit indicates the sign of the result of the most recent arithmetic, logic, or data transfer operation. A 0 indicates a positive result, and a 1 indicates a negative result.

5. *Interrupt (I)*: The I-bit is used to indicate whether or not the INTERRUPT structure has been masked off. A 1 in this CCR bit turns off the INTER-RUPT structure and a 0 turns it on.

6. *Half-carry (H)*: The H-bit indicates whether or not a CARRY occurred from the least significant half-byte (nibble) in the most recent arithmetic operation. This CCR bit is normally used with BCD arithmetic.

The Stack Pointer (SP) Register

The stack pointer register is used by the MC6800 to keep track of its last-in, first-out (LIFO) stack. The *SP* contains a number that points to (indexes) an area of the memory designated as the stack. When a program directs the MC6800 to place information on its stack, the MC6800 stores a byte of data at the location indexed by the *SP* and then decrements the *SP*. Whenever data are removed from the stack, this sequence is reversed: The stack pointer is first incremented, and then the data are removed. This allows the MC6800 in effect to reuse the same area of memory for its LIFO stack.

The Program Counter (PC)

The program counter is used by the microprocessor to find the next step in a program. The address of the next step of a program is always maintained in the *PC*. When the MC6800 retrieves an instruction from the memory, it automatically adds one to the *PC* so that the next instruction is retrieved from the next memory location in sequence.

THE INSTRUCTION SET

There are three basic forms of instructions used by the MC6800: 1-, 2-, and 3-byte instructions. Fig. 12-21 shows the three forms of instructions and the type of information usually placed in each.

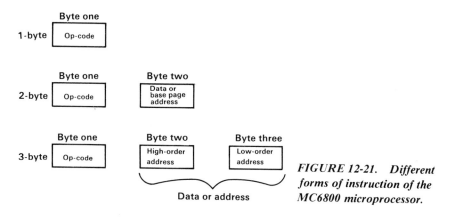

FIGURE 12-21. *Different forms of instruction of the MC6800 microprocessor.*

In all three forms the first byte is always an op-code (operation code). The op-code indicates which command the MC6800 is to execute. In the 1-byte instruction, no other information besides the op-code is required.

In the 2-byte instructions, the second byte is always an 8-bit number used as data or as an address in the first 256 bytes of the memory (known as the base page). For example, the LDAA $42 instruction is a 2-byte instruction with an op-code of $96 ($96 is the machine code for the mnemonic code LDAA) and an address of $42. This command loads accumulator A from memory location $42 in the base page. To specify data, the instruction would be LDAA #$42. (If the # sign is placed in front of a hexadecimal quantity, it indicates that the quantity represents data and not an address.) The LDAA #$42 instruction will load a data value of $42 into accumulator A.

The 3-byte commands contain either data or a memory address in the second and third bytes. An example of a 3-byte command is LDAA $1000. The first byte is the op-code for $B6, and it is followed by the 16-bit memory address $1000. This command loads accumulator A with the number stored in memory location $1000. With the MC6800, whenever an address or data are placed in a 3-byte command, the most significant eight bits (high-order portion) immediately follow the op-code, and the least significant eight bits (low-order portion) follow the most significant eight bits. The command LDAA $1000 is encoded into machine language and stored as B6-10-00.

MC6800 Instruction Types

The complete set of instructions for the MC6800 is provided in Fig. 12-22 along with notes about the addressing, operation, and status of the CCR bits. The main groups of instructions and operations are discussed below.

ACCUMULATOR AND MEMORY OPERATIONS

Included are Arithmetic Logic, Data Test and Data Handling instructions.

OPERATIONS	MNEMONIC	IMMED OP	~	#	DIRECT OP	~	#	INDEX OP	~	#	EXTND OP	~	#	IMPLIED OP	~	#	BOOLEAN/ARITHMETIC OPERATION (All register labels refer to contents)	H	I	N	Z	V	C
Add	ADDA	8B	2	2	9B	3	2	AB	5	2	BB	4	3				A + M → A	:	●	:	:	:	:
	ADDB	CB	2	2	DB	3	2	EB	5	2	FB	4	3				B + M → B	:	●	:	:	:	:
Add Acmltrs	ABA													1B	2	1	A + B → A	:	●	:	:	:	:
Add with Carry	ADCA	89	2	2	99	3	2	A9	5	2	B9	4	3				A + M + C → A	:	●	:	:	:	:
	ADCB	C9	2	2	D9	3	2	E9	5	2	F9	4	3				B + M + C → B	:	●	:	:	:	:
And	ANDA	84	2	2	94	3	2	A4	5	2	B4	4	3				A · M → A	●	●	:	:	R	●
	ANDB	C4	2	2	D4	3	2	E4	5	2	F4	4	3				B · M → B	●	●	:	:	R	●
Bit Test	BITA	85	2	2	95	3	2	A5	5	2	B5	4	3				A · M	●	●	:	:	R	●
	BITB	C5	2	2	D5	3	2	E5	5	2	F5	4	3				B · M	●	●	:	:	R	●
Clear	CLR							6F	7	2	7F	6	3				00 → M	●	●	R	S	R	R
	CLRA													4F	2	1	00 → A	●	●	R	S	R	R
	CLRB													5F	2	1	00 → B	●	●	R	S	R	R
Compare	CMPA	81	2	2	91	3	2	A1	5	2	B1	4	3				A − M	●	●	:	:	:	:
	CMPB	C1	2	2	D1	3	2	E1	5	2	F1	4	3				B − M	●	●	:	:	:	:
Compare Acmltrs	CBA													11	2	1	A − B	●	●	:	:	:	:
Complement, 1's	COM							63	7	2	73	6	3				M̄ → M	●	●	:	:	R	S
	COMA													43	2	1	Ā → A	●	●	:	:	R	S
	COMB													53	2	1	B̄ → B	●	●	:	:	R	S
Complement, 2's (Negate)	NEG							60	7	2	70	6	3				00 − M → M	●	●	:	:	①	②
	NEGA													40	2	1	00 − A → A	●	●	:	:	①	②
	NEGB													50	2	1	00 − B → B	●	●	:	:	①	②
Decimal Adjust, A	DAA													19	2	1	Converts Binary Add. of BCD Characters into BCD Format	●	●	:	:	:	③
Decrement	DEC							6A	7	2	7A	6	3				M − 1 → M	●	●	:	:	4	●
	DECA													4A	2	1	A − 1 → A	●	●	:	:	4	●
	DECB													5A	2	1	B − 1 → B	●	●	:	:	4	●
Exclusive OR	EORA	88	2	2	98	3	2	A8	5	2	B8	4	3				A ⊕ M → A	●	●	:	:	R	●
	EORB	C8	2	2	D8	3	2	E8	5	2	F8	4	3				B ⊕ M → B	●	●	:	:	R	●
Increment	INC							6C	7	2	7C	6	3				M + 1 → M	●	●	:	:	⑤	●
	INCA													4C	2	1	A + 1 → A	●	●	:	:	⑤	●
	INCB													5C	2	1	B + 1 → B	●	●	:	:	⑤	●
Load Acmltr	LDAA	86	2	2	96	3	2	A6	5	2	B6	4	3				M → A	●	●	:	:	R	●
	LDAB	C6	2	2	D6	3	2	E6	5	2	F6	4	3				M → B	●	●	:	:	R	●
Or, Inclusive	ORAA	8A	2	2	9A	3	2	AA	5	2	BA	4	3				A + M → A	●	●	:	:	R	●
	ORAB	CA	2	2	DA	3	2	EA	5	2	FA	4	3				B + M → B	●	●	:	:	R	●
Push Data	PSHA													36	4	1	A → M$_{SP}$, SP − 1 → SP	●	●	●	●	●	●
	PSHB													37	4	1	B → M$_{SP}$, SP − 1 → SP	●	●	●	●	●	●
Pull Data	PULA													32	4	1	SP + 1 → SP, M$_{SP}$ → A	●	●	●	●	●	●
	PULB													33	4	1	SP + 1 → SP, M$_{SP}$ → B	●	●	●	●	●	●
Rotate Left	ROL							69	7	2	79	6	3				M	●	●	:	:	⑥	:
	ROLA													49	2	1	A	●	●	:	:	⑥	:
	ROLB													59	2	1	B	●	●	:	:	⑥	:
Rotate Right	ROR							66	7	2	76	6	3				M	●	●	:	:	⑥	:
	RORA													46	2	1	A	●	●	:	:	⑥	:
	RORB													56	2	1	B	●	●	:	:	⑥	:
Shift Left, Arithmetic	ASL							68	7	2	78	6	3				M	●	●	:	:	⑥	:
	ASLA													48	2	1	A	●	●	:	:	⑥	:
	ASLB													58	2	1	B	●	●	:	:	⑥	:
Shift Right, Arithmetic	ASR							67	7	2	77	6	3				M	●	●	:	:	⑥	:
	ASRA													47	2	1	A	●	●	:	:	⑥	:
	ASRB													57	2	1	B	●	●	:	:	⑥	:
Shift Right, Logic	LSR							64	7	2	74	6	3				M	●	●	R	:	⑥	:
	LSRA													44	2	1	A	●	●	R	:	⑥	:
	LSRB													54	2	1	B	●	●	R	:	⑥	:
Store Acmltr	STAA				97	4	2	A7	6	2	B7	5	3				A → M	●	●	:	:	R	●
	STAB				D7	4	2	E7	6	2	F7	5	3				B → M	●	●	:	:	R	●
Subtract	SUBA	80	2	2	90	3	2	A0	5	2	B0	4	3				A − M → A	●	●	:	:	:	:
	SUBB	C0	2	2	D0	3	2	E0	5	2	F0	4	3				B − M → B	●	●	:	:	:	:
Subtract Acmltrs	SBA													10	2	1	A − B → A	●	●	:	:	:	:
Subtr. with Carry	SBCA	82	2	2	92	3	2	A2	5	2	B2	4	3				A − M − C → A	●	●	:	:	:	:
	SBCB	C2	2	2	D2	3	2	E2	5	2	F2	4	3				B − M − C → B	●	●	:	:	:	:
Transfer Acmltrs	TAB													16	2	1	A → B	●	●	:	:	R	●
	TBA													17	2	1	B → A	●	●	:	:	R	●
Test, Zero or Minus	TST							6D	7	2	7D	6	3				M − 00	●	●	:	:	R	R
	TSTA													4D	2	1	A − 00	●	●	:	:	R	R
	TSTB													5D	2	1	B − 00	●	●	:	:	R	R
																		H	I	N	Z	V	C

LEGEND:

OP Operation Code (Hexadecimal);
~   Number of MPU Cycles;
#   Number of Program Bytes;
+   Arithmetic Plus;
−   Arithmetic Minus;
·   Boolean AND;
M$_{SP}$ Contents of memory location pointed to be Stack Pointer;

+   Boolean Inclusive OR;
⊙   Boolean Exclusive OR;
M̄   Complement of M;
→   Transfer Into;
0   Bit = Zero;
00 Byte = Zero.

Note — Accumulator addressing mode instructions are included in the column for IMPLIED addressing

CONDITION CODE SYMBOLS:

H   Half carry from bit 3;
I   Interrupt mask
N   Negative (sign bit)
Z   Zero (byte)
V   Overflow, 2's complement
C   Carry from bit 7
R   Reset Always
S   Set Always
:   Test and set if true, cleared otherwise
●   Not Affected

INDEX REGISTER AND STACK POINTER INSTRUCTIONS

POINTER OPERATIONS	MNEMONIC	IMMED OP	~	#	DIRECT OP	~	#	INDEX OP	~	#	EXTND OP	~	#	IMPLIED OP	~	#	BOOLEAN/ARITHMETIC OPERATION	H	I	N	Z	V	C
Compare Index Reg	CPX	8C	3	3	9C	4	2	AC	6	2	BC	5	3				$X_H - M, X_L - (M+1)$	•	•	①	:	②	•
Decrement Index Reg	DEX													09	4	1	$X - 1 \to X$	•	•	•	:	•	•
Decrement Stack Pntr	DES													34	4	1	$SP - 1 \to SP$	•	•	•	•	•	•
Increment Index Reg	INX													08	4	1	$X + 1 \to X$	•	•	•	:	•	•
Increment Stack Pntr	INS													31	4	1	$SP + 1 \to SP$	•	•	•	•	•	•
Load Index Reg	LDX	CE	3	3	DE	4	2	EE	6	2	FE	5	3				$M \to X_H, (M+1) \to X_L$	•	•	③	:	R	•
Load Stack Pntr	LDS	8E	3	3	9E	4	2	AE	6	2	BE	5	3				$M \to SP_H, (M+1) \to SP_L$	•	•	③	:	R	•
Store Index Reg	STX				DF	5	2	EF	7	2	FF	6	3				$X_H \to M, X_L \to (M+1)$	•	•	③	:	R	•
Store Stack Pntr	STS				9F	5	2	AF	7	2	BF	6	3				$SP_H \to M, SP_L \to (M+1)$	•	•	③	:	R	•
Indx Reg → Stack Pntr	TXS													35	4	1	$X - 1 \to SP$	•	•	•	•	•	•
Stack Pntr → Indx Reg	TSX													30	4	1	$SP + 1 \to X$	•	•	•	•	•	•

① (Bit N) Test: Sign bit of most significant (MS) byte of result = 1?
② (Bit V) Test: 2's complement overflow from subtraction of ms bytes?
③ (Bit N) Test: Result less than zero? (Bit 15 = 1)

JUMP AND BRANCH INSTRUCTIONS

OPERATIONS	MNEMONIC	RELATIVE OP	~	#	INDEX OP	~	#	EXTND OP	~	#	IMPLIED OP	~	#	BRANCH TEST	H	I	N	Z	V	C
Branch Always	BRA	20	4	2										None	•	•	•	•	•	•
Branch If Carry Clear	BCC	24	4	2										$C = 0$	•	•	•	•	•	•
Branch If Carry Set	BCS	25	4	2										$C = 1$	•	•	•	•	•	•
Branch If = Zero	BEQ	27	4	2										$Z = 1$	•	•	•	•	•	•
Branch If ≥ Zero	BGE	2C	4	2										$N \oplus V = 0$	•	•	•	•	•	•
Branch If > Zero	BGT	2E	4	2										$Z + (N \oplus V) = 0$	•	•	•	•	•	•
Branch If Higher	BHI	22	4	2										$C + Z = 0$	•	•	•	•	•	•
Branch If ≤ Zero	BLE	2F	4	2										$Z + (N \oplus V) = 1$	•	•	•	•	•	•
Branch If Lower Or Same	BLS	23	4	2										$C + Z = 1$	•	•	•	•	•	•
Branch If < Zero	BLT	2D	4	2										$N \oplus V = 1$	•	•	•	•	•	•
Branch If Minus	BMI	2B	4	2										$N = 1$	•	•	•	•	•	•
Branch If Not Equal Zero	BNE	26	4	2										$Z = 0$	•	•	•	•	•	•
Branch If Overflow Clear	BVC	28	4	2										$V = 0$	•	•	•	•	•	•
Branch If Overflow Set	BVS	29	4	2										$V = 1$	•	•	•	•	•	•
Branch If Plus	BPL	2A	4	2										$N = 0$	•	•	•	•	•	•
Branch To Subroutine	BSR	8D	8	2											•	•	•	•	•	•
Jump	JMP				6E	4	2	7E	3	3				See Special Operations	•	•	•	•	•	•
Jump To Subroutine	JSR				AD	8	2	BD	9	3					•	•	•	•	•	•
No Operation	NOP										01	2	1	Advances Prog. Cntr Only	•	•	•	•	•	•
Return From Interrupt	RTI										3B	10	1		•	•	•	①	•	•
Return From Subroutine	RTS										39	5	1	See Special Operations	•	•	•	•	•	•
Software Interrupt	SWI										3F	12	1		•	•	•	•	•	•
Wait for Interrupt *	WAI										3E	9	1		•	②	•	•	•	•

*WAI puts Address Bus, R/W, and Data Bus in the three-state mode while VMA is held low.

① (All) Load Condition Code Register from Stack. (See Special Operations)
② (Bit I) Set when interrupt occurs. If previously set, a Non Maskable Interrupt is required to exit the wait state.

CONDITION CODE REGISTER INSTRUCTIONS

OPERATIONS	MNEMONIC	IMPLIED OP	~	#	BOOLEAN OPERATION	H	I	N	Z	V	C
Clear Carry	CLC	0C	2	1	$0 \to C$	•	•	•	•	•	R
Clear Interrupt Mask	CLI	0E	2	1	$0 \to I$	•	R	•	•	•	•
Clear Overflow	CLV	0A	2	1	$0 \to V$	•	•	•	•	R	•
Set Carry	SEC	0D	2	1	$1 \to C$	•	•	•	•	•	S
Set Interrupt Mask	SEI	0F	2	1	$1 \to I$	•	S	•	•	•	•
Set Overflow	SEV	0B	2	1	$1 \to V$	•	•	•	•	S	•
Acmltr A → CCR	TAP	06	2	1	$A \to CCR$	①					
CCR → Acmltr A	TPA	07	2	1	$CCR \to A$	•	•	•	•	•	•

R = Reset
S = Set
• = Not affected
① (ALL) Set according to the contents of Accumulator A.

CONDITION CODE REGISTER NOTES: (Bit set if test is true and cleared otherwise)

1 (Bit V) Test: Result = 10000000?
2 (Bit C) Test: Result = 00000000?
3 (Bit C) Test: Decimal value of most significant BCD Character greater than nine? (Not cleared if previously set.)
4 (Bit V) Test: Operand = 10000000 prior to execution?
5 (Bit V) Test: Operand = 01111111 prior to execution?
6 (Bit V) Test: Set equal to result of N⊕C after shift has occurred.

FIGURE 12-22. Instruction set of the MC6800 microprocessor (Courtesy of Motorola, Inc.).

1. *Accumulator and memory operations*: The accumulator operations include addition, subtraction, decrement, increment, etc. The instructions specify which accumulator (*A* or *B*) is to be used for a particular operation. The memory operations include clearing, shifting, testing, etc.

2. *Index register and stack pointer instructions*: The index register (*IX*) instructions allow the index register to be loaded with information, incremented, decremented, and stored in the memory. The stack pointer instructions allow the *SP* value to be loaded, stored, and transferred to and from the *IX*.

3. *JUMP and BRANCH instructions*: The JUMP and BRANCH instructions are used to modify the flow of the program. The BRANCH instructions allow the programmer to modify the flow of a program by loading the *PC* with a new number. The JUMP instructions are mainly used for branching to a subroutine, and the RETURN instructions are used for returning from a subroutine.

4. *Condition code register instructions*: These instructions modify condition code bits. They also allow the contents of the CCR to be transferred to accumulator *A* and vice versa.

Addressing Modes

The MC6800 can address data in six different modes: implied, immediate, direct, indexed, extended, and relative. These modes allow data to be addressed: in a register, in a memory location, immediately following an op-code, or through an index register. The six addressing modes are explained below.

1. *Implied addressing*: Implied instructions are 1-byte instructions that contain all of the necessary information for the MC6800. Examples of implied addressing include: CLRA, which clears accumulator *A*; DECA, which decrements accumulator *A*; TAB, which copies accumulator *A* into accumulator *B*; and TSTA, which modifies the CCR so that it reflects the contents of accumulator *A*.

2. *Immediate addressing*: There are two forms of immediate instructions: 2-byte and 3-byte. Most of the 2-byte immediate instructions use an 8-bit number as data or an address. The ADDA #$55 instruction is a 2-byte immediate instruction that adds a $55 to accumulator *A*. The 3-byte immediate instructions load the 16-bit *SP* or index register. The LDX #$9999 instruction would store a $9999 into the index register.

3. *Direct addressing*: Direct addressing uses the base page of the memory: that is, it addresses the first 256 bytes, numbered from $0000 to $00FF. Since the first two hexadecimal digits are always zeros in the base page, only the last two digits of the address are coded in the direct addressing

mode. The instruction for loading accumulator *B* with the contents of memory location $0066 is LDAB $66. This command is coded into machine language as a D6-66.

4. *Extended addressing*: Extended addressing works exactly like direct addressing, except that the address is four hexadecimal digits in width. If accumulator *B* is to be loaded from memory location $1033, then the LDAB $1033 instruction is used. This instruction is coded into machine language as F6-10-33.

5. *Indexed addressing*: In this mode, the number stored in the *IX* points to the data. This instruction takes two forms: one includes an "offset" address, and the other does not. If the LDAA X instruction is executed, the data stored at the memory location currently indicated by the *IX* are loaded into accumulator *A*. If an offset is added to the instruction, the value of the offset is added to the *IX* value and data are referenced at the new location. For example, if the *IX* indicates location $1000 and the LDAB 8,X instruction is executed, then the contents of memory location $1008 will be loaded into accumulator *B*. The *IX* is offset by eight to generate the memory address.

6. *Relative addressing*: Relative addressing is used in the BRANCH and JUMP subroutine instructions. Instead of storing the 16-bit address of the location to be branched to, the instruction stores an 8-bit signed displacement value. If the instruction BRA $03 is executed, then the MC6800 branches to three locations beyond the address of the next sequential instruction in the memory. If a BRA $FE instruction is executed, the computer will get hung up because the branch transfers it to the address of the next instruction minus two bytes (since FE is the hex value of −2). That means the computer will return to the BRA $FE instruction. This creates an infinite wait loop.

Example 12-2. Suppose that the data in memory locations $1000 through $1004 are to be added together and the sum is to be stored in memory location $1005.

Solution This procedure is listed in symbolic assembly language in the MC6800 program that follows. The meaning of each command is explained on the right.

```
START: LDX $1000    Index the data at $1000
       LDAA X       Load data at $1000 into accumulator A
       ADDA 1,X     Add contents of $1001 to accumulator A
       ADDA 2,X     Add contents of $1002 to accumulator A
       ADDA 3,X     Add contents of $1003 to accumulator A
       ADDA 4,X     Add contents of $1004 to accumulator A
       STAA 5,X     Save sum at $1005
       WAI          Wait
```

INTRODUCTION TO THE Z80® MICROPROCESSOR

The Z80® microprocessor (see Figs. 12-23 and 12-24) is a popular 8-bit microprocessor found in such applications as home computer systems, automobile fuel-injection systems, and automated bank tellers. This section will introduce this device for those of you who will continue your education on this micropro-

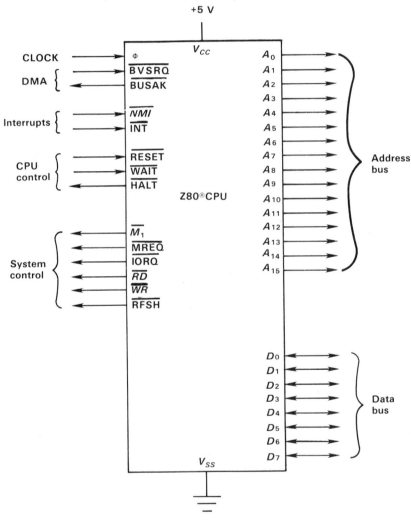

FIGURE 12-23. *Pinout of the Z80® microprocessor.*

FIGURE 12-24. The Z80® microprocessor. (Reproduced by permission © 1981 Zilog, Inc. This material shall not be reproduced without the written consent of Zilog, Inc.)

cessor. If you plan your future studies on the MC6800 from Motorola or the 8085A from Intel, please refer to one of the preceding sections.

MEMORY AND I/O ORGANIZATION

The memory and I/O are organized as two separate areas in many Z80® microprocessor-based systems. Fig. 12-25 illustrates the directly addressable areas for the memory and I/O in such a system. Notice that the memory is numbered 0000H–FFFFH and the I/O space is numbered 00H–FFH (H is used to designate a hexadecimal quantity). Each location in the memory and I/O space contains an 8-bit number. This means that the Z80® microprocessor can address 64k bytes of memory and 256 different 8-bit I/O locations or devices.

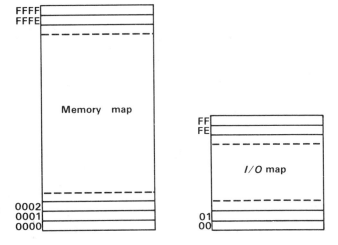

FIGURE 12-25. Memory and I/O maps of the Z80® microprocessor.

Fig. 12-26 shows how a memory system based on the Z80® microprocessor would be set up for a dot-matrix printer. The bottom area of the memory contains the program stored in a ROM. The ROM also contains tables for converting ASCII code to dot-matrix code and diagnostic programs for troubleshooting the printer. The RAM is used for the stack and also to store temporary information used by the printer such as tab positions.

Notice that there are only a few I/O devices used in a dot-matrix printer: the print head, the head-positioning motor, the paper-advance solenoid, and a serial interface to the computer system. Port 13H positions the print head and advances the paper, port 12H feeds the print head with coded data, and ports 10H and 11H control the serial interface.

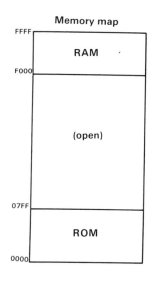

FIGURE 12-26. *Memory and I/O maps of a dot-matrix printer based on the Z80® microprocessor.*

MEMORY AND I/O CONTROL SIGNALS

The memory and *I/O* control signals of the Z80® microprocessor are READ (RD), WRITE (WR), MEMORY REQUEST (MREQ), and *I/O* REQUEST (IORQ). With these four signals and an address from the address bus, it is possible to control any *I/O* or memory device.

If the MREQ signal is at a logic 0 level and the IORQ is a logic 1, then a memory device will be accessed by the Z80® microprocessor; if the conditions on these two lines are reversed, then an *I/O* device will be accessed. When the Z80® is to write to a memory location, a 0 is placed on WR and MREQ and a 1 is placed on the RD and IORQ. See Table 12-5 for the complete truth table for these control signals.

THE PROGRAMMING MODEL OF THE Z80® MICROPROCESSOR

Fig. 12-27 shows the internal programming model of the Z80® microprocessor. This structure must be learned completely to ensure successful machine or assembly language programming.

The General-purpose Registers

This microprocessor contains six 8-bit general-purpose registers: *B, C, D, E, H,* and *L.* General-purpose registers are used to hold intermediate results in computations or in any way that the programmer sees fit.

TABLE 12-5 THE TRUTH TABLE FOR THE Z80® MICROPROCESSOR CONTROL SIGNALS.

$\overline{MREQ}$	$\overline{IORQ}$	$\overline{RD}$	$\overline{WR}$	Function
0	0	0	0	—
0	0	0	1	—
0	0	1	0	—
0	0	1	1	—
0	1	0	0	—
0	1	0	1	MEMORY READ
0	1	1	0	MEMORY WRITE
0	1	1	1	—
1	0	0	0	—
1	0	0	1	*I/O* READ
1	0	1	0	*I/O* WRITE
1	0	1	1	—
1	1	0	0	—
1	1	0	1	—
1	1	1	0	—
1	1	1	1	—

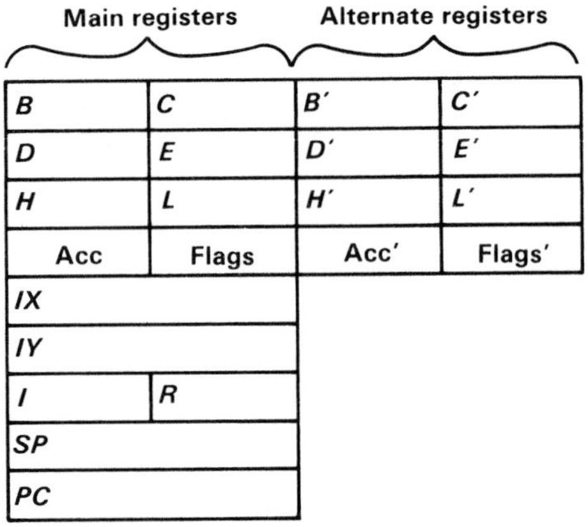

Main registers		Alternate registers	
B	C	B′	C′
D	E	D′	E′
H	L	H′	L′
Acc	Flags	Acc′	Flags′
IX			
IY			
I	R		
SP			
PC			

FIGURE 12-27. Internal register structure (programming model) of the Z80® microprocessor.

In addition to functioning as six general-purpose 8-bit registers, they can also be grouped into pairs of registers (*BC, DE,* and *HL*). Register pairs are general-purpose registers capable of holding a 16-bit number. A number in a register pair is used as data or to point to (index) data in the memory.

The Accumulator Register (A)

The accumulator register is a very important register in all microprocessors. The Z80® microprocessor contains one accumulator register designed to receive the results from almost all 8-bit arithmetic or logic operations. The only exceptions to this are for instructions used to increment (add one to) or decrement (subtract one from) the contents of any register or register pair. Whenever a number is added, subtracted, ANDed, ORed, or Exclusive ORed, the answer will end up in the accumulator.

The Flag Register (F)

The flag register was introduced as the condition code register. This register indicates various conditions of the result of an arithmetic or logic operation. It does *not* measure the value of a particular register or memory location.

Fig. 12-28 shows the flag bit positions of the Z80® microprocessor. A description of each flag bit follows.

1. *Carry (C)*: The carry flag indicates whether or not a CARRY from the most significant bit resulted from the most recent addition, or whether a BORROW resulted from the most recent subtraction. A 1 in the carry flag indicates a CARRY or BORROW, and a 0 indicates no CARRY or BORROW. (All logic operations clear this bit position, and it is not affected by increment or decrement instructions.)

2. *Add/subtract (N)*: This special flag bit indicates whether the most recent arithmetic instruction is an ADD or a SUBTRACT. The ADD instruction clears this bit, and SUBTRACT sets it. The N-flag bit is used by the DAA instruction in correcting the result of a BCD addition or subtraction.

3. *Parity/overflow (P/V)*: Logic instructions use this bit as a parity flag. The parity flag indicates the parity of the result of an arithmetic or logic operation. If the P-flag bit contains a 1, then the parity—which is the count of the number of 1s, expressed as even or odd—is even. If the P-flag bit contains a 0, then the result is odd parity.

    Arithmetic instructions use this bit as an overflow flag. If the result of an 8-bit addition or subtraction exceeds 127 or -128, then the overflow flag will be set.

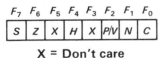

X = Don't care

FIGURE 12-28. Flag bits in the flag register of the Z80® microprocessor.

4. *Half-carry (H)*: The *H*-flag bit indicates whether or not a CARRY occurred from the least significant half-byte (nibble) in the most recent arithmetic or logic operation. This flag is cleared by all logic operations except AND, which sets it. This bit is not affected by increment or decrement instructions.

5. *Zero (Z)*: The zero flag indicates whether or not the result of the most recent arithmetic or logic operation is zero. A 1 in the *Z*-flag indicates zero, and a 0 indicates a nonzero condition.

6. *Sign (S)*: The *S*-flag indicates the sign of the result of the most recent arithmetic or logic operation. A 0 indicates a positive result, and a 1 indicates a negative result.

The Alternate Register Set

In addition to the general-purpose registers, the Z80® microprocessor has an alternate set of registers. This alternate set can be traded with the main register set by using the EX AF,AF' command, which exchanges the main and alternate accumulator and flag register, or the EXX command, which exchanges *BC* with *BC'*, *DE* with *DE'*, and *HL* with *HL'*. This feature is very useful in subroutines.

The INTERRUPT Vector Register (I)

The INTERRUPT vector register is used in INTERRUPT mode two to form the address of the INTERRUPT service subroutine.

The Memory REFRESH Register (R)

The memory REFRESH register is used when the Z80® microprocessor is connected to a system that uses dynamic memory. The REFRESH address, which is held in *R,* is output on address on pins A_0 through A_7 after each op-code retrieval. After each REFRESH operation, *R* is automatically incremented for the next REFRESH cycle.

The Index Registers (IX and IY)

The Z80® microprocessor contains two internal index registers that index the memory. If *IX* is loaded with a 1000H and the LD B,(IX) instruction is executed, the contents of memory location 1000H will be loaded into general-purpose register *B*.

The Stack Pointer Register (SP)

The stack pointer register is used by the Z80® microprocessor to keep track of its last-in, first-out (LIFO) stack. The *SP* contains a number that points to (indexes) an area of the memory designated as the stack. Whenever a program directs the Z80® microprocessor to place information on its stack, the Z80® microprocessor decrements the *SP*, stores a byte of data in the memory location indexed by the *SP*, again decrements the *SP*, and places a second byte of data in the new memory location indexed by the *SP*. When data are removed from the stack, this sequence is reversed, with the stack pointer incremented instead of decremented. This allows the Z80® microprocessor in effect to reuse the same area of memory for its LIFO stack.

The Program Counter (PC)

The program counter is used by the microprocessor to find the next step in a program. The address of the next step of a program is always maintained in the *PC*. When the Z80® microprocessor retrieves an instruction from the memory, it automatically adds one to the *PC* so that the next instruction is always fetched from the next memory location in sequence.

THE INSTRUCTION SET

There are four basic forms of instructions used by the Z80® microprocessor: 1-, 2-, 3-, and 4-byte instructions. Fig. 12-29 shows the four forms of instructions and the type of information used in each format.

In all four forms the first byte is an op-code (operation code). The op-code indicates which command is to be executed by the Z80® microprocessor. In the 1-byte instruction, no other information besides the op-code is required.

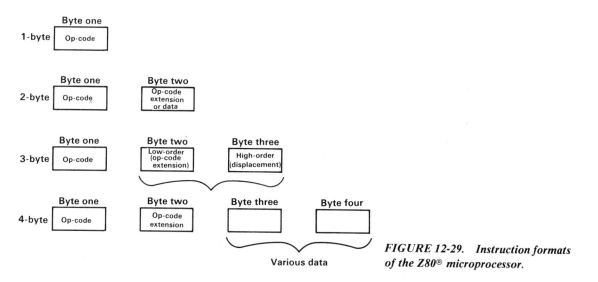

FIGURE 12-29. *Instruction formats of the Z80® microprocessor.*

In the 2-byte instructions, the second byte usually contains an 8-bit value of data (D_8) to be used when the op-code is executed or a displacement value, which will be discussed along with relative addressing. The ADD A,08H instruction, for example, is a 2-byte instruction with an op-code of C6H (in machine language) and data of 08H. This command adds 08H to the accumulator value and returns the result to the accumulator.

The 3-byte commands contain either data or a memory address in the second and third bytes. In the LD A,1000H command, for example, the first byte is an op-code of 3AH followed by the 16-bit memory address 1000H. This command would load the accumulator with the same number as that stored in memory location 1000H. With the Z80® microprocessor, whenever an address or data are placed in a 3-byte command, the least significant eight bits (low-order portion) immediately follow the op-code, and the most significant eight bits (high-order portion) follow the least significant eight bits. The LD A,1000H command is encoded into machine language and stored as 3A-00-10.

Four-byte instructions have two op-codes stored in the first and second bytes. The remaining two bytes are used for displacement or for other fixed or variable information.

Z80® Microprocessor Instruction Types

The complete set of instructions for the Z80® microprocessor is listed in Table 12-6. Following is an explanation of the main instruction categories.

1. *8-bit LOAD instructions*: The 8-bit LOAD instructions allow data to be loaded into the general-purpose registers, the accumulator, and the memory location indexed by *BC, DE, HL, IX,* or *IY*. In addition, they allow data to be loaded from any memory location into any register, as indexed by *HL, IX,* or *IY*.

2. *16-bit LOAD instructions*: The 16-bit LOAD instructions allow information to be placed into the register pairs, *IX,* and *IY*. The PUSH and POP instructions allow register pair data, the accumulator, and the flag byte to be transferred to and from the stack.

3. *EXCHANGE, BLOCK TRANSFER, and SEARCH instructions*: These instructions are extremely important because they allow entire blocks of data to be transferred from one part of the memory to another, allow blocks of memory to be searched for a particular number, and allow the main registers to be exchanged with the alternate registers. All the BLOCK TRANSFER and SEARCH instructions use *HL* and *DE* to point to the data in the memory and us *BC* as a counter. The counter is used to specify how many bytes are to be transferred or searched.

4. *8-bit arithmetic and logic instructions*: These instructions are used to add, add with CARRY, subtract, subtract with BORROW, AND, OR, Exclu-

sive OR, compare, increment, and decrement data. The data used by these instructions can be in any register or any memory location indexed by *HL*, *IX*, or *IY*.

5. *General-purpose arithmetic and CPU control instructions*: These instructions include commands to complement the accumulator and carry flag, twos complement the accumulator, and control the INTERRUPT structure of the Z80® microprocessor.

6. *ROTATE instructions*: ROTATEs allow any register or memory location to be rotated either to the right or to the left. There are also SHIFT instructions for both arithmetic and logical SHIFT operations.

7. *Bit SET, RESET, and TEST instructions*: This grouping allows any bit position of any register or memory location to be set, reset, or tested. When a bit is tested, a copy of it is placed in the zero flag.

8. *JUMP instructions*: The JUMP instructions allow the flow of a program to be controlled. The JUMPs accomplish this by conditionally or unconditionally modifying the contents of the *PC*. Whenever the contents of the program counter are changed, the flow of the program changes and continues at another point.

9. *CALL and RETURN instructions*: The CALL and RETURN instructions are used with subroutines and INTERRUPT service subroutines. The CALL initiates a subroutine and the RETURN returns from a subroutine.

10. *Input and output instructions*: These instructions transfer data to and from external *I/O* devices. In most cases the data are transferred from or to the accumulator register.

TABLE 12-6 THE Z80® MICROPROCESSOR INSTRUCTION SET.

Machine code	Mnemonic	Description
8-bit LOAD instructions		
01DDDSSS	LD r1,r2	Contents of r2 are copied into r1.
00DDD110-nn	LD r,nn	Data nn are loaded into r.
DD-01DDD110-dd	LD r, (IX+dd)	Data indexed by *IX* plus dd are copied into r.
FD-01DDD110-dd	LD r, (IY+dd)	Data indexed by *IY* plus dd are copied into r.
DD-01110DDD-dd	LD (IX+dd),r	Data in r are copied into the location indexed by *IX* plus dd.
FD-01110DDD-dd	LD (IY+dd),r	Data in r are copied into the location indexed by *IY* plus dd.
DD-36-dd-nn	LD (IX+dd),nn	Data nn are loaded into the location indexed by *IX* plus dd.
FD-36-dd-nn	LD (IY+dd),nn	Data nn are loaded into the location indexed by *IY* plus dd.
0A	LD A,(BC)	Data indexed by *BC* are copied into the Acc.
1A	LD A,(DE)	Data indexed by *DE* are copied into the Acc.
3A-ll-hh	LD A,hhll	Data addressed by hhll are copied into the Acc.
02	LD (BC),A	Data in Acc are copied into the location indexed by *BC*.
12	LD (DE),A	Data in Acc are copied into the location indexed by *DE*.
32-ll-hh	LD (hhll),A	Data in Acc are copied into hhll.

TABLE 12-6 (continued).

Machine code	Mnemonic	Description
ED-57	LD A,I	Data in *I* are copied into Acc.
ED-5F	LD A,R	Data in *R* are copied into Acc.
ED-47	LD I,A	Data in Acc are copied into *I*.
ED-4F	LD R,A	Data in Acc are copied into *R*.

16-bit LOAD instructions

00pp0001-ll-hh	LD pp,hhll	Data hhll are copied into register pair pp.
DD-21-ll-hh	LD IX,hhll	Data hhll are copied into *IX*.
FD-21-ll-hh	LD IY,hhll	Data hhll are copied into *IY*.
2A-ll-hh	LD HL,(hhll)	Data in hhll are copied into *HL*.
ED-01pp1011-ll-hh	LD pp,(hhll)	Data in hhll are copied into register pair pp.
DD-2A-ll-hh	LD IX,(hhll)	Data in hhll are copied into *IX*.
FD-2A-ll-hh	LD IY,(hhll)	Data in hhll are copied into *IY*.
22-ll-hh	LD (hhll),HL	Data in *HL* are copied into hhll and hhll+1.
ED-01pp0011-ll-hh	LD (hhll),pp	Data in register pair pp are copied into hhll and hhll+1.
DD-22-ll-hh	LD (hhll),IX	Data in *IX* are copied into hhll and hhll+1.
FD-22-ll-hh	LD (hhll),IY	Data in *IY* are copied into hhll and hhll+1.
F9	LD SP,HL	Data in *HL* are copied into *SP*.
DD-F9	LD SP,IX	Data in *IX* are copied into *SP*.
FD-F9	LD SP,IY	Data in *IY* are copied into *SP*.
11qq0101	PUSH qq	Data in register pair qq are pushed on the stack.
DD-E5	PUSH IX	Data in *IX* are pushed on the stack.
FD-E5	PUSH IY	Data in *IY* are pushed on the stack.
11qq0001	POP qq	Data are removed from the stack and placed in register pair qq.
DD-E1	POP IX	Data are removed from the stack and placed into *IX*.
FD-E1	POP IY	Data are removed from the stack and placed into *IY*.

EXCHANGE, BLOCK TRANSFER, and SEARCH instructions.

EB	EX DE,HL	Exchange *DE* with *HL*.
08	EX AF,AF'	Exchange *AF* with *AF'*.
D9	EXX	Exchange *BC*, *DE*, and *HL* with *BC'*, *DE'*, and *HL'*.
E3	EX (SP),HL	Exchange data indexed by *SP* with *HL*.
DD-E3	EX (SP),IX	Exchange data indexed by *SP* with *IX*.
FD-E3	EX (SP),IY	Exchange data indexed by *SP* with *IY*.
ED-A0	LDI	Data indexed by *HL* are transferred to location indexed by *DE*, *BC* = count.
ED-B0	LDIR	Same as LDI, except it is repeated until *BC* = 0.
ED-A8	LDD	Same as LDI, except *HL* and *DE* are decremented instead of incremented.
ED-B8	LDDR	Same as LDD, except it is repeated until *BC* = 0.
ED-A1	CPI	Acc. is compared with the data indexed by *HL*, *BC* = count.
ED-B1	CPIR	Same as CPI, except it is repeated until *BC* = 0.
ED-A9	CPD	Same as CPI, except *HL* is decremented instead of incremented.
ED-B9	CPDR	Same as CPD, except it is repeated until *BC* = 0.

TABLE 12-6 *(continued).*

Machine code	Mnemonic	Description
8-bit arithmetic and logic instructions.		
10000DDD	ADD A,r	Acc. plus r; answer to Acc.
C6-nn	ADD A,nn	Acc. plus nn; answer to Acc.
DD-86-dd	ADD A,(IX+dd)	Acc. plus data indexed by *IX* plus dd; answer to Acc.
FD-86-dd	ADD A,(IY+dd)	Acc. plus data indexed by *IY* plus dd; answer to Acc.
10001DDD	ADC A,r	Acc. plus r plus CARRY; answer to Acc.
CE-nn	ADC A,nn	Acc. plus nn plus CARRY; answer to Acc.
DD-8E-dd	ADC A,(IX+dd)	Acc. plus data indexed by (*IX* plus dd) plus CARRY; answer to Acc.
FD-8E-dd	ADC A,(IY+dd)	Acc. plus data indexed by (*IY* plus dd) plus CARRY; answer to Acc.
10010DDD	SUB A,r	Acc. minus r; answer to Acc.
D6-nn	SUB A,nn	Acc. minus nn; answer to Acc.
DD-96-dd	SUB A,(IX+dd)	Acc. minus data indexed by *IX* plus dd; answer in Acc.
FD-96-dd	SUB A,(IY+dd)	Acc. minus data indexed by *IY* plus dd; answer in Acc.
10011DDD	SBC A,r	Acc. minus r minus CARRY; answer to Acc.
DE-nn	SBC A,nn	Acc. minus nn minus CARRY; answer to Acc.
DD-9E-dd	SBC A,(IX+dd)	Acc. minus data indexed by (*IX* plus dd) minus CARRY; answer to Acc.
FD-9E-dd	SBC A,(IY+dd)	Acc. minus data indexed by (*IY* plus dd) minus CARRY; answer to Acc.
10100DDD	AND r	Acc. AND r; answer to Acc.
E6-nn	AND nn	Acc. AND nn; answer to Acc.
DD-A6-dd	AND (IX+dd)	Acc. AND data indexed by *IX* plus dd; answer to Acc.
FD-A6-dd	AND (IY+dd)	Acc. AND data indexed by *IY* plus dd; answer to Acc.
10110DDD	OR r	Acc. OR r; answer to Acc.
F6-nn	OR nn	Acc. OR nn; answer to Acc.
DD-B6-dd	OR (IX+dd)	Acc. OR data indexed by *IX* plus dd; answer to Acc.
FD-B6-dd	OR (IY+dd)	Acc. OR data indexed by *IY* plus dd; answer to Acc.
10101DDD	XOR r	Acc. Exclusive OR r; answer to Acc.
EE-nn	XOR nn	Acc. Exclusive OR nn; answer to Acc.
DD-AE-dd	XOR (IX+dd)	Acc. Exclusive OR data indexed by *IX* plus dd; answer to Acc.
FD-AE-dd	XOR (IY+dd)	Acc. Exclusive OR data indexed by *IY* plus dd; answer to Acc.
10111DDD	CP r	Acc. minus r; answer to flags.
FE-nn	CP nn	Acc. minus nn; answer to flags.
DD-BE-dd	CP (IX+dd)	Acc. minus data indexed by *IX* plus dd; answer to flags.
FD-BE-dd	CP (IY+dd)	Acc. minus data indexed by *IY* plus dd; answer to flags.
00DDD100	INC r	Add one to r.
DD-34-dd	INC (IX+dd)	Add one to data indexed by *IX* plus dd.
FD-34-dd	INC (IY+dd)	Add one to data indexed by *IY* plus dd.
00DDD101	DEC r	Subtract one from r.
DD-35-dd	DEC (IX+dd)	Subtract one from data indexed by *IX* plus dd.
FD-35-dd	DEC (IY+dd)	Subtract one from data indexed by *IY* plus dd.
General-purpose arithmetic and CPU control instructions.		
27	DAA	Decimal accumulator adjust.
2F	CPL	Invert Acc.
ED-44	NEG	Twos complement Acc.
3F	CCF	Complement CARRY.
37	SCF	Set CARRY.

TABLE 12-6 (continued).

Machine code	Mnemonic	Description
00	NOP	No operation.
76	HALT	Halt until a RESET or INTERRUPT.
F3	DI	Disable INTERRUPTS.
FB	EI	Enable INTERRUPTS.
ED-46	IM 0	INTERRUPT mode 0.
ED-56	IM 1	INTERRUPT mode 1.
ED-5E	IM 2	INTERRUPT mode 2.

16-bit arithmetic instructions.

Machine code	Mnemonic	Description
00pp1001	ADD HL,pp	*HL* plus pp; answer to *HL*.
ED-01pp1010	ADC HL,pp	*HL* plus pp plus CARRY; answer to *HL*.
ED-01pp0010	SBC HL,pp	*HL* minus pp minus CARRY; answer to *HL*.
DD-00jj1001	ADD IX,jj	*IX* plus jj; answer to *IX*.
FD-00kk1001	ADD IY,kk	*IY* plus kk; answer to *IY*.
00pp0011	INC pp	Add one to pp.
DD-23	INC IX	Add one to *IX*.
FD-23	INC IY	Add one to *IY*.
00pp1011	DEC pp	Subtract one from pp.
DD-2B	DEC IX	Subtract one from *IX*.
FD-2B	DEC IY	Subtract one from *IY*.

ROTATE and SHIFT instructions.

Machine code	Mnemonic	Description
07	RLCA	Rotate Acc. left.
17	RLA	Rotate Acc. and CARRY left.
0F	RRCA	Rotate Acc. right.
1F	RRA	Rotate Acc. and CARRY right.
CB-00000DDD	RLC r	Rotate r left.
DD-CB-dd-06	RLC (IX+dd)	Rotate data indexed by *IX* plus dd left.
FD-CB-dd-06	RLC (IY+dd)	Rotate data indexed by *IY* plus dd left.
CB-00010DDD	RL r	Rotate r and CARRY left.
DD-CB-dd-16	RL (IX+dd)	Rotate data indexed by *IX* plus dd left through CARRY.
FD-CB-dd-16	RL (IY+dd)	Rotate data indexed by *IY* plus dd left through CARRY.
CB-00001DDD	RRC r	Rotate r right.
DD-CB-dd-0E	RRC (IX+dd)	Rotate data indexed by *IX* plus dd right.
FD-CB-dd-0E	RRC (IY+dd)	Rotate data indexed by *IY* plus dd right.
CB-00011DDD	RR r	Rotate r right through CARRY.
DD-CB-dd-1E	RR (IX+dd)	Rotate data indexed by *IX* plus dd right through CARRY.
FD-CB-dd-1E	RR (IY+dd)	Rotate data indexed by *IY* plus dd right through CARRY.
CB-00100DDD	SLA r	Shift r left.
DD-CB-dd-26	SLA (IX+dd)	Shift data indexed by *IX* plus dd left.
FD-CB-dd-26	SLA (IY+dd)	Shift data indexed by *IY* plus dd left.
CB-00101DDD	SRA r	Shift arithmetic right r.

TABLE 12-6 (continued).

Machine code	Mnemonic	Description
DD-CB-dd-2E	SRA (IX+dd)	Shift arithmetic right data indexed by *IX* plus dd.
FD-CB-dd-2E	SRA (IY+dd)	Shift arithmetic right data indexed by *IY* plus dd.
CB-00111DDD	SRL r	Shift r right.
DD-CB-dd-3E	SRL (IX+dd)	Shift right data indexed by *IX* plus dd.
FD-CB-dd-3E	SRL (IY+dd)	Shift right data indexed by *IY* plus dd.
ED-6F	RLD	See programming manual.
ED-67	RRD	See programming manual.

Bit SET, RESET and TEST instructions.

CB-01bbbDDD	BIT bbb,r	Test bit position bbb of r.
DD-CB-dd-01bbb110	BIT bbb,(IX+dd)	Test bit position bbb of data indexed by *IX* plus dd.
FD-CB-dd-01bbb110	BIT bbb,(IY+dd)	Test bit position bbb of data indexed by *IY* plus dd.
CB-11bbbDDD	SET bbb,r	Set bit position bbb of r.
DD-CB-dd-11bbb110	SET bbb,(IX+dd)	Set bit position bbb of data indexed by *IX* plus dd.
FD-CB-dd-11bbb110	SET bbb,(IY+dd)	Set bit position bbb of data indexed by *IY* plus dd.
CB-10bbbDDD	RES bbb,r	Reset bit position bbb of r.
DD-CB-dd-10bbb110	RES bbb,(IX+dd)	Reset bit position bbb of data indexed by *IX* plus dd.
FD-CB-dd-10bbb110	RES bbb,(IY+dd)	Reset bit position bbb of data indexed by *IY* plus dd.

JUMP instructions

C3-ll-hh	JP hhll	Jump to hhll.
11ccc010-ll-hh	JP cc,hhll	Jump on condition ccc true.
18-dd	JP dd	Jump to dd plus *PC*.
38-dd	JP C,dd	Jump CARRY SET to dd plus *PC*.
30-dd	JP NC,dd	Jump CARRY RESET to dd plus *PC*.
28-dd	JP Z,dd	Jump zero to dd plus *PC*.
20-dd	JP NZ,dd	Jump not zero to dd plus *PC*.
E9	JP (HL)	Jump to location indexed by *HL*.
DD-E9	JP (IX)	Jump to location indexed by *IX*.
FD-E9	JP (IY)	Jump to location indexed by *IY*.
10-dd	DJNZ, dd	Decrement B and if not zero add dd to *PC*.

CALL and RETURN instructions.

CD-ll-hh	CALL hhll	Call the subroutine at hhll.
11ccc100-ll-hh	CALL ccc,hhll	Call the subroutine at hhll if condition ccc is true.
C9	RET	Return from a subroutine.
11ccc000	RET ccc	Return from a subroutine if condition ccc is true.
ED-4D	RETI	Return from an INTERRUPT.
ED-45	RETN	Return from a nonmaskable INTERRUPT.
11ttt111	RST ttt	See note ttt for subroutine CALL address.

TABLE 12-6 (continued).

Machine code	Mnemonic	Description

Input and output instructions.

DB-nn	IN A,(nn)	Input data from I/O device nn to Acc.
ED-01DDD000	IN r,(C)	Input data from I/O device (C) to r.
ED-A2	INI	B=B-1, (C) = I/O port number, data stored at HL indexed location, $HL=HL+1$.
ED-B2	INIR	Same as INI, except it repeats until $B=0$.
ED-AA	IND	Same as INI, except $HL=HL-1$.
ED-BA	INDR	Same as IND except it repeats until $B=0$.
D3-nn	OUT (nn),A	Output Acc. to device nn.
ED-01DDD001	OUT (C),r	Output r to the device addressed by C.
ED-A3	OUTI	Same as INI, except output.
ED-B3	OTIR	Same as INIR, except output.
ED-A8	OUTD	Same as IND, except output.
ED-B8	OTDR	Same as INDR, except output.

Notes:

DDD = destination register
SSS = source register
(r) = register A, B, C, D, E, H, L, and (HL)
(nn) = 8-bit number
(dd) = 8-bit signed offset address with a range of +127 to -128
(hhll) = 16-bit data or address
(DDD or SSS) = B=000, C=001, D=010, E=011, H=100, L=101, (HL)=110, and A=111
(pp) = BC=00, DE=01, HL=10, and SP=11
(qq) = BC=00, DE=01, HL=10, and AF=11
(jj) = BC=00, DE=01, IX=10, and SP=11
(kk) = BC=00, DE=01, IY=10, and SP=11
(rrr) = bit position 0=000, 1=001, 2=010, 3=011, 4=100, 5=101, 6=110, and 7=111
(ccc) = NZ not zero = 000, Z zero = 001, NC no CARRY = 010, C CARRY = 011,
    PO parity odd = 100, PE parity even = 101, P plus = 110, M minus = 111
(ttt) = 000=0000H, 001=0008H, 010=0010H, 011=0018H, 100=0020H, 101=0028H, 110=0030H, and 111=0038H

ADDRESSING MODES

The Z80® microprocessor can address data in six different modes. Data can be addressed:

1.   following an op-code
2.   directly in a memory location
3.   in a register
4.   through an index register

5. through an index register with an offset

6. in relation to the program counter

The six ways of addressing data are explained below.

1. *Immediate addressing*: This method of addressing data is one of the easiest to understand. In immediate mode addressing, the data (nn) used in an instruction immediately follow the op-code. Looking at Table 12-6, note that the symbolic or mnemonic op-code for the SUBTRACT IMMEDIATE instruction is SUB A,nn. To subtract 12H from the contents of the accumulator, you would use the SUB A,12H instruction.

2. *Direct addressing*: This form of addressing data requires that the op-code be followed by the memory address of the data. There are quite a few instructions of this type listed in various categories of Table 12-6. An example is the LD (hh11),A instruction, which is the mnemonic code for loading the memory with the accumulator value. If the LD (2000H),A instruction were executed, a copy of the data in the accumulator would be loaded into memory location 2000H.

3. *Register addressing*: Register addressing is used in the bulk of the arithmetic, logic, and LOAD instructions. The key to understanding register addressing is remembering that DDD is the destination register and SSS is the source register. This is indicated at the end of the instruction list in Table 12-6 as a note. Suppose you wanted to copy or load the number in the *B*-register into the *D*-register. This is accomplished by using the LD r1,r2 instruction and by coding DDD with 010 and SSS with 000 (the codes for the *D*-register and the *B*-register, respectively). The symbolic form of this command is LD D,B. Zilog symbolic code requires that the destination be listed first, followed by the source.

4. *Indexed addressing*: This type of addressing is probably the hardest to understand because the address of the data is not stored with the op-code. The address of the data is stored in one of the register pairs, *BC, DE,* or *HL.* If a LD (HL),A instruction is executed by the Z80® microprocessor, the contents of the accumulator (*A*) are copied into the memory location indicated by the *HL* pair. The *HL* pair usually addresses or indexes the memory. The *BC* and *DE* pairs only index memory for the LOAD instructions.

5. *Indexed addressing with an offset (dd)*: If either the *IX* or *IY* register indexes memory, a displacement is added to the register before data are indexed. For example, if *IX* contains a 1000H and the LD A,(IX + 10H) instruction is executed, the contents of memory location 1010H will be loaded into the accumulator. The addresses 1000H and 10H are added to form the address for this instruction.

6. *Relative addressing*: Some of the JUMP instructions are the only ones that use this form of addressing. If a JP 04H instruction is entered, the machine will jump to four locations beyond the address of the next instruction. If the next instruction is stored at location 1010H, then the computer will jump to 1014H. This occurs because the machine adds the displacement value (04H) to the contents of the program counter (1010H). A backward jump occurs if the displacement is negative. For example, if the address of the next instruction is 100CH and you wish to jump to location 1000H, then you would use the displacement value of −0CH (0F4H).

Example 12-3. Suppose the data in memory locations 1000H through 1004H are to be added, and the sum is to be stored in memory location 1005H.

Solution This procedure is listed in symbolic assembly language in the program that follows. The symbolic language is explained on the right.

```
START:  LD IX,1000H       Point to 1000H with IX
        LD A,(IX+00H)      Get first byte
        ADD A,(IX+01H)     Add 1001H
        ADD A,(IX+02H)     Add 1002H
        ADD A,(IX+03H)     Add 1003H
        ADD A,(IX+04H)     Add 1004H
        LD (IX+05H),A      Store 1005H
        HALT               Stop program
```

SUMMARY

1 Microprocessors are digital sequential logic elements whose function can be changed by changing the program.

2 A computer is a machine that solves problems by manipulating binary data.

3 A computer system is composed of four parts: CPU, memory, *I/O*, and auxiliary memory.

4 The CPU in a computer system performs data transfer, arithmetic, and logic, and makes decisions.

5 The memory in a computer system stores programs and data manipulated by the programs.

6 The auxiliary memory in a computer system stores large volumes of data.

7   The input/output equipment in a computer system allows the machine to transfer data to the user or other machines.

8   A typical microprocessor is constructed with an ALU, a register array, an instruction register, and a control logic circuit.

9   The ALU performs all the arithmetic and logic operations in a microprocessor. The information operated on by the ALU is in the form of signed or unsigned numbers.

10   The register array contains an accumulator, which is used to accumulate answers after an arithmetic or logic operation; a general-purpose set of registers; a stack pointer; a condition code register; and a program counter.

11   The bus structure of most modern microprocessor-based systems is composed of three buses: the data bus, the control bus, and the address bus.

12   The fixed-word-width microprocessor, the single-component microprocessor, and the bit-slice microprocessor are the three types of microcomputers in production.

13   Microprocessors find application in video games, microwave ovens, automobile fuel-injection systems, printers, home computers, and almost any other modern electrical product.

14   Microprocessor trainers teach the operation, instruction set, and programming of a microprocessor. A typical trainer consists of a keyboard, LED displays, memory, and some form of I/O in addition to the keyboard and LED displays.

15   Programming consists of writing a program, which is a group of instructions arranged in a logical sequence, in either assembly language or a high-level language.

16   The first step in writing a program is to place your thoughts on paper in the form of a flowchart. Once the flowchart is accurately constructed, the programmer can concentrate on developing the code for the program from the flowchart.

QUESTIONS AND PROBLEMS

1   What is a microprocessor?

2   What is a computer?

3   What is a program?

KEY TERMS

accumulator (A or Acc)

address bus

ALU

ASCII

assembler

assembly language

auxiliary memory

BASIC

bit-slice microprocessor

bus

compiler

computer

condition code register (bits)

control bus

CPU

data bus

displacement

fixed-word-width microprocessor

flag register (bits)

flowchart

I/O

instruction register (I)

instructions

interpreter

machine language

microcomputer

microprocessor

microprocessor trainer

nibble

op-code

parity

program

register array

single-component microcomputer

4 Draw the block diagram of a computer system.

5 What are the main functions of the CPU in a computer, and what does the acronym CPU indicate?

6 What are the main functions of the memory in a computer system?

7 What types of memory devices are normally labeled "auxiliary memory" in a computer system?

8 List some common I/O devices.

9 What are the main elements of a microprocessor?

10 Compare the speed of the microprocessor with that of the main-frame computer system.

11 Draw a block diagram of a microprocessor.

12 Convert the following decimal numbers into unsigned binary 8-bit numbers for use with an 8-bit microprocessor: 12, 55, 66, 127, 100, and 88.

13 Convert the following decimal numbers into signed binary 8-bit numbers for use with an 8-bit microprocessor: +33, −59, +99, −127, −30, and +77.

14 What is an accumulator and where would it be found in a microprocessor?

15 What is the purpose of the program counter and why is it called a counter?

16 What are the condition code bits and how does the microprocessor use them?

17 List the three buses most often found in a microprocessor-based computer system.

18 Define the purpose of the three buses indicated in problem 17.

19 What is the difference between a fixed-word-width microprocessor and a bit-slice microprocessor?

20 Who developed the first microprocessor?

21 When did the microprocessor age begin and what initiated it?

22 List eight examples of fixed-word-width microprocessors.

23 Draw the block diagram of a single-component microcomputer.

24 Where would a single-component microcomputer be applied?

25 Describe the component parts of a microprocessor trainer and the purpose of each part.

26 Briefly describe each of the following terms: machine language, assembly language, interpretive language, and compiled language.

27 What is the first step in writing a program?

28 Design a flowchart that explains how to get off a garbage truck safely, pick up a trash can and dump it into the truck, compact the garbage, and safely reboard the truck. (Don't forget that before the garbage can be dumped into the truck, the compactor may need to be operated.)

29 Develop a flowchart that will determine whether you will get an A, B, C, D, or F for this course. The data should include your test scores, homework scores, quizzes, etc. and should include the weight of each factor.

APPENDIX A: THE NEW UNIFORM LOGIC SYMBOLS

Uniform logic symbols were first published in 1972 by the International Electrotechnical Commission (IEC) in publication number 117—15. Shortly thereafter, the Institute of Electrical and Electronics Engineers, Inc. (IEEE), published these symbols in *IEEE/ANSI Standard Y32. 14—1973*. Since 1973, however, the uniform logic symbols have all but been ignored by the users and manufacturers of digital electronic components. Today they are beginning to become standard, as witnessed by their use in most of the new manufacturer's data books on digital components. It is for this reason that this appendix is included in this text. It is expected that they will appear in all of the illustrations in the next edition of this text as they become the standard rather than the exception.

NAME	OLD(MIL-SPEC-806)	NEW (UNIFORM LOGIC SYMBOL)
AND	$\begin{smallmatrix}A\\B\end{smallmatrix}$ ⟩— T	$\begin{smallmatrix}A\\B\end{smallmatrix}$ — & — T
OR	$\begin{smallmatrix}A\\B\end{smallmatrix}$ ⟩— T	$\begin{smallmatrix}A\\B\end{smallmatrix}$ — ≥1 — T
NOT	A —▷∘— T A —∘▷— T	A — 1 ▷ T A ▷ 1 ▷ T
NAND	$\begin{smallmatrix}A\\B\end{smallmatrix}$ ⟩∘— T	$\begin{smallmatrix}A\\B\end{smallmatrix}$ — & ▷— T
NOR	$\begin{smallmatrix}A\\B\end{smallmatrix}$ ⟩∘— T	$\begin{smallmatrix}A\\B\end{smallmatrix}$ — ≥1 ▷— T
AND	$\begin{smallmatrix}A\\B\end{smallmatrix}$ ∘⟩∘— T	$\begin{smallmatrix}A\\B\end{smallmatrix}$ ▷ ≥1 ▷— T
OR	$\begin{smallmatrix}A\\B\end{smallmatrix}$ ∘⟩∘— T	$\begin{smallmatrix}A\\B\end{smallmatrix}$ ▷ & ▷— T
NAND	$\begin{smallmatrix}A\\B\end{smallmatrix}$ ∘∘⟩— T	$\begin{smallmatrix}A\\B\end{smallmatrix}$ ▷ ≥1 — T

NAME	OLD (MIL-SPEC-806)	NEW (UNIFORM LOGIC SYMBOL)
NOR	$\begin{matrix} A \\ B \end{matrix}$ ⟶ T	$\begin{matrix} A \\ B \end{matrix}$ ⟶ [&] ⟶ T
Exclusive OR	$\begin{matrix} A \\ B \end{matrix}$ ⟶⊕⟶ T	$\begin{matrix} A \\ B \end{matrix}$ ⟶ [=1] ⟶ T
Exclusive NOR	$\begin{matrix} A \\ B \end{matrix}$ ⟶⊕∘⟶ T	$\begin{matrix} A \\ B \end{matrix}$ ⟶ [=1] ⟶ T
Schmitt Trigger NAND	$\begin{matrix} A \\ B \end{matrix}$ ⟶[⎍]∘⟶ T	$\begin{matrix} A \\ B \end{matrix}$ ⟶ [& ⎍] ⟶ T
Open-collector NAND	$\begin{matrix} A \\ B \end{matrix}$ ⟶∘⟶ T	$\begin{matrix} A \\ B \end{matrix}$ ⟶ [& ◊] ⟶ T
Pulse-Triggered Flip-flop	$\overline{PRESET}$ — J / Q / CLK / K / Q̄ — $\overline{CLR}$	J S 7 / C / K R 7
Edge-Triggered Flip-flop	$\overline{PRESET}$ — J / Q / ▷CLK / K / Q̄ — $\overline{CLR}$	J S / C / K R

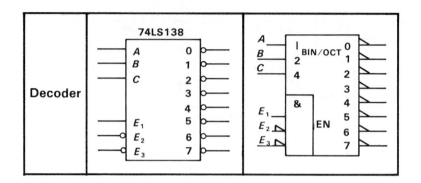

Decoder

APPENDIX B: DATA SHEETS

schematic (each gate)

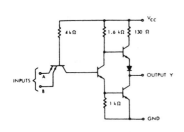

NOTE: Component values shown are nominal.

W FLAT PACKAGE
(TOP VIEW)

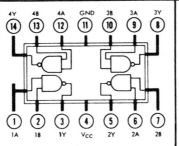

J OR N DUAL-IN-LINE PACKAGE
(TOP VIEW)

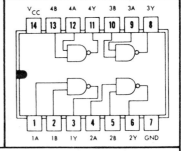

positive logic: $Y = \overline{AB}$

recommended operating conditions

	MIN	NOM	MAX	UNIT
Supply Voltage V_{CC}: SN5400 Circuits	4.5	5	5.5	V
SN7400 Circuits	4.75	5	5.25	V
Normalized Fan-Out From Each Output, N			10	
Operating Free-Air Temperature Range, T_A: SN5400 Circuits	−55	25	125	°C
SN7400 Circuits	0	25	70	°C

Courtesy of Texas Instruments Incorporated.

421

electrical characteristics over recommended operating free-air temperature (unless otherwise noted)

	PARAMETER	TEST FIGURE	TEST CONDITIONS[†]		MIN	TYP[‡]	MAX	UNIT
$V_{in(1)}$	Logical 1 input voltage required at both input terminals to ensure logical 0 level at output	1			2			V
$V_{in(0)}$	Logical 0 input voltage required at either input terminal to ensure logical 1 level at output	2					0.8	V
$V_{out(1)}$	Logical 1 output voltage	2	V_{CC} = MIN, I_{load} = −400 µA	V_{in} = 0.8 V,	2.4	3.3		V
$V_{out(0)}$	Logical 0 output voltage	1	V_{CC} = MIN, I_{sink} = 16 mA	V_{in} = 2 V,		0.22	0.4	V
$I_{in(0)}$	Logical 0 level input current (each input)	3	V_{CC} = MAX,	V_{in} = 0.4 V			−1.6	mA
$I_{in(1)}$	Logical 1 level input current (each input)	4	V_{CC} = MAX,	V_{in} = 2.4 V			40	µA
			V_{CC} = MAX,	V_{in} = 5.5 V			1	mA
I_{OS}	Short-circuit output current[§]	5	V_{CC} = MAX	SN5400	−20		−55	mA
				SN7400	−18		−55	
$I_{CC(0)}$	Logical 0 level supply current	6	V_{CC} = MAX,	V_{in} = 5 V		12	22	mA
$I_{CC(1)}$	Logical 1 level supply current	6	V_{CC} = MAX,	V_{in} = 0		4	8	mA

switching characteristics, V_{CC} = 5 V, T_A = 25°C, N = 10

	PARAMETER	TEST FIGURE	TEST CONDITIONS		MIN	TYP	MAX	UNIT
t_{pd0}	Propagation delay time to logical 0 level	65	C_L = 15 pF,	R_L = 400 Ω		7	15	ns
t_{pd1}	Propagation delay time to logical 1 level	65	C_L = 15 pF,	R_L = 400 Ω		11	22	ns

[†] For conditions shown as MIN or MAX, use the appropriate value specified under recommended operating conditions for the applicable device type.

[‡] All typical values are at V_{CC} = 5 V, T_A = 25°C.

[§] Not more than one output should be shorted at a time.

intel

2716*
16K (2K x 8) UV ERASABLE PROM

- **Fast Access Time**
  - 2716-1: 350 ns Max.
  - 2716-2: 390 ns Max.
  - 2716:   450 ns Max.
  - 2716-5: 490 ns Max.
  - 2716-6: 650 ns Max.
- **Single +5V Power Supply**
- **Low Power Dissipation**
  - **Active Power: 525 mW Max.**
  - **Standby Power: 132 mW Max.**

- **Pin Compatible to Intel 2732A EPROM**
- **Simple Programming Requirements**
  - **Single Location Programming**
  - **Programs with One 50 ms Pulse**
- **Inputs and Outputs TTL Compatible During Read and Program**
- **Completely Static**

The Intel® 2716 is a 16,384-bit ultraviolet erasable and electrically programmable read-only memory (EPROM). The 2716 operates from a single 5-volt power supply, has a static standby mode, and features fast single-address location programming. It makes designing with EPROMs faster, easier and more economical.

The 2716, with its single 5-volt supply and with an access time up to 350 ns, is ideal for use with the newer high-performance +5V microprocessors such as Intel's 8085 and 8086. A selected 2716-5 and a 2716-6 are available for slower speed applications. The 2716 is also the first EPROM with a static standby mode which reduces the power dissipation without increasing access time. The maximum active power dissipation is 525 mW while the maximum standby power dissipation is only 132 mW, a 75% savings.

The 2716 has the simplest and fastest method yet devised for programming EPROMs—single-pulse, TTL-level programming. No need for high voltage pulsing because all programming controls are handled by TTL signals. Program any location at any time—either individually, sequentially or at random, with the 2716's single-address location programming. Total programming time for all 16,384 bits is only 100 seconds.

*Part(s) also available in extended temperature range for Military and Industrial grade applications.

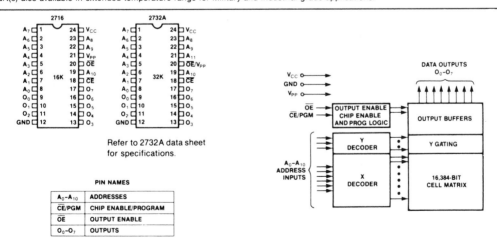

Refer to 2732A data sheet for specifications.

PIN NAMES

A_0-A_{10}	ADDRESSES
$\overline{CE}$/PGM	CHIP ENABLE/PROGRAM
$\overline{OE}$	OUTPUT ENABLE
O_0-O_7	OUTPUTS

Figure 1. Pin Configuration

Figure 2. Block Diagram

SEPTEMBER 1981
AFN-00811B

Courtesy of Intel Corporation.

DEVICE OPERATION

The five modes of operation of the 2716 are listed in Table 1. It should be noted that all inputs for the five modes are at TTL levels. The power supplies required are a +5V V_{CC} and a V_{PP}. The V_{PP} power supply must be at 25V during the three programming modes, and must be at 5V in the other two modes.

Read Mode

The 2716 has two control functions, both of which must be logically satisfied in order to obtain data at the outputs. Chip Enable ($\overline{CE}$) is the power control and should be used for device selection. Output Enable ($\overline{OE}$) is the output control and should be used to gate data to the output pins, independent of device selection. Assuming that addresses are stable, address access time (t_{ACC}) is equal to the delay from $\overline{CE}$ to output (t_{CE}). Data is available at the outputs t_{OE} after the falling edge of $\overline{OE}$, assuming that $\overline{CE}$ has been low and addresses have been stable for at least $t_{ACC}-t_{OE}$.

Standby Mode

The 2716 has a standby mode which reduces the active power dissipation by 75%, from 525 mW to 132 mW. The 2716 is placed in the standby mode by applying a TTL high signal to the $\overline{CE}$ input. When in standby mode, the outputs are in a high impedence state, independent of the $\overline{OE}$ input.

Output OR-Tieing

Because 2716s are usually used in larger memory arrays, Intel has provided a 2-line control function that accomodates this use of multiple memory connections. The two-line control function allows for:

a) the lowest possible memory power dissipation, and

b) complete assurance that output bus contention will not occur.

To most efficiently use these two control lines, it is recommended that $\overline{CE}$ (pin 18) be decoded and used as the primary device selecting function, while $\overline{OE}$ (pin 20) be made a common connection to all devices in the array and connected to the READ line from the system control bus. This assures that all deselected memory devices are in their low-power standby modes and that the output pins are only active when data is desired from a particular memory device.

Programming

Initially, and after each erasure, all bits of the 2716 are in the "1" state. Data is introduced by selectively programming "0's" into the desired bit locations. Although only "0's" will be programmed, both "1's" and "0's" can be presented in the data word. The only way to change a "0" to a "1" is by ultraviolet light erasure.

The 2716 is in the programming mode when the V_{PP} power supply is at 25V and $\overline{OE}$ is at V_{IH}. The data to be programmed is applied 8 bits in parallel to the data output pins. The levels required for the address and data inputs are TTL.

When the address and data are stable, a 50 msec, active-high, TTL program pulse is applied to the $\overline{CE}$/PGM input. A program pulse must be applied at each address location to be programmed. You can program any location at any time—either individually, sequentially, or at random. The program pulse

Table 1. Mode Selection

Pins Mode	$\overline{CE}$/PGM (18)	$\overline{OE}$ (20)	V_{PP} (21)	V_{CC} (24)	Outputs (9–11, 13–17)
Read	V_{IL}	V_{IL}	+5	+5	D_{OUT}
Standby	V_{IH}	Don't Care	+5	+5	High Z
Program	Pulsed V_{IL} to V_{IH}	V_{IH}	+25	+5	D_{IN}
Program Verify	V_{IL}	V_{IL}	+25	+5	D_{OUT}
Program Inhibit	V_{IL}	V_{IH}	+25	+5	High Z

has a maximum width of 55 msec. The 2716 must not be programmed with a DC signal applied to the $\overline{CE}$/PGM input.

Programming of multiple 2716s in parallel with the same data can be easily accomplished due to the simplicity of the programming requirements. Like inputs of the paralleled 2716s may be connected together when they are programmed with the same data. A high-level TTL pulse applied to the $\overline{CE}$/PGM input programs the paralleled 2716s.

Program Inhibit

Programming of multiple 2716s in parallel with different data is also easily accomplished. Except for $\overline{CE}$/PGM, all like inputs (including $\overline{OE}$) of the parallel 2716s may be common. A TTL-level program pulse applied to a 2716's $\overline{CE}$/PGM input with V_{PP} at 25V will program that 2716. A low-level $\overline{CE}$/PGM input inhibits the other 2716 from being programmed.

Program Verify

A verify should be performed on the programmed bits to determine that they were correctly programmed. The verify may be performed with V_{PP} at 25V. Except during programming and program verify, V_{PP} must be at 5V.

ERASURE CHARACTERISTICS

The erasure characteristics of the 2716 are such that erasure begins to occur when exposed to light with wavelengths shorter than approximately 4000 Angstroms (Å). It should be noted that sunlight and certain types of fluorescent lamps have wavelengths in the 3000–4000 Å range. Data show that constant exposure to room-level fluorescent lighting could erase the typical 2716 in approximately 3 years, while it would take approximately 1 week to cause erasure when exposed to direct sunlight. If the 2716 is to be exposed to these types of lighting conditions for extended periods of time, opaque labels are available from Intel which should be placed over the 2716 window to prevent unintentional erasure.

The recommended erasure procedure (see Data Catalog PROM/ROM Programming Instruction Section) for the 2716 is exposure to shortwave ultraviolet light which has a wavelength of 2537 Angstroms (Å). The integrated dose (i.e., UV intensity X exposure time) for erasure should be a minimum of 15 W-sec/cm^2. The erasure time with this dosage is approximately 15 to 20 minutes using an ultraviolet lamp with a 1200 μW/cm^2 power rating. The 2716 should be placed within 1 inch of the lamp tubes during erasure. Some lamps have a filter on their tubes which should be removed before erasure.

ABSOLUTE MAXIMUM RATINGS*

Temperature Under Bias .... $-10°C$ to $+80°C$
Storage Temperature .... $-65°C$ to $+125°C$
All Input or Output Voltages with
   Respect to Ground .... $+6V$ to $-0.3V$
V_{PP} Supply Voltage with Respect
   to Ground During Program .... $+26.5V$ to $-0.3V$

*NOTICE: Stresses above those listed under "Absolute Maximum Ratings" may cause permanent damage to the device. This is a stress rating only and functional operation of the device at these or any other conditions above those indicated in the operational sections of this specification is not implied. Exposure to absolute maximum rating conditions for extended periods may affect device reliability.

DC AND AC OPERATING CONDITIONS DURING READ

	2716	2716-1	2716-2	2716-5	2716-6
Temperature Range	0°C–70°C	0°C–70°C	0°C–70°C	0°C–70°C	0°C–70°C
V_{CC} Power Supply[1,2]	5V ±5%	5V ±10%	5V ±5%	5V ±5%	5V ±5%
V_{PP} Power Supply[2]	V_{CC}	V_{CC}	V_{CC}	V_{CC}	V_{CC}

READ OPERATION
D.C. AND OPERATING CHARACTERISTICS

Symbol	Parameter	Limits			Units	Test Conditions
		Min.	Typ.[3]	Max.		
I_{LI}	Input Load Current			10	μA	$V_{IN} = 5.25V$
I_{LO}	Output Leakage Current			10	μA	$V_{OUT} = 5.25V$
I_{PP1}[2]	V_{PP} Current			5	mA	$V_{PP} = 5.25V$
I_{CC1}[2]	V_{CC} Current (Standby)		10	25	mA	$\overline{CE} = V_{IH}, \overline{OE} = V_{IL}$
I_{CC2}[2]	V_{CC} Current (Active)		57	100	mA	$\overline{OE} = \overline{CE} = V_{IL}$
V_{IL}	Input Low Voltage	-0.1		0.8	V	
V_{IH}	Input High Voltage	2.0		$V_{CC} + 1$	V	
V_{OL}	Output Low Voltage			0.45	V	$I_{OL} = 2.1$ mA
V_{OH}	Output High Voltage	2.4			V	$I_{OH} = -400 \mu A$

TYPICAL CHARACTERISTICS

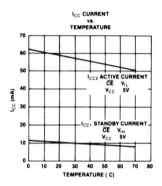

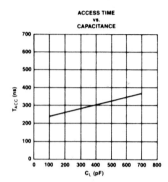

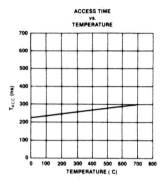

A.C. CHARACTERISTICS

Symbol	Parameter	Limits (ns)										Test Conditions
		2716		2716-1		2716-2		2716-5		2716-6		
		Min.	Max.	Min.	Max.	Min.	Max.	Min.	Max.	Min.	Max.	
t_{ACC}	Address to Output Delay		450		350		390		450		450	$\overline{CE} = \overline{OE} = V_{IL}$
t_{CE}	$\overline{CE}$ to Output Delay		450		350		390		490		650	$\overline{OE} = V_{IL}$
t_{OE}[4]	Output Enable to Output Delay		120		120		120		160		200	$\overline{CE} = V_{IL}$
t_{DF}[4]	Output Enable High to Output Float	0	100	0	100	0	100	0	100	0	100	$\overline{CE} = V_{IL}$
t_{OH}	Output Hold from Addresses, $\overline{CE}$ or $\overline{OE}$ Whichever Occurred First	0		0		0		0		0		$\overline{CE} = \overline{OE} = V_{IL}$

CAPACITANCE[4] (T_A = 25°C, f = 1 MHz)

Symbol	Parameter	Typ.	Max.	Units	Test Conditions
C_{IN}	Input Capacitance	4	6	pF	$V_{IN} = 0V$
C_{OUT}	Output Capacitance	8	12	pF	$V_{OUT} = 0V$

A.C. TEST CONDITIONS

Output Load .... 1 TTL gate and
                          C_L = 100 pF
Input Rise and Fall Times .... ≤20 ns
Input Pulse Levels .... 0.8V to 2.2V
Timing Measurement Reference Level:
    Inputs .... 1V and 2V
    Outputs .... 0.8V and 2V

A.C. WAVEFORMS[1]

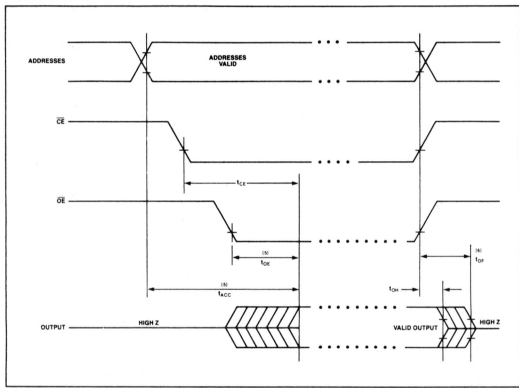

NOTES:
1. V_{CC} must be applied simultaneously or before V_{PP} and removed simultaneously or after V_{PP}.
2. V_{PP} may be connected to V_{CC} except during programming. The supply current would then be the sum of I_{CC} and I_{PP1}.
3. Typical values are for $T_A = 25°C$ and nominal supply voltages.
4. This parameter is only sampled and is not 100% tested.
5. $\overline{OE}$ may be delayed up to $t_{ACC}-t_{OE}$ after the falling edge of $\overline{CE}$ without impact on t_{ACC}.
6. t_{DF} is specified from $\overline{OE}$ or $\overline{CE}$, whichever occurs first.

PROGRAMMING CHARACTERSITICS[1]

D.C. PROGRAMMING CHARACTERISTICS (T_A = 25°C ±5°C, V_{CC}[2] = 5V ±5%, V_{PP}[2,3] = 25V ±1V)

Symbol	Parameter	Min.	Typ.	Max.	Units	Test Conditions
I_{LI}	Input Current (for Any Input)			10	μA	V_{IN} = 5.25V/0.45
I_{PP1}	V_{PP} Supply Current			5	mA	$\overline{CE}$/PGM = V_{IL}
I_{PP2}	V_{PP} Supply Current During Programming Pulse			30	mA	$\overline{CE}$/PGM = V_{IH}
I_{CC}	V_{CC} Supply Current			100	mA	
V_{IL}	Input Low Level	−0.1		0.8	V	
V_{IH}	Input High Level	2.0		V_{CC}+1	V	

A.C. PROGRAMMING CHARACTERISTICS (T_A = 25°C ±5°C, V_{CC}[2] = 5V ±5%, V_{PP}[2,3] = 25V ±1V)

Symbol	Parameter	Min.	Typ.	Max.	Units	Test Conditions
t_{AS}	Address Setup Time	2			μs	
t_{OES}	$\overline{OE}$ Setup Time	2			μs	
t_{DS}	Data Setup Time	2			μs	
t_{AH}	Address Hold Time	2			μs	
t_{OEH}	$\overline{OE}$ Hold Time	2			μs	
t_{DH}	Data Hold Time	2			μs	
t_{DF}	Output Enable to Output Float Delay	0		200	ns	$\overline{CE}$/PGM = V_{IL}
t_{OE}	Output Enable to Output Delay			200	ns	$\overline{CE}$/PGM = V_{IL}
t_{PW}	Program Pulse Width	45	50	55	ms	
t_{PRT}	Program Pulse Rise Time	5			ns	
t_{PFT}	Program Pulse Fall Time	5			ns	

A.C. CONDITIONS OF TEST

V_{CC} . . . . 5V ± 5%
V_{PP} . . . . 25V ± 1V
Input Rise and Fall Times (10% to 90%) . . . . 20 ns

Input Pulse Levels . . . . 0.8 to 2.2V
Input Timing Reference Level . . . . 1V and 2V
Output Timing Reference Level . . . . 0.8V and 2V

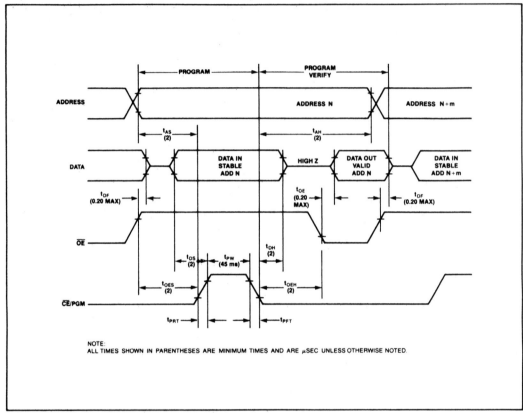

PROGRAMMING WAVEFORMS (V_{PP} = 25V ±1V, V_{CC} = 5V ±5%)

NOTE:
ALL TIMES SHOWN IN PARENTHESES ARE MINIMUM TIMES AND ARE μSEC UNLESS OTHERWISE NOTED.

NOTES:
1. Intel's standard product warranty applies only to devices programmed to specifications described herein.
2. V_{CC} must be applied simultaneously or before V_{PP} and removed simultaneously or after V_{PP}. The 2716 must not be inserted into or removed from a board with V_{PP} at 25 ±1V to prevent damage to the device.
3. The maximum allowable voltage which may be applied to the V_{PP} pin during programming is +26V. Care must be taken when switching the V_{PP} supply to prevent overshoot exceeding this 26V maximum specification.

National Semiconductor

CD4001BM/CD4001BC Quad 2-Input NOR Buffered B Series Gate
CD4011BM/CD4011BC Quad 2-Input NAND Buffered B Series Gate

General Description

These quad gates are monolithic complementary MOS (CMOS) integrated circuits constructed with N- and P- channel enhancement mode transistors. They have equal source and sink current capabilities and conform to standard B series output drive. The devices also have buffered outputs which improve transfer characteristics by providing very high gain.

All inputs are protected against static discharge with diodes to V_{DD} and V_{SS}.

Features

- Low power TTL compatibility   fan out of 2 driving 74L or 1 driving 74LS

- 5 V—10 V—15 V parametric ratings

- Symmetrical output characteristics

- Maximum input leakage 1 µA at 15 V over full temperature range

Schematic and Connection Diagrams

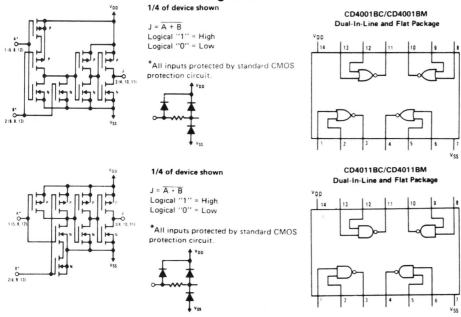

1/4 of device shown

$J = \overline{A + B}$
Logical "1" = High
Logical "0" = Low

*All inputs protected by standard CMOS protection circuit.

CD4001BC/CD4001BM
Dual-In-Line and Flat Package

1/4 of device shown

$J = \overline{A \cdot B}$
Logical "1" = High
Logical "0" = Low

*All inputs protected by standard CMOS protection circuit.

CD4011BC/CD4011BM
Dual-In-Line and Flat Package

TOP VIEW

Courtesy of National Semiconductor Corporation. Authorization has been given by National Semiconductor Corporation, Santa Clara, California, world leaders in Digital Acquisition Circuitry.

Absolute Maximum Ratings (Notes 1 and 2)

Voltage at Any Pin	$-0.5V$ to $V_{DD} + 0.5V$
Package Dissipation	500 mW
V_{DD} Range	$-0.5\ V_{DC}$ to $+18\ V_{DC}$
Storage Temperature	$-65^\circ C$ to $+150^\circ C$
Lead Temperature (Soldering, 10 seconds)	$300^\circ C$

Operating Conditions

Operating V_{DD} Range	$3\ V_{DC}$ to $15\ V_{DC}$
Operating Temperature Range	
CD4001BM, CD4011BM	$-55^\circ C$ to $+125^\circ C$
CD4001BC, CD4011BC	$-40^\circ C$ to $+85^\circ C$

DC Electrical Characteristics CD4001BM, CD4011BM (Note 2)

PARAMETER		CONDITIONS	−55 C		+25 C			+125 C		UNITS
			MIN	MAX	MIN	TYP	MAX	MIN	MAX	
I_{DD}	Quiescent Device Current	$V_{DD} = 5V$		0.25		0.004	0.25		7.5	μA
		$V_{DD} = 10V$		0.50		0.005	0.50		15	μA
		$V_{DD} = 15V$		1.0		0.006	1.0		30	μA
V_{OL}	Low Level Output Voltage	$V_{DD} = 5V$		0.05		0	0.05		0.05	V
		$V_{DD} = 10V$ $\}$ $|I_O| < 1\mu A$		0.05		0	0.05		0.05	V
		$V_{DD} = 15V$ $\}$		0.05		0	0.05		0.05	V
V_{OH}	High Level Output Voltage	$V_{DD} = 5V$	4.95		4.95	5		4.95		V
		$V_{DD} = 10V$ $\}$ $|I_O| < 1\mu A$	9.95		9.95	10		9.95		V
		$V_{DD} = 15V$ $\}$	14.95		14.95	15		14.95		V
V_{IL}	Low Level Input Voltage	$V_{DD} = 5V$, $V_O = 4.5V$		1.5		2	1.5		1.5	V
		$V_{DD} = 10V$, $V_O = 9.0V$		3.0		4	3.0		3.0	V
		$V_{DD} = 15V$, $V_O = 13.5V$		4.0		6	4.0		4.0	V
V_{IH}	High Level Input Voltage	$V_{DD} = 5V$, $V_O = 0.5V$	3.5		3.5	3		3.5		V
		$V_{DD} = 10V$, $V_O = 1.0V$	7.0		7.0	6		7.0		V
		$V_{DD} = 15V$, $V_O = 1.5V$	11.0		11.0	9		11.0		V
I_{OL}	Low Level Output Current	$V_{DD} = 5V$, $V_O = 0.4V$	0.64		0.51	0.88		0.36		mA
		$V_{DD} = 10V$, $V_O = 0.5V$	1.6		1.3	2.25		0.9		mA
		$V_{DD} = 15V$, $V_O = 1.5V$	4.2		3.4	8.8		2.4		mA
I_{OH}	High Level Output Current	$V_{DD} = 5V$, $V_O = 4.6V$	−0.64		−0.51	−0.88		−0.36		mA
		$V_{DD} = 10V$, $V_O = 9.5V$	−1.6		−1.3	−2.25		−0.9		mA
		$V_{DD} = 15V$, $V_O = 13.5V$	−4.2		−3.4	−8.8		−2.4		mA
I_{IN}	Input Current	$V_{DD} = 15V$, $V_{IN} = 0V$		−0.10		-10^{-5}	−0.10		−1.0	μA
		$V_{DD} = 15V$, $V_{IN} = 15V$		0.10		10^{-5}	0.10		1.0	μA

Note 1: "Absolute Maximum Ratings" are those values beyond which the safety of the device cannot be guaranteed. Except for "Operating Temperature Range" they are not meant to imply that the devices should be operated at these limits. The table of "Electrical Characteristics" provides conditions for actual device operation.

Note 2: All voltages measured with respect to V_{SS} unless otherwise specified.

DC Electrical Characteristics CD4001BC, CD4011BC (Note 2)

	PARAMETER	CONDITIONS	40°C MIN	40°C MAX	+25°C MIN	+25°C TYP	+25°C MAX	+85°C MIN	+85°C MAX	UNITS
I_{DD}	Quiescent Device Current	$V_{DD} = 5V$		1		0.004	1		7.5	μA
		$V_{DD} = 10V$		2		0.005	2		15	μA
		$V_{DD} = 15V$		4		0.006	4		30	μA
V_{OL}	Low Level Output Voltage	$V_{DD} = 5V$ ⎫		0.05		0	0.05		0.05	V
		$V_{DD} = 10V$ ⎬ $|I_O| < 1μA$		0.05		0	0.05		0.05	V
		$V_{DD} = 15V$ ⎭		0.05		0	0.05		0.05	V
V_{OH}	High Level Output Voltage	$V_{DD} = 5V$ ⎫	4.95		4.95	5		4.95		V
		$V_{DD} = 10V$ ⎬ $|I_O| < 1μA$	9.95		9.95	10		9.95		V
		$V_{DD} = 15V$ ⎭	14.95		14.95	15		14.95		V
V_{IL}	Low Level Input Voltage	$V_{DD} = 5V$, $V_O = 4.5V$		1.5		2	1.5		1.5	V
		$V_{DD} = 10V$, $V_O = 9.0V$		3.0		4	3.0		3.0	V
		$V_{DD} = 15V$, $V_O = 13.5V$		4.0		6	4.0		4.0	V
V_{IH}	High Level Input Voltage	$V_{DD} = 5V$, $V_O = 0.5V$	3.5		3.5	3		3.5		V
		$V_{DD} = 10V$, $V_O = 1.0V$	7.0		7.0	6		7.0		V
		$V_{DD} = 15V$, $V_O = 1.5V$	11.0		11.0	9		11.0		V
I_{OL}	Low Level Output Current	$V_{DD} = 5V$, $V_O = 0.4V$	0.52		0.44	0.88		0.36		mA
		$V_{DD} = 10V$, $V_O = 0.5V$	1.3		1.1	2.25		0.9		mA
		$V_{DD} = 15V$, $V_O = 1.5V$	3.6		3.0	8.8		2.4		mA
I_{OH}	High Level Output Current	$V_{DD} = 5V$, $V_O = 4.6V$	−0.52		−0.44	−0.88		−0.36		mA
		$V_{DD} = 10V$, $V_O = 9.5V$	−1.3		−1.1	−2.25		−0.9		mA
		$V_{DD} = 15V$, $V_O = 13.5V$	−3.6		−3.0	−8.8		−2.4		mA
I_{IN}	Input Current	$V_{DD} = 15V$, $V_{IN} = 0V$		−0.30		-10^{-5}	−0.30		−1.0	μA
		$V_{DD} = 15V$, $V_{IN} = 15V$		0.30		10^{-5}	0.30		1.0	μA

AC Electrical Characteristics CD4001BC, CD4001BM

$T_A = 25°C$, Input t_r; $t_f = 20$ ns. $C_L = 50$ pF, $R_L = 200k$. Typical temperature coefficient is 0.3%/°C.

	PARAMETER	CONDITIONS	TYP	MAX	UNITS
t_{PHL}	Propagation Delay Time, High-to-Low Level	$V_{DD} = 5V$	120	250	ns
		$V_{DD} = 10V$	50	100	ns
		$V_{DD} = 15V$	35	70	ns
t_{PLH}	Propagation Delay Time, Low-to-High Level	$V_{DD} = 5V$	110	250	ns
		$V_{DD} = 10V$	50	100	ns
		$V_{DD} = 15V$	35	70	ns
t_{THL}, t_{TLH}	Transition Time	$V_{DD} = 5V$	90	200	ns
		$V_{DD} = 10V$	50	100	ns
		$V_{DD} = 15V$	40	80	ns
C_{IN}	Average Input Capacitance	Any Input	5	7.5	pF
C_{PD}	Power Dissipation Capacity	Any Gate	14		pF

AC Electrical Characteristics CD4011BC, CD4011BM

T_A = 25°C, Input t_r; t_f = 20 ns. C_L = 50 pF, R_L = 200k. Typical Temperature Coefficient is 0.3%/°C.

	PARAMETER	CONDITIONS	TYP	MAX	UNITS
t_{PHL}	Propagation Delay, High-to-Low Level	V_{DD} = 5V	120	250	ns
		V_{DD} = 10V	50	100	ns
		V_{DD} = 15V	35	70	ns
t_{PLH}	Propagation Delay, Low-to-High Level	V_{DD} = 5V	85	250	ns
		V_{DD} = 10V	40	100	ns
		V_{DD} = 15V	30	70	ns
t_{THL}, t_{TLH}	Transition Time	V_{DD} = 5V	90	200	ns
		V_{DD} = 10V	50	100	ns
		V_{DD} = 15V	40	80	ns
C_{IN}	Average Input Capacitance	Any Input	5	7.5	pF
C_{PD}	Power Dissipation Capacity	Any Gate	14		pF

Typical Performance Characteristics

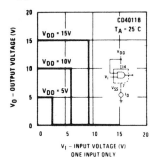

FIGURE 1. Typical Transfer Characteristics

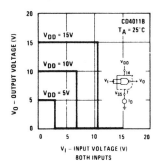

FIGURE 2. Typical Transfer Characteristics

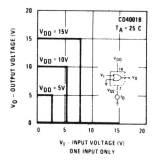

FIGURE 3. Typical Transfer Characteristics

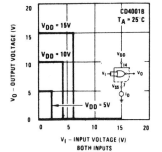

FIGURE 4. Typical Transfer Characteristics

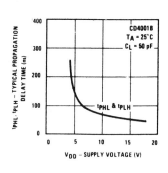

FIGURE 5

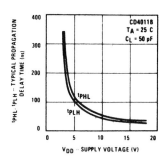

FIGURE 6

Typical Performance Characteristics (Cont'd.)

FIGURE 7

FIGURE 8

FIGURE 9

FIGURE 10

FIGURE 11

FIGURE 12

FIGURE 13

FIGURE 14

National Semiconductor

CD4049M/CD4049C Hex Inverting Buffer
CD4050BM/CD4050BC Hex Non-Inverting Buffer

General Description

These hex buffers are monolithic complementary MOS (CMOS) integrated circuits constructed with N- and P-channel enhancement mode transistors. These devices feature logic level conversion using only one supply voltage (V_{DD}). The input signal high level (V_{IH}) can exceed the V_{DD} supply voltage when these devices are used for logic level conversions. These devices are intended for use as hex buffers, CMOS to DTL/TTL converters, or as CMOS current drivers, and at $V_{DD} = 5.0$V, they can drive directly two DTL/TTL loads over the full operating temperature range.

Features

- Wide supply voltage range   3.0 V to 15 V
- Direct drive to 2 TTL loads at 5.0 V over full temperature range
- High source and sink current capability
- Special input protection permits input voltages greater than V_{DD}

Applications

- CMOS hex inverter/buffer
- CMOS to DTL/TTL hex converter
- CMOS current "sink" or "source" driver
- CMOS high-to-low logic level converter

Connection Diagrams

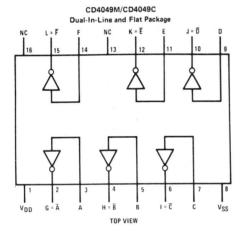

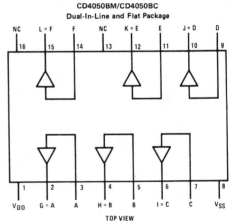

Schematic Diagrams

Absolute Maximum Ratings

(Notes 1 and 2)

V_{DD} Supply Voltage	−0.5V to +18V
V_{IN} Input Voltage	−0.5V to +18V
V_{OUT} Voltage at Any Output Pin	−0.5V to V_{DD} +0.5V
T_S Storage Temperature Range	−65°C to +150°C
P_D Package Dissipation	500 mW
T_L Lead Temperature (Soldering, 10 seconds)	300°C

Recommended Operating Conditions

(Note 2)

V_{DD} Supply Voltage	3V to 15V
V_{IN} Input Voltage	0V to 15V
V_{OUT} Voltage at Any Output Pin	0 to V_{DD}
T_A Operating Temperature Range	
CD4049M, CD4050BM	−55°C to +125°C
CD4049C, CD4050BC	−40°C to +85°C

DC Electrical Characteristics CD4049C/CD4050BC (Note 2)

PARAMETER		CONDITIONS	−40°C		25°C			85°C		UNITS
			MIN	MAX	MIN	TYP	MAX	MIN	MAX	
I_{DD}	Quiescent Device Current	V_{DD} = 5V		4		0.03	4.0		30	μA
		V_{DD} = 10V		8		0.05	8.0		60	μA
		V_{DD} = 15V		16		0.07	16.0		120	μA
V_{OL}	Low Level Output Voltage	$V_{IH} = V_{DD}$, $V_{IL} = 0V$, $|I_O| < 1 \mu A$								
		V_{DD} = 5V		0.05	0	0.05			0.05	V
		V_{DD} = 10V		0.05	0	0.05			0.05	V
		V_{DD} = 15V		0.05	0	0.05			0.05	V
V_{OH}	High Level Output Voltage	$V_{IH} = V_{DD}$, $V_{IL} = 0V$, $|I_O| < 1 \mu A$								
		V_{DD} = 5V	4.95		4.95	5		4.95		V
		V_{DD} = 10V	9.95		9.95	10		9.95		V
		V_{DD} = 15V	14.95		14.95	15		14.95		V
V_{IL}	Low Level Input Voltage (CD4050BC Only)	$|I_O| < 1 \mu A$								
		V_{DD} = 5V, V_O = 0.5V		1.5		2.25	1.5		1.5	V
		V_{DD} = 10V, V_O = 1V		3.0		4.5	3.0		3.0	V
		V_{DD} = 15V, V_O = 1.5V		4.0		6.75	4.0		4.0	V
V_{IL}	Low Level Input Voltage (CD4049C Only)	$|I_O| < 1 \mu A$								
		V_{DD} = 5V, V_O = 4.5V		1.0		1.5	1.0		1.0	V
		V_{DD} = 10V, V_O = 9V		2.0		2.5	2.0		2.0	V
		V_{DD} = 15V, V_O = 13.5V		3.0		3.5	3.0		3.0	V
V_{IH}	High Level Input Voltage (CD4050BC Only)	$|I_O| < 1 \mu A$								
		V_{DD} = 5V, V_O = 4.5V	3.5		3.5	2.75		3.5		V
		V_{DD} = 10V, V_O = 9V	7.0		7.0	5.5		7.0		V
		V_{DD} = 15V, V_O = 13.5V	11.0		11.0	8.25		11.0		V
V_{IH}	High Level Input Voltage (CD4049C Only)	$|I_O| < 1 \mu A$								
		V_{DD} = 5V, V_O = 0.5V	4.0		4.0	3.5		4.0		V
		V_{DD} = 10V, V_O = 1V	8.0		8.0	7.5		8.0		V
		V_{DD} = 15V, V_O = 1.5V	12.0		12.0	11.5		12.0		V
I_{OL}	Low Level Output Current (Note 3)	$V_{IH} = V_{DD}$, $V_{IL} = 0V$								
		V_{DD} = 5V, V_O = 0.4V	4.6		4.0	5		3.2		mA
		V_{DD} = 10V, V_O = 0.5V	9.8		8.5	12		6.8		mA
		V_{DD} = 15V, V_O = 1.5V	29		25	40		20		mA
I_{OH}	High Level Output Current (Note 3)	$V_{IH} = V_{DD}$, $V_{IL} = 0V$								
		V_{DD} = 5V, V_O = 4.6V	−1.0		−0.9	−1.6		−0.72		mA
		V_{DD} = 10V, V_O = 9.5V	−2.1		−1.9	−3.6		−1.5		mA
		V_{DD} = 15V, V_O = 13.5V	−7.1		−6.2	−12		−5		mA
I_{IN}	Input Current	V_{DD} = 15V, V_{IN} = 0V	−0.3		−0.3	-10^{-5}			−1.0	μA
		V_{DD} = 15V, V_{IN} = 15V	0.3		0.3	10^{-5}			1.0	μA

DC Electrical Characteristics CD4049C/CD4050BC (Note 2)

PARAMETER		CONDITIONS	−40°C		25°C			85°C		UNITS
			MIN	MAX	MIN	TYP	MAX	MIN	MAX	
I_{DD}	Quiescent Device Current	V_{DD} = 5V		4		0.03	4.0		30	μA
		V_{DD} = 10V		8		0.05	8.0		60	μA
		V_{DD} = 15V		16		0.07	16.0		120	μA
V_{OL}	Low Level Output Voltage	$V_{IH} = V_{DD}, V_{IL} = 0V,$ $|I_O| < 1\,\mu A$								
		V_{DD} = 5V		0.05		0	0.05		0.05	V
		V_{DD} = 10V		0.05		0	0.05		0.05	V
		V_{DD} = 15V		0.05		0	0.05		0.05	V
V_{OH}	High Level Output Voltage	$V_{IH} = V_{DD}, V_{IL} = 0V,$ $|I_O| < 1\,\mu A$								
		V_{DD} = 5V	4.95		4.95	5		4.95		V
		V_{DD} = 10V	9.95		9.95	10		9.95		V
		V_{DD} = 15V	14.95		14.95	15		14.95		V
V_{IL}	Low Level Input Voltage (CD4050BC Only)	$|I_O| < 1\,\mu A$								
		V_{DD} = 5V, V_O = 0.5V		1.5		2.25	1.5		1.5	V
		V_{DD} = 10V, V_O = 1V		3.0		4.5	3.0		3.0	V
		V_{DD} = 15V, V_O = 1.5V		4.0		6.75	4.0		4.0	V
V_{IL}	Low Level Input Voltage (CD4049C Only)	$|I_O| < 1\,\mu A$								
		V_{DD} = 5V, V_O = 4.5V		1.0		1.5	1.0		1.0	V
		V_{DD} = 10V, V_O = 9V		2.0		2.5	2.0		2.0	V
		V_{DD} = 15V, V_O = 13.5V		3.0		3.5	3.0		3.0	V
V_{IH}	High Level Input Voltage (CD4050BC Only)	$|I_O| < 1\,\mu A$								
		V_{DD} = 5V, V_O = 4.5V	3.5		3.5	2.75		3.5		V
		V_{DD} = 10V, V_O = 9V	7.0		7.0	5.5		7.0		V
		V_{DD} = 15V, V_O = 13.5V	11.0		11.0	8.25		11.0		V
V_{IH}	High Level Input Voltage (CD4049C Only)	$|I_O| < 1\,\mu A$								
		V_{DD} = 5V, V_O = 0.5V	4.0		4.0	3.5		4.0		V
		V_{DD} = 10V, V_O = 1V	8.0		8.0	7.5		8.0		V
		V_{DD} = 15V, V_O = 1.5V	12.0		12.0	11.5		12.0		V
I_{OL}	Low Level Output Current (Note 3)	$V_{IH} = V_{DD}, V_{IL} = 0V$								
		V_{DD} = 5V, V_O = 0.4V	4.6		4.0	5		3.2		mA
		V_{DD} = 10V, V_O = 0.5V	9.8		8.5	12		6.8		mA
		V_{DD} = 15V, V_O = 1.5V	29		25	40		20		mA
I_{OH}	High Level Output Current (Note 3)	$V_{IH} = V_{DD}, V_{IL} = 0V$								
		V_{DD} = 5V, V_O = 4.6V	−1.0		−0.9	−1.6		−0.72		mA
		V_{DD} = 10V, V_O = 9.5V	−2.1		−1.9	−3.6		−1.5		mA
		V_{DD} = 15V, V_O = 13.5V	−7.1		−6.2	−12		−5		mA
I_{IN}	Input Current	V_{DD} = 15V, V_{IN} = 0V	−0.3		−0.3	-10^{-5}			−1.0	μA
		V_{DD} = 15V, V_{IN} = 15V	0.3		0.3	10^{-5}			1.0	μA

AC Electrical Characteristics CD4049M/CD4049C

$T_A = 25°C$, $C_L = 50$ pF, $R_L = 200$k, $t_r = t_f = 20$ ns, unless otherwise specified.

	PARAMETER	CONDITIONS	MIN	TYP	MAX	UNITS
t_{PHL}	Propagation Delay Time High-to-Low Level	$V_{DD} = 5V$		30	65	ns
		$V_{DD} = 10V$		20	40	ns
		$V_{DD} = 15V$		15	30	ns
t_{PLH}	Propagation Delay Time Low-to-High Level	$V_{DD} = 5V$		45	85	ns
		$V_{DD} = 10V$		25	45	ns
		$V_{DD} = 15V$		20	35	ns
t_{THL}	Transition Time High-to-Low Level	$V_{DD} = 5V$		30	60	ns
		$V_{DD} = 10V$		20	40	ns
		$V_{DD} = 15V$		15	30	ns
t_{TLH}	Transition Time Low-to-High Level	$V_{DD} = 5V$		60	120	ns
		$V_{DD} = 10V$		30	55	ns
		$V_{DD} = 15V$		25	45	ns
C_{IN}	Input Capacitance	Any Input		15	22.5	pF

AC Electrical Characteristics CD4050BM/CD4050BC

$T_A = 25°C$, $C_L = 50$ pF, $R_L = 200$k, $t_r = t_f = 20$ ns, unless otherwise specified.

	PARAMETER	CONDITIONS	MIN	TYP	MAX	UNITS
t_{PHL}	Propagation Delay Time High-to-Low Level	$V_{DD} = 5V$		60	110	ns
		$V_{DD} = 10V$		25	55	ns
		$V_{DD} = 15V$		20	30	ns
t_{PLH}	Propagation Delay Time Low-to-High Level	$V_{DD} = 5V$		60	120	ns
		$V_{DD} = 10V$		30	55	ns
		$V_{DD} = 15V$		25	45	ns
t_{THL}	Transition Time High-to-Low Level	$V_{DD} = 5V$		30	60	ns
		$V_{DD} = 10V$		20	40	ns
		$V_{DD} = 15V$		15	30	ns
t_{TLH}	Transition Time Low-to-High Level	$V_{DD} = 5V$		60	120	ns
		$V_{DD} = 10V$		30	55	ns
		$V_{DD} = 15V$		25	45	ns
C_{IN}	Input Capacitance	Any Input		5	7.5	pF

Switching Time Waveforms

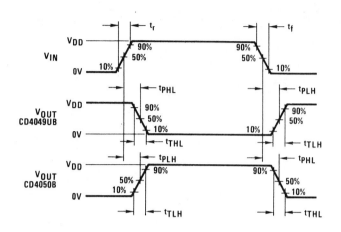

Typical Applications

CMOS to TTL or CMOS at a Lower V$_{DD}$

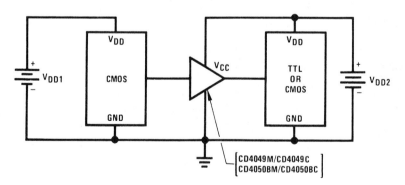

Note: V$_{DD1}$ ≥ V$_{DD2}$

Note: In the case of the CD4049M/CD4049C
the output drive capability increases with increasing
input voltage. E.g., If V$_{DD1}$= 10V the CD4049M/
CD4049C could drive 4 TTL loads.

National Semiconductor

CD4051BM/CD4051BC Single 8-Channel Analog Multiplexer/Demultiplexer
CD4052BM/CD4052BC Dual 4-Channel Analog Multiplexer/Demultiplexer
CD4053BM/CD4053BC Triple 2-Channel Analog Multiplexer/Demultiplexer

General Description

These analog multiplexers/demultiplexers are digitally controlled analog switches having low "ON" impedance and very low "OFF" leakage currents. Control of analog signals up to $15\,V_{p-p}$ can be achieved by digital signal amplitudes of 3-15 V. For example, if $V_{DD} = 5\,V$, $V_{SS} = 0\,V$ and $V_{EE} = -5\,V$, analog signals from $-5\,V$ to $+5\,V$ can be controlled by digital inputs of 0-5 V. The multiplexer circuits dissipate extremely low quiescent power over the full $V_{DD} - V_{SS}$ and $V_{DD} - V_{EE}$ supply voltage ranges, independent of the logic state of the control signals. When a logical "1" is present at the inhibit input terminal all channels are "OFF".

CD4051BM/CD4051BC is a single 8-channel multiplexer having three binary control inputs. A, B, and C, and an inhibit input. The three binary signals select 1 of 8 channels to be turned "ON" and connect the input to the output.

CD4052BM/CD4052BC is a differential 4-channel multiplexer having two binary control inputs, A and B, and an inhibit input. The two binary input signals select 1 or 4 pairs of channels to be turned on and connect the differential analog inputs to the differential outputs.

CD4053BM/CD4053BC is a triple 2-channel multiplexer having three separate digital control inputs, A, B, and C, and an inhibit input. Each control input selects one of a pair of channels which are connected in a single-pole double-throw configuration.

Features

- Wide range of digital and analog signal levels: digital 3-15 V, analog to $15\,V_{p-p}$
- Low "ON" resistance: $80\,\Omega$ (typ.) over entire $15\,V_{p-p}$ signal-input range for $V_{DD} - V_{EE} = 15\,V$
- High "OFF" resistance: channel leakage of $\pm 10\,pA$ (typ.) at $V_{DD} - V_{EE} = 10\,V$
- Logic level conversion for digital addressing signals of 3-15 V ($V_{DD} - V_{SS} = 3$-15 V) to switch analog signals to $15\,V_{p-p}$ ($V_{DD} - V_{EE} = 15\,V$)
- Matched switch characteristics: $\Delta R_{ON} = 5\,\Omega$ (typ.) for $V_{DD} - V_{EE} = 15\,V$
- Very low quiescent power dissipation under all digital-control input and supply conditions: $1\,\mu W$ (typ.) at $V_{DD} - V_{SS} = V_{DD} - V_{EE} = 10\,V$
- Binary address decoding on chip

Connection Diagrams

CD4051BM/CD4051BC

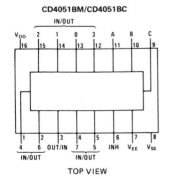

TOP VIEW

CD4052BM/CD4052BC

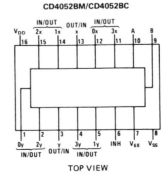

TOP VIEW

CD4053BM/CD4053BC

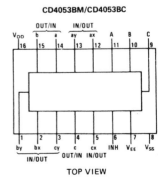

TOP VIEW

TTL
MSI

TYPES SN54LS138, SN54LS139, SN54S138, SN54S139,
SN74LS138, SN74LS139, SN74S138, SN74S139
DECODERS/DEMULTIPLEXERS

BULLETIN NO. DL-S 7611804, DECEMBER 1972–REVISED OCTOBER 1976

- **Designed Specifically for High-Speed:**
  Memory Decoders
  Data Transmission Systems

- **'S138 and 'LS138 3-to-8-Line Decoders**
  Incorporate 3 Enable Inputs to Simplify
  Cascading and/or Data Reception

- **'S139 and 'LS139 Contain Two Fully**
  Independent 2-to-4-Line Decoders/
  Demultiplexers

- **Schottky Clamped for High Performance**

TYPE	TYPICAL PROPAGATION DELAY (3 LEVELS OF LOGIC)	TYPICAL POWER DISSIPATION
'LS138	22 ns	32 mW
'S138	8 ns	245 mW
'LS139	22 ns	34 mW
'S139	7.5 ns	300 mW

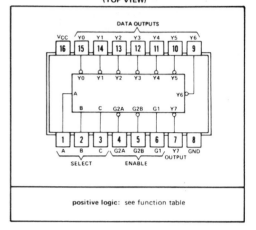

SN54LS138, SN54S138 . . . J OR W PACKAGE
SN74LS138, SN74S138 . . . J OR N PACKAGE
(TOP VIEW)

positive logic: see function table

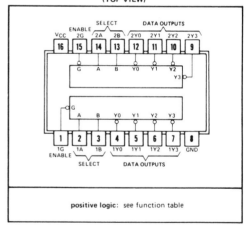

SN54LS139, SN54S139 . . . J OR W PACKAGE
SN74LS139, SN74S139 . . . J OR N PACKAGE
(TOP VIEW)

positive logic: see function table

description

These Schottky-clamped TTL MSI circuits are designed to be used in high-performance memory-decoding or data-routing applications requiring very short propagation delay times. In high-performance memory systems these decoders can be used to minimize the effects of system decoding. When employed with high-speed memories utilizing a fast-enable circuit the delay times of these decoders and the enable time of the memory are usually less than the typical access time of the memory. This means that the effective system delay introduced by the Schottky-clamped system decoder is negligible.

The 'LS138 and 'S138 decode one-of-eight lines dependent on the conditions at the three binary select inputs and the three enable inputs. Two active-low and one active-high enable inputs reduce the need for external gates or inverters when expanding. A 24-line decoder can be implemented without external inverters and a 32-line decoder requires only one inverter. An enable input can be used as a data input for demultiplexing applications.

The 'LS139 and 'S139 comprise two individual two-line-to-four-line decoders in a single package. The active-low enable input can be used as a data line in demultiplexing applications.

All of these decoders/demultiplexers feature fully buffered inputs each of which represents only one normalized Series 54LS/74LS load ('LS138, 'LS139) or one normalized Series 54S/74S load ('S138, 'S139) to its driving circuit. All inputs are clamped with high-performance Schottky diodes to suppress line-ringing and simplify system design. Series 54LS and 54S devices are characterized for operation over the full military temperature range of −55°C to 125°C; Series 74LS and 74S devices are characterized for 0°C to 70°C industrial systems.

functional block diagrams and logic

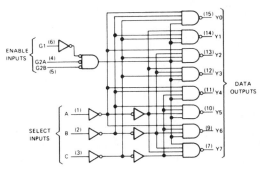

'LS138, 'S138

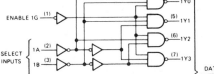

'LS139, 'S139

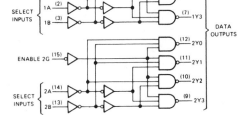

'LS138, 'S138 FUNCTION TABLE

INPUTS					OUTPUTS							
ENABLE		SELECT										
G1	G2*	C	B	A	Y0	Y1	Y2	Y3	Y4	Y5	Y6	Y7
X	H	X	X	X	H	H	H	H	H	H	H	H
L	X	X	X	X	H	H	H	H	H	H	H	H
H	L	L	L	L	L	H	H	H	H	H	H	H
H	L	L	L	H	H	L	H	H	H	H	H	H
H	L	L	H	L	H	H	L	H	H	H	H	H
H	L	L	H	H	H	H	H	L	H	H	H	H
H	L	H	L	L	H	H	H	H	L	H	H	H
H	L	H	L	H	H	H	H	H	H	L	H	H
H	L	H	H	L	H	H	H	H	H	H	L	H
H	L	H	H	H	H	H	H	H	H	H	H	L

*G2 = G2A + G2B

H = high level, L = low level, X = irrelevant

'LS139, 'S139 (EACH DECODER/DEMULTIPLEXER) FUNCTION TABLE

INPUTS			OUTPUTS			
ENABLE	SELECT					
G	B	A	Y0	Y1	Y2	Y3
H	X	X	H	H	H	H
L	L	L	L	H	H	H
L	L	H	H	L	H	H
L	H	L	H	H	L	H
L	H	H	H	H	H	L

H = high level, L = low level, X = irrelevant

schematics of inputs and outputs

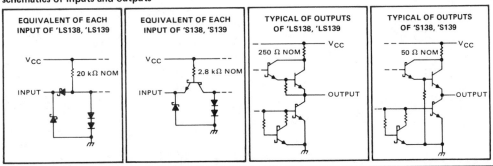

REVISED DECEMBER 1980

absolute maximum ratings over operating free-air temperature range (unless otherwise noted)

Supply voltage, V_{CC} (see Note 1) . . . . 7 V
Input voltage . . . . 7 V
Operating free-air temperature range: SN54LS138, SN54LS139 Circuits . . . . −55°C to 125°C
SN74LS138, SN74LS139 Circuits . . . . 0°C to 70°C
Storage temperature range . . . . −65°C to 150°C

NOTE 1: Voltage values are with respect to network ground terminal.

recommended operating conditions

	SN54LS138 SN54LS139			SN74LS138 SN74LS139			UNIT
	MIN	NOM	MAX	MIN	NOM	MAX	
Supply voltage, V_{CC}	4.5	5	5.5	4.75	5	5.25	V
High-level output current, I_{OH}			−400			−400	μA
Low-level output current, I_{OL}			4			8	mA
Operating free-air temperature, T_A	−55		125	0		70	°C

electrical characteristics over recommended operating free-air temperature range (unless otherwise noted)

PARAMETER		TEST CONDITIONS[†]		SN54LS138 SN54LS139			SN74LS138 SN74LS139			UNIT
				MIN	TYP[‡]	MAX	MIN	TYP[‡]	MAX	
V_{IH}	High-level input voltage			2			2			V
V_{IL}	Low-level input voltage					0.7			0.8	V
V_{IK}	Input clamp voltage	V_{CC} = MIN, I_I = −18 mA				−1.5			−1.5	V
V_{OH}	High-level output voltage	V_{CC} = MIN, V_{IH} = 2 V, V_{IL} = $V_{IL max}$, I_{OH} = −400 μA		2.5	3.4		2.7	3.4		V
V_{OL}	Low-level output voltage	V_{CC} = MIN, V_{IH} = 2 V, V_{IL} = $V_{IL max}$	I_{OL} = 4 mA		0.25	0.4		0.25	0.4	V
			I_{OL} = 8 mA					0.35	0.5	
I_I	Input current at maximum input voltage	V_{CC} = MAX, V_I = 7 V				0.1			0.1	mA
I_{IH}	High-level input current	V_{CC} = MAX, V_I = 2.7 V				20			20	μA
I_{IL}	Low-level input current	V_{CC} = MAX, V_I = 0.4 V				−0.4			−0.4	mA
I_{OS}	Short circuit output current §	V_{CC} = MAX	'LS138	−20		−100	−20		−100	mA
			'LS139	−6		−40	−5		−42	
I_{OS}	Supply current	V_{CC} = MAX, Outputs enabled and open	'LS138		6.3	10		6.3	10	mA
			'LS139		6.8	11		6.8	11	

[†]For conditions shown as MIN or MAX, use the appropriate value specified under recommended operating conditions for the applicable device type.
[‡]All typical values are at V_{CC} = 5 V, T_A = 25°C.
§ Not more than one output should be shorted at a time.

switching characteristics, V_{CC} = 5 V, T_A = 25°C

PARAMETER[¶]	FROM (INPUT)	TO (OUTPUT)	LEVELS OF DELAY	TEST CONDITIONS	SN54LS138 SN74LS138			SN54LS139 SN74LS139			UNIT
					MIN	TYP	MAX	MIN	TYP	MAX	
t_{PLH}	Binary Select	Any	2	C_L = 15 pF, R_L = 2 kΩ, See Note 2		13	20		13	20	ns
t_{PHL}						27	41		22	33	ns
t_{PLH}			3			18	27		18	29	ns
t_{PHL}						26	39		25	38	ns
t_{PLH}	Enable	Any	2			12	18		16	24	ns
t_{PHL}						21	32		21	32	ns
t_{PLH}			3			17	26				ns
t_{PHL}						25	38				ns

[¶] t_{PLH} ≡ propagation delay time, low-to-high-level output; t_{PHL} ≡ propagation delay time, high-to-low-level output.
NOTE 2: Load circuits and waveforms are shown on page 3-11.

absolute maximum ratings over operating free-air temperature range (unless otherwise noted)

Supply voltage, V_{CC} (see Note 1) . . . . 7 V

Input voltage . . . . 5.5 V

Operating free-air temperature range: SN54S138, SN54S139 Circuits . . . . $-55°$C to $125°$C

SN74S138, SN74S139 Circuits . . . . $0°$C to $70°$C

Storage temperature range . . . . $-65°$C to $150°$C

NOTE 1: Voltage values are with respect to network ground terminal.

recommended operating conditions

	SN54S138 SN54S139			SN74S138 SN74S139			UNIT
	MIN	NOM	MAX	MIN	NOM	MAX	
Supply voltage, V_{CC}	4.5	5	5.5	4.75	5	5.25	V
High-level output current, I_{OH}			-1			-1	mA
Low-level output current, I_{OL}			20			20	mA
Operating free-air temperature, T_A	-55		125	0		70	$°$C

electrical characteristics over recommended operating free-air temperature range (unless otherwise noted)

PARAMETER	TEST CONDITIONS[†]		SN54S138 SN74S138			SN54S139 SN74S139			UNIT
			MIN	TYP[‡]	MAX	MIN	TYP[‡]	MAX	
V_{IH} High-level input voltage			2			2			V
V_{IL} Low-level input voltage					0.8			0.8	V
V_{IK} Input clamp voltage	V_{CC} = MIN, I_I = -18 mA				-1.2			-1.2	V
V_{OH} High-level output voltage	V_{CC} = MIN, V_{IH} = 2 V,	SN54S'	2.5	3.4		2.5	3.4		V
	V_{IL} = 0.8 V, I_{OH} = -1 mA	SN74S'	2.7	3.4		2.7	3.4		
V_{OL} Low-level output voltage	V_{CC} = MIN, V_{IH} = 2 V, V_{IL} = 0.8 V, I_{OL} = 20 mA				0.5			0.5	V
I_I Input current at maximum input voltage	V_{CC} = MAX, V_I = 5.5 V				1			1	mA
I_{IH} High-level input current	V_{CC} = MAX, V_I = 2.7 V				50			50	μA
I_{IL} Low-level input current	V_{CC} = MAX, V_I = 0.5 V				-2			-2	mA
I_{OS} Short-circuit output current[§]	V_{CC} = MAX		-40		-100	-40		-100	mA
I_{CC} Supply current	V_{CC} = MAX, Outputs enabled and open			49	74		60	90	mA

[†]For conditions shown as MIN or MAX, use the appropriate value specified under recommended operating conditions for the applicable device type.

[‡]All typical values are at V_{CC} = 5 V, T_A = $25"$C.

[§]Not more than one output should be shorted at a time, and duration of the short-circuit test should not exceed one second.

switching characteristics, V_{CC} = 5 V, T_A = $25°$C

PARAMETER[¶]	FROM (INPUT)	TO (OUTPUT)	LEVELS OF DELAY	TEST CONDITIONS	SN54S138, SN74S138			SN54S139 SN74S139			UNIT
					MIN	TYP	MAX	MIN	TYP	MAX	
t_{PLH}	Binary select	Any	2			4.5	7		5	7.5	ns
t_{PHL}						7	10.5		6.5	10	
t_{PLH}			3	C_L = 15 pF, R_L = 280 Ω, See Note 3		7.5	12		7	12	ns
t_{PHL}						8	12		8	12	
t_{PLH}	Enable	Any	2			5	8		5	8	ns
t_{PHL}						7	11		6.5	10	
t_{PLH}			3			7	11				ns
t_{PHL}						7	11				

[¶] t_{PLH} $\equiv$ propagation delay time, low-to-high-level output

t_{PHL} $\equiv$ propagation delay time, high-to-low-level output

NOTE 3: Load circuits and waveforms are shown on page 3-10.

- Cascading Circuitry Provided Internally

- Synchronous Operation

- Individual Preset to Each Flip-Flop

- Fully Independent Clear Input

- Typical Maximum Input Count Frequency . . . 32 MHz

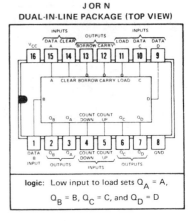

**J OR N
DUAL-IN-LINE PACKAGE (TOP VIEW)**

logic: Low input to load sets Q_A = A,
Q_B = B, Q_C = C, and Q_D = D

description

These monolithic circuits are synchronous reversible (up/down) counters having a complexity of 55 equivalent gates. The SN54192 and SN74192 are BCD counters and the SN54193 and SN74193 are 4-bit binary counters. Synchronous operation is provided by having all flip-flops clocked simultaneously so that the outputs change coincidently with each other when so instructed by the steering logic. This mode of operation eliminates the output counting spikes which are normally associated with asynchronous (ripple-clock) counters.

The outputs of the four master-slave flip-flops are triggered by a low-to-high-level transition of either count (clock) input. The direction of counting is determined by which count input is pulsed while the other count input is high.

All four counters are fully programmable; that is, the outputs may be preset to any state by entering the desired data at the data inputs while the load input is low. The output will change to agree with the data inputs independently of the count pulses. This feature allows the counters to be used as modulo-N dividers by simply modifying the count length with the preset inputs.

A clear input has been provided which forces all outputs to the low level when a high level is applied. The clear function is independent of the count and load inputs. An input buffer has been placed on the clear, count, and load inputs to lower the drive requirements to one normalized Series 54/74 load. This is important when the output of the driving circuitry is somewhat limited.

These counters were designed to be cascaded without the need for external circuitry. Both borrow and carry outputs are available to cascade both the up- and down-counting functions. The borrow output produces a pulse equal in width to the count-down input when the counter underflows. Similarly, the carry output produces a pulse equal in width to the count-up input when an overflow condition exists. The counters can then be easily cascaded by feeding the borrow and carry outputs to the count-down and count-up inputs respectively of the succeeding counter.

Power dissipation is typically 325 milliwatts for either the decade or binary version. Maximum input count frequency is typically 32 megahertz and is guaranteed to be 25 MHz minimum. All inputs are buffered and represent only one normalized Series 54/74 load. Input clamping diodes are provided to minimize transmission-line effects and thereby simplify system design. The SN54192 and SN54193 are characterized for operation over the full military temperature range of $-55°C$ to $125°C$; the SN74192 and SN74193 are characterized for operation from $0°C$ to $70°C$.

absolute maximum ratings over operating free-air temperature range (unless otherwise noted)

Supply voltage V_{CC} (see Note 1) . . . . 7 V
Input voltage (see Note 1) . . . . 5.5 V
Operating free-air temperature range: SN54192 and SN54193 Circuits . . . . $-55°C$ to $125°C$
                                  SN74192 and SN74193 Circuits . . . . $0°C$ to $70°C$
Storage temperature range . . . . $-65°C$ to $150°C$

NOTE 1: Voltage values are with respect to network ground terminal.

recommended operating conditions

	SN54192, SN54193			SN74192, SN74193			UNIT
	MIN	NOM	MAX	MIN	NOM	MAX	
Supply voltage V_{CC}	4.5	5	5.5	4.75	5	5.25	V
Normalized fan-out from each output, N			10			10	
Input count frequency, f_{count}	0		25	0		25	MHz
Width of any input pulse, t_w	20			20			ns
Data setup time, t_{setup} (see Figure 7 and Note 2)	20			20			ns
Data hold time, t_{hold} (see Note 3)	0			0			ns
Operating free-air temperature range, T_A	−55	25	125	0	25	70	°C

NOTES: 2. Setup time is the interval immediately preceding the positive-going edge of the load pulse during which interval the data to be recognized must be maintained at the input to ensure its recognition.
3. Hold time is the interval immediately following the positive-going edge of the load pulse during which interval the data to be recognized must be maintained at the input to ensure its recognition.

electrical characteristics over recommended operating free-air temperature range (unless otherwise noted)

	PARAMETER	TEST FIGURE	TEST CONDITIONS[†]	SN54192, SN54193			SN74192, SN74193			UNIT
				MIN	TYP[‡]	MAX	MIN	TYP[‡]	MAX	
V_{IH}	High-level input voltage	1 and 2		2			2			V
V_{IL}	Low-level input voltage	1 and 2				0.8			0.8	V
V_{OH}	High-level output voltage	1	$V_{CC} = MIN$, $V_{IH} = 2$ V, $V_{IL} = 0.8$ V, $I_{OH} = -400 \mu A$	2.4			2.4			V
V_{OL}	Low-level output voltage	2	$V_{CC} = MIN$, $V_{IH} = 2$ V, $V_{IL} = 0.8$ V, $I_{OL} = 16$ mA			0.4			0.4	V
I_{IH}	High-level input current	3	$V_{CC} = MAX$, $V_I = 2.4$ V			40			40	μA
			$V_{CC} = MAX$, $V_I = 5.5$ V			1			1	mA
I_{IL}	Low-level input current	4	$V_{CC} = MAX$, $V_I = 0.4$ V			−1.6			−1.6	mA
I_{OS}	Short-circuit output current[§]	5	$V_{CC} = MAX$	−20		−65	−18		−65	mA
I_{CC}	Supply current	6	$V_{CC} = MAX$		65	89		65	102	mA

[†]For conditions shown as MIN or MAX, use the appropriate value specified under recommended operating conditions for the applicable device type.
[‡] All typical values are at $V_{CC} = 5$ V, $T_A = 25°C$.
[§] Not more than one output should be shorted at a time.

switching characteristics, $V_{CC} = 5$ V, $T_A = 25°C$, N = 10

PARAMETER[¶]	FROM INPUT	TO OUTPUT	TEST FIGURE	TEST CONDITIONS	MIN	TYP	MAX	UNIT
f_{max}					25	32		MHz
t_{setup}			7			14	20	ns
t_{PLH}	Count-up	Carry	8			17	26	ns
t_{PHL}						16	24	
t_{PLH}	Count-down	Borrow	8	$C_L = 15$ pF, $R_L = 400 \Omega$		16	24	ns
t_{PHL}						16	24	
t_{PLH}	Either Count	Q	8			25	38	ns
t_{PHL}						31	47	
t_{PLH}	Load	Q	7			27	40	ns
t_{PLH}						29	40	
t_{PHL}	Clear	Q	7			22	35	ns

[¶] f_{max} = maximum clock frequency
t_{PLH} = propagation delay time, low-to-high-level output
t_{PHL} = propagation delay time, high-to-low-level output

SN54192, SN74192 DECADE COUNTER

functional block diagram

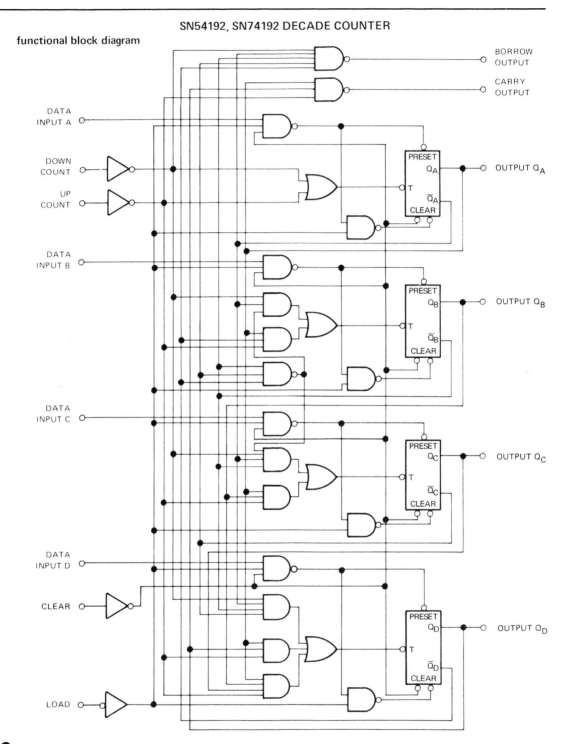

SN54193, SN74193 BINARY COUNTER

functional block diagram

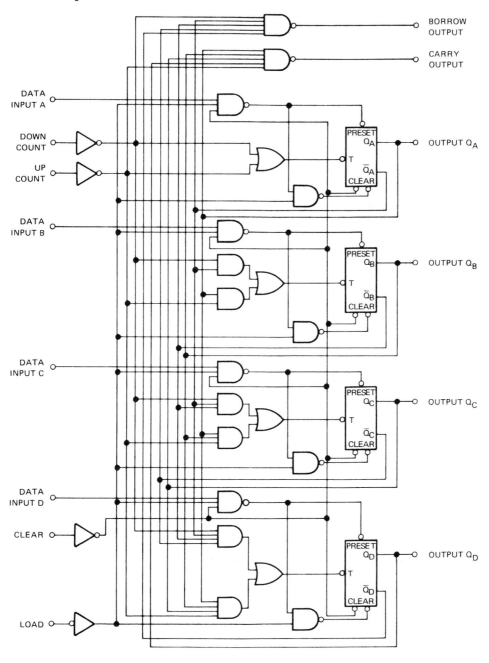

SN54192, SN74192 DECADE COUNTERS

typical clear, load, and count sequences

Illustrated below is the following sequence:

1. Clear outputs to zero.
2. Load (pr set) to BCD seven.
3. Count up to eight, nine, carry, zero, one, and two.
4. Count down to one, zero, borrow, nine, eight, and seven.

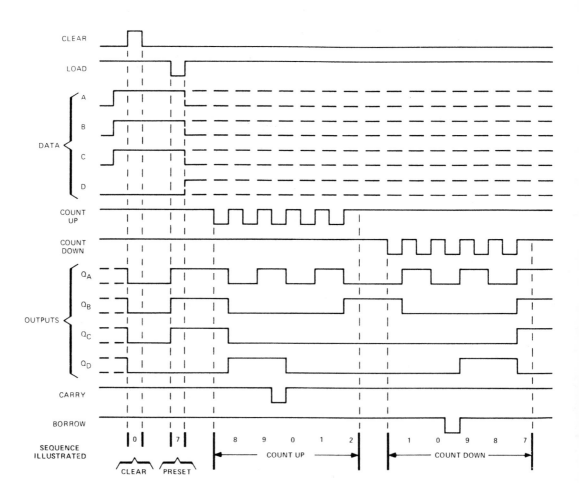

NOTES: A. Clear overrides load, data, and count inputs.

         B. When counting up, count-down input must be high; when counting down, count up input must be high.

SN54193, SN74193 BINARY COUNTERS

typical clear, load, and count sequences

Illustrated below is the following sequence:

1. Clear outputs to zero.
2. Load (preset) to BCD thirteen.
3. Count up to fourteen, fifteen, carry, zero, one, and two.
4. Count down to one, zero, borrow, fifteen, fourteen, and thirteen.

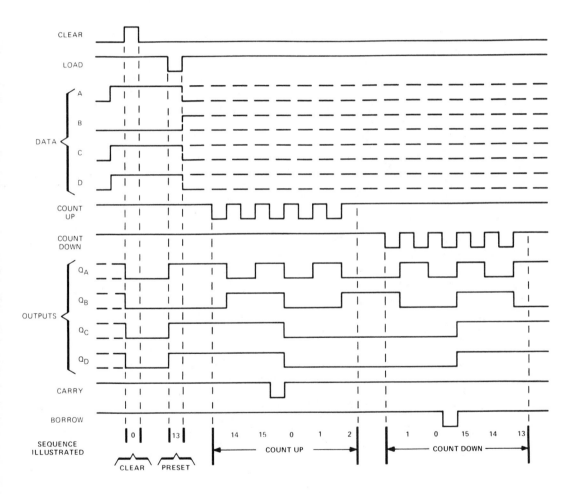

NOTES: A. Clear overrides load, data, and count inputs.
B. When counting up, count-down input must be high; when counting down, count-up input must be high.

APPENDIX C: GLOSSARY

Access Time.   The delay in a memory between the time the address is applied and valid data are obtained at the output.

Accumulator.   A register in a microprocessor that holds the results of the operations performed by the ALU.

Accuracy.   The percentage of error between the expected and actual outputs of an ADC or DAC.

Active Pullup.   A MOSFET that represents a fixed resistance value and is connected to V_{DD}.

ADC.   *See* Analog-to-Digital Converter.

Address.   The memory location of a certain piece of data or connection to an I/O device.

Address Bus.   The group of signals that make up the address.

ALU.   The Arithmetic and Logic Unit, which performs the arithmetic and logic operations in a CPU or microprocessor.

Analog.   Having an infinite number or a continuum of values. (*See also* Digital.)

Analog-to-Digital Converter (ADC).   A circuit that converts an analog voltage into a binary or BCD number.

Analog Switch.   A transmission gate configured to switch analog signals instead of digital signals. In effect, it is a digitally controlled switch.

453

AND Gate. One of the three primary gates, its output is 1 if and only if all of its inputs are 1.

ASCII (American Standard Code for Information Interchange). A computer code used to encode alphabetic and numeric characters.

Assembler. A program that converts symbolic machine codes (op-codes) into binary machine codes.

Assembly Language. The symbolic programming language used by the assembler.

Astable Multivibrator. A circuit that has no stable states and, as a result, produces a digital squarewave at its output.

Asynchronous Counter. A counter whose outputs do not change in synchronization.

Asynchronous Latch. A basic storage device that can store a logic 1 or 0 without the use of a CLOCK.

Auxiliary Memory. Additional memory in a computer system, such as disk, tape, or magnetic bubble memory, that is not directly addressable by the CPU.

BASIC. A high-level computer language commonly found in microcomputer systems.

BCD (Binary Coded Decimal). A coding system in which four bits are used to represent a digit in the decimal number system (0–9). The remaining six 4-bit binary values (hex A–F) are disallowed.

BCD Counter. A counter that counts in BCD order.

Bidirectional Current Driver. A device that passes current in both directions through a selection line in a magnetic core memory.

Bilateral Switch. A transmission gate designed so that the terminals can be used as either inputs or outputs.

Binary. The base two number system.

Bipolar Transistor. A transistor that uses the diode effect at junctions for control.

Binary Weighted Ladder. A DAC composed of a series of resistors, each doubling the value of the one before.

Biquinary Counter. A counter constructed from a modulo-2 and a modulo-5 counter, which generates a binary-quinary counting sequence.

Bit. An abbreviation for binary digit, a bit represents a single binary character (1 or 0).

Bit-slice Microprocessor. A microprocessor designed so that its word size may be expanded.

Block Mark. A mark indicating the starting point of a block of data stored on a magnetic tape.

Boolean Algebra. A branch of mathematics invented by George Boole to express logic functions in mathematical terms.

Bubble Generator. A device found in a magnetic bubble memory that creates new magnetic bubbles.

Buffer. A noninverting gate used to isolate parts of a digital circuit or to provide additional drive.

Bus. A group of digital signals. (*See also* Address Bus *and* Data Bus.)

Bypass Capacitor. A capacitor that decouples an integrated circuit from the power supply, which helps to eliminate switching noise.

Byte. A group of bits, most often eight, that is used to represent a character or number in a computer system.

Byte-wide®. Memory in which a chip stores its data in 8-bit groups.

Charge Packet. The storage element in a charge-coupled device (CCD).

Charge Pump. A circuit used to charge a capacitor in a power supply or a ramp generator.

Chevron Propagating Elements. Chevron-shaped elements in a magnetic bubble memory that are used to move bubbles.

Circuit Simplification. A systematic technique whereby digital circuits are reduced to use the fewest number of gates to implement a certain function.

CLEAR. An asynchronous input that will set the output Q of a flip-flop at a logic 0 level. (*See also* PRESET.)

CLOCK. An input to a digital circuit used for timing.

CMOS (Complementary Metal Oxide Semiconductor). A semiconductor characterized by complementary transistor pairs and extremely low power consumption.

Combinational Logic. Simple switching circuits which contain no memory. Their outputs depend only upon the present inputs.

Comparator. A circuit that compares two analog input voltages and produces a digital output indicating the larger input voltage.

Compiler. A program that converts a high-level language program into a binary machine language program.

Complement. The inverse or opposite. In the binary number system, a 1 is the complement of a 0 and vice versa.

Complementary Pair. A pair of MOSFETs, one a *p*-channel and one an *n*-channel. A bipolar complementary pair is an NPN-PNP combination, although this is not as common.

Completely Covered. With reference to a Karnaugh map, this indicates a condition in which all implicants have been enclosed in at least one group (even though the group may contain only one implicant).

Computer. A machine designed to solve problems by manipulating binary information with a program.

Condition Code Register. A register that holds information about the operating condition of a microprocessor.

Contact Bounce. Whenever a mechanical switch is changed to a new position, the metal contacts physically bounce, creating noise in a digital system.

Contact Bounce Eliminator. An electronic circuit (or microcomputer program) that removes mechanical contact bounces from the electrical signal produced by a switch.

Continuous ADC. *See* Tracking ADC.

Control Bus. The bus in a computer system that is used to control the memory and *I/O*.

Conversion Time. The amount of time required to convert an analog voltage to a digital output in an ADC, or a digital input to an analog voltage in a DAC.

Core Bit Plane. A storage location for each bit position in the binary word stored in a magnetic core memory.

CPU (Central Processing Unit). The part of a computer system which directs the operation of the memory, *I/O*, and ancillary memory.

Cross-coupled Latch. *See* Asynchronous Latch.

CRT (Cathode Ray Tube). A television "picture tube" designed specifically for use as a display device.

Current Limit Resistor. In a circuit with a constant voltage-drop device (such as a diode), this resistor is used to limit the maximum current flowing through the device.

Current Mode Logic (CML). *See* Emitter-Coupled Logic.

Cyclic Redundancy Check (CRC). A method employed to test the accuracy of data stored on a magnetic disk. It is usually a combination of shifts and Exclusive ORs.

D-type Flip-flop. A special flip-flop that transfers the data present at the *D* input to the output at the time of the GATE or CLOCK pulse.

DAC. *See* Digital-to-Analog Converter.

Data. The information either stored or dynamically acted upon in a digital system. When data are stored, the storage location is called the address.

Data Bus. The bus used to transfer information into or out of the CPU, microprocessor, memory, or *I/O* in a computer system.

Decade Counter. A counter that has ten states.

DeMorgan's Theorem. A principle in Boolean algebra that defines the complementing of a function.

Demultiplexer. A digital means of converting a given set of inputs to a larger set of outputs.

Destructive Read(out). Whenever data are read from a core memory, the contents of the memory are destroyed. This is destructive read, since the data must be rewritten to remain stored for future access.

Differential Inputs. Inputs that are subtracted from each other, supplied to an amplifier, and amplified.

Digital. Having a finite number of values (usually two). (*See also* Analog.)

Digital-to-Analog Converter (DAC). A circuit that converts a binary or BCD number to an analog voltage.

Diode Logic (DL). A logic circuit based on switching diodes.

Directory. The directory on a magnetic disk or tape indicates the name and relative position of programs and data stored on the disk or tape.

Disallowed State. A digital state which the circuit can never assume. Either the circuit is incapable of assuming this state or it has been designed so that this state cannot be reached.

Disk-Operating System (DOS). The program in a system that is used to control transfer of data and programs to or from a disk. Control Program for Microprocessors (CP/M), written by Digital Research, Inc., is an example of a common DOS.

Displacement. The distance that the data or an instruction is offset from the current contents of the program counter.

Don't Care. Denoted by the symbol "X," this is a state in which the value doesn't matter: it can be either a 1 or a 0 and have the same result.

Double-density. A storage technique that employs MFM recording to place about 200k bytes of data on a 5¼-inch mini-floppy diskette.

Down Counter. A type of counter that counts in descending order.

Drain. One of the three terminals on a MOSFET. The drain is the MOS equivalent of the emitter on a bipolar transistor.

Dynamic RAM. Random Access Memory that uses a capacitor as the storage element. Since the charge can leak off, the dynamic RAM must be periodically refreshed.

ECL. *See* Emitter-Coupled Logic.

Edge-triggered Flip-flop. A flip-flop whose outputs will change in response to the inputs only on the positive or negative edge of the CLOCK pulse.

EEPROM (Electrically Erasable Programmable Read Only Memory). Memory that can be programmed and erased incircuit and is nonvolatile. An EEPROM is erased with electrical pulses.

Electromagnetic Interference (EMI). Noise that enters or exits a system as a radiowave. Noise exiting a digital system is usually generated by CLOCK signals.

Emitter-Coupled Logic (ECL). A bipolar logic family noted for its fast operating speed.

EPROM (Erasable Programmable Read Only Memory). Memory that can be programmed or erased and remains nonvolatile. EPROM is erased by exposure to UV light.

Essential Prime Implicant. On a Karnaugh map, an implicant which is included in the minimum number of groups that will completely cover the map.

Exclusive NOR. A 2-input logic gate whose output is a 1 only when its inputs are identical.

Exclusive OR. A 2-input logic gate whose output is a 1 only when its inputs are a 1 and a 0.

Fanout. The number of input connections that can be driven from a given output.

FCC (Federal Communications Commission). The government body that regulates the radiowave spectrum.

FIFO (First In, First Out). A buffer in which the first information put in is the first available for recall. A shift register is a FIFO buffer.

Fixed-word-width Microprocessor. A microprocessor designed to work with a word of fixed width. The word width is usually 4, 8, 16, or 32 bits.

Flag Register. *See* Condition Code Register.

Flash Converter. *See* Instantaneous ADC.

Flexible Disk (Diskette). *See* Floppy Disk.

Flip-flop. A device that can store one binary bit of information.

Floating Gate. A type of MOSFET in which the gate is not connected to any other part of the transistor and is floating in the insulator material. The floating gate is used to store data in the EPROM and EEPROM memories.

Floppy Disk (Diskette). A magnetic recording device constructed from a thin disk of plastic about the size of a 45 rpm record. (The standard size is 8 inches in diameter.)

Flow Chart. A chart that delineates the sequential steps and alternate "routes" in a program.

Frequency Modulation (FM). A modulation technique used to encode data for magnetic disk memory systems.

Frequency Shift Keying (FSK). A modulation method where digital information is encoded into two different frequencies.

Gate. 1. A logic circuit, generally having two states, whose output is some function of its input. 2. The control input on a MOSFET.

Gate Array. A semicustom logic circuit containing many digital function blocks. The functional blocks are interconnected by the semiconductor manufacturer to specifications supplied by the user.

Gated Latch. A flip-flop whose outputs will change in response to the inputs only if the GATE input is active.

Glitch.   An electrical noise spike.

Hard Disk.   A magnetic storage medium constructed from a platter of aluminum that is coated with magnetic oxide.

Head Crash.   A condition that occurs when the flying head in a hard disk drive hits the surface of the disk. This is usually caused by a power failure or vibration and generally ruins the disk and head.

Hexadecimal (Hex).   The base sixteen number system. The digits in the hex system are: 0,1,2,3,4,5,6,7,8,9,A,B,C,D,E,F.

Hub.   The center hole in a floppy disk, which is used to spin the disk.

Hunting.   The pattern used by a tracking ADC of oscillating around the value of an input voltage.

Hysteresis.   The difference between two voltage levels, such as the two trip points in a Schmitt trigger or the two magnetic polarities in a magnetic core.

Hysteresis Curve.   A plot of current versus flux density that describes a magnetic core's operation. It can also describe the operation of a Schmitt trigger or other electronic circuit in terms of two parameters such as input and output voltage.

IC (Integrated Circuit).   A device in which several electronic devices, generally transistors, are integrated on a single piece of semiconductor material, and which performs a specific function. (*See also* SSI, MSI, LSI.)

ID Field.   A section on a disk that identifies the track, sector, and head number.

IIL.   *See* Integrated Injection Logic.

Implicant.   On a Karnaugh map, a 1 (for a Sum of Products map) or a 0 (for a Product of Sums map) which is grouped with similar values in order to minimize the function.

Index Hole.   The hole used by the disk drive electronics to detect the beginning of a track.

Inherent Error.   The error that occurs in an ADC because of the comparator's inability to resolve the difference between certain voltages.

Inhibit Amplifier.   An amplifier used to prevent, or inhibit, the writing of data into a core of a magnetic core memory.

Input/Output Equipment (I/O).   Equipment that is attached to a computer system for the purpose of transferring data into and out of the computer.

Instantaneous ADC.   The quickest type of analog-to-digital converter, which is often called a "flash converter" because of its speed.

Instructions.   The steps in a program that direct the operation of the computer system.

Instruction Register.   A register used to hold the op-code so that the microprocessor can determine which instruction to execute.

Integrated Circuit. *See* IC.

Integrated Injection Logic. A bipolar logic family noted for its fairly high speed and low power consumption.

Interface. A circuit that connects two circuits in a compatible manner in terms of input and output power requirements.

Interpreter. A program that converts a high-level language into a pseudo-machine language. Interpreters work on single high-order instructions. (*See also* Compiler.)

INVERTER. One of the three primary gates, the INVERTER, or NOT gate, is a single-input gate whose output is the opposite, or inverse, of the input.

Johnson Counter. A special form of shift register that divides the input frequency by twice the number of flip-flops in the counter.

Junction Capacitance. The capacitance exhibited by the junction area of a diode.

Karnaugh Map. A graphical technique for minimizing digital combinational circuits.

Keeper. A device used to reduce the loss of magnetic energy in a permanent magnet.

Large-Scale Integration (LSI). A microelectronic IC containing between 100 and 1000 logic gates.

Latch. A device that can store a logic 1 or logic 0 by "latching" onto the input information.

LCD (Liquid Crystal Display). A type of display in which electric fields are used to move crystals into patterns which block light and thus form characters. The light for an LCD must be provided from an external source, as it is not a light-emitting display.

LED (Light-Emitting Diode). A special diode that, when energized, emits light.

LIFO (Last In, First Out). A buffer in which the last data input are the first available at the output. A stack in a microprocessor system is a LIFO buffer.

Load Transistor. The dynamic pullup in an NMOS digital circuit.

LSI. *See* Large-Scale Integration.

Machine Language. The most basic programming language, in which commands are entered in binary, usually coded in hexadecimal.

Magnetic Bubbles. Magnetized areas on a substrate that, when viewed under an electron microscope, appear as bubbles.

Magnetic Core. A doughnut-shaped magnet used to store digital information.

Magneto-resistive Detector. A device that detects bubbles in a magnetic bubble memory device. Its resistance changes whenever a magnetic bubble passes near it.

Major Loop. A structure in a magnetic bubble memory that is used to enter or extract information to or from the bubble memory device.

Master-slave Flip-flop. A flip-flop constructed from two gated latches. The master accepts input data on the logic 1 level of the CLOCK, and the slave accepts data from the master on the logic 0 level of the CLOCK.

Maxterm. A "1" value for a circuit in Sum of Products standard form.

Mean Time Between Failures (MTBF). The average amount of time before a piece of equipment fails to operate properly.

Medium-Scale Integration (MSI). A microelectronic integrated circuit containing between 11 and 99 logic gates.

Memory Element. The storage portion of a sequential logic circuit, usually a flip-flop.

Merged-Transistor Logic (MTL). *See* Integrated Injection Logic.

Microcomputer. A computer system constructed with a microprocessor.

Micro-floppy Disk (Diskette). The latest version of the floppy disk, which is scaled down to $3\frac{1}{2}$ inches in diameter.

Microprocessor. The modern, miniaturized version of a CPU.

Microprocessor Trainer. A programming aide which provides direct control over a microprocessor through its use of the machine language of the microprocessor.

Mini-floppy Disk (Diskette). The most common version of the floppy disk, which is scaled down to $5\frac{1}{4}$ inches in diameter.

Minor Loop. A structure used to store information in a magnetic bubble memory device.

Minterm. A "0" value for a circuit in Product of Sums standard form.

Modified Frequency Modulation (MFM). A modulation technique that encodes digital data for storage on a magnetic disk.

Modulus. The number of states in a synchronous circuit.

Modulus Counter. A counter whose modulus is any integer value.

Monostable Multivibrator. A device whose output will become active for a predetermined amount of time after it is triggered.

MOSFET (Metal Oxide Semiconductor Field Effect Transistor). A transistor controlled by electric fields.

MSI. *See* Medium-Scale Integration.

NAND. An AND gate followed by an INVERTER. Its output is 0 only when all inputs are 1.

n-channel. A MOSFET whose channel is formed from *n*-type semiconductor material. (*See also p*-channel.)

Nibble. A half-byte, usually four bits.

NMOS. A logic family constructed with n-channel MOS transistors.

Noise Immunity. The circuit's resistance to noise, measured as the difference between the input and output logic voltage levels.

Noise Margin. *See* Noise Immunity.

Nonreturn-to-Zero (NRZ). A recording technique that fully saturates a magnetic tape in one or the other direction.

Nonvolatile. Memory that is capable of retaining data after the power supply voltage is removed.

NOR. An OR gate followed by an INVERTER. Its output is 1 only if all inputs are 0.

NOT. *See* INVERTER.

NOVRAM ®. Nonvolatile RAM.

Number System. A system of numbering using a specific base. The decimal, or base ten, number system is the most common. The number of individual digits (including zero) is equal to the base.

One-shot. *See* Monostable Multivibrator.

Op-code. A binary number that indicates which operation is to be performed by a computer system.

Open-collector Logic. A form of TTL in which the collector resistor is not included in the IC and must be provided externally.

Open-drain Logic. The MOS equivalent of open-collector logic.

OR Gate. One of the three primary gates, its output is a 0 only when all inputs are 0s.

PAL (Programmable Array Logic). A semicustom logic device that is programmed by the semiconductor manufacturer based on the customer's circuit requirements. (*See also* Gate array.)

Parity. A count of the number of 1s in the result of an arithmetic or logic operation, expressed as even or odd.

p-channel. A MOSFET whose channel is formed from p-type semiconductor material. (*See also* n-channel.)

PLA (Programmable Logic Array). *See* PAL.

PMOS. A logic family constructed with p-channel MOS transistors.

Present State–Next State Diagram. A diagram that illustrates the present and next states of a synchronous system.

PRESET. An asynchronous input that will set the output Q of a flip-flop to a logic 1 level. (*See also* CLEAR.)

Priority Encoder. A digital circuit that can determine which input is occuring and assign a priority to it.

Product of Sums. A standard form for expressing combinational logic equations, it is written as the ANDing of OR functions. Example: $(A+B)(A+B+D)$ $(C+D) = Y$.

Program. An organized set of instructions stored in a computer system's memory in sequential order.

Programmable Counter. A counter whose counting sequence can be programmed.

Programmable Modulus Counter. A counter with a LOAD input used to program its modulus.

PROM (Programmable Read Only Memory). A type of ROM that can be programmed by the user, but only once.

Propagating Element. A means used in magnetic bubble memory devices to move bubbles across the surface of the device.

Propagation Delay Time. The amount of time required for a digital signal to pass through a digital circuit.

Pulldown Resistor. A resistor connected to ground that is used to set a logic 0 level on an input.

Pullup Resistor. A resistor connected to the power supply voltage that is used to set a logic 1 level on an input.

Quantizing Error. *See* Inherent Error.

Race. A condition that results when the output of a gated latch is connected to its input. The latch "races" or oscillates toward its final state, but never reaches it.

RAM (Random Access Memory). A type of memory used for fast, temporary storage of data in a digital system.

Ramp/Slope ADC. A type of analog-to-digital converter that uses a voltage ramp to determine the value of the analog input.

READ/RECORD Head. A device that converts a current flow into a magnetic field and that is used to impress information on the surface of a magnetic disk or tape. It also translates the magnetic energy on the tape or disk back to a voltage.

Redundant. With reference to Karnaugh maps, terms in an equation which are unnecessary. A circuit is not completely reduced if it contains redundant terms.

Refresh. The process of rewriting data periodically to a dynamic RAM.

Register. A collection of flip-flops that holds a multiple-digit binary or BCD number.

Register Array. A grouping of registers that is usually found in a microprocessor.

Replicator/Annihilator. A circuit used to replicate (copy) and annihilate (destroy) magnetic bubbles.

Resolution. The number of input bits in an ADC or DAC, or the reciprocal of the number of input combinations, expressed as a percentage.

Ring Counter. A special shift register that counts. The length of the counting sequence is determined by the number of flip-flops in the ring counter.

Ring Oscillator. An oscillator constructed from INVERTERs in series. Its output is connected to its input, so it never reaches a stable state.

Ripple Counter. *See* Asynchronous Counter.

ROM (Read Only Memory). A type of memory in which the data are permanently stored. ROM is nonvolatile and cannot be reprogrammed.

R-2R Ladder. A DAC that uses only two resistance values, R and $2R$.

SAR/ADC. *See* Successive Approximation Register ADC.

Schmitt Trigger. A special circuit that has hysteresis and is used either to produce a squarewave or to reduce noise.

Schottky-barrier Diode. A diode with an extremely low forward bias voltage and low storage capacitance.

Schottky-clamped Transistor. A transistor with a Schottky diode connected from base to collector. This prevents the base region from saturating completely.

Sector. A pie-shaped portion of a track on the surface of a magnetic disk.

Select Lines. Digital signals used to select a magnetic core in a magnetic core memory system.

Sense Amplifier. A device used to amplify a low-level current or voltage and produce a digital output.

Sense Winding. The wire threaded through all the cores in a bit plane of a magnetic core memory. It is used to sense a change in the polarity of the magnetic field in a selected core.

Sequential Logic. A logic circuit that uses present data to generate future data. Sequential logic circuits exhibit memory.

Settling Time. The amount of time required for the outputs of a latch to "settle" to their final state.

Seven-segment Display. A format for displaying numerals and some letters. If all seven segments are lit, the numeral 8 is displayed.

Shift Register. A register that is capable of shifting data either to the left or to the right.

Silicon Gate MOSFET. A MOSFET whose gate is synthesized from silicon instead of the aluminum used in most MOSFETs.

Simultaneous ADC. *See* Instantaneous ADC.

Single-component Microcomputer. A microprocessor that includes memory and I/O in the same IC package. (Also called a single-chip microcomputer.)

Single Density. A storage technique used when data are stored on a magnetic disk using FM. On a 5¼-inch disk, 100k bytes of data are usually stored in this format.

Single-shot. *See* Monostable Multivibrator.

Sink Current. Current that flows, or "sinks," to ground in a digital circuit.

Solid State Relay. A transistorized equivalent of an electromechanical relay.

Source. One of the three terminals on a MOSFET. The source is equivalent to the collector on a bipolar transistor.

Source Current. Current that flows from, or is "sourced" by, the power supply in a digital system.

SSI (Small-Scale Integration). A microelectronic IC containing nine or fewer logic gates.

State Transition Diagram. A flowchart or map of the operation of a synchronous system.

Static RAM (SRAM). RAM that stores data for as long as power is applied and needs no refreshing.

Steering Network. A network in a magnetic core memory used to steer the current flow through the select lines in one of two directions.

Stepper Motor. A special type of motor that turns in predefined steps in response to digital pulse.

STROBE. A pulse used to indicate or gate a unique event.

Successive Approximation Register ADC (SAR/ADC). A type of analog-to-digital converter which uses a comparator to determine each bit position of the digital representation of the analog input.

Sum of Products. A standard form for writing combinational logic equations as an ORing of ANDs. Example: $AB + ABC + AD = Y$.

Switch Bounce. *See* Contact Bounce.

Synchronous Counter. A counter whose outputs change in synchronization because the CLOCK inputs are in parallel.

T-bar Propagating Element. A device in a magnetic bubble memory that is constructed from T- and bar-shaped elements, which are used to move the bubbles.

Thin-film Permalloy. A very thin coating of material that conducts magnetic energy but does not retain a magnetic field.

Threshold Voltage. The voltage between gate and source on a MOSFET that will turn the device on.

Toggle. The complementing of the output of a flip-flop.

Totem-pole. A special output circuit configuration found in most TTL devices, consisting of pullup and pulldown transistors.

Track. A band of magnetic digital data that is created on the surface of a magnetic tape or disk by the positioning of a READ/WRITE head.

Tracking ADC. An analog-to-digital converter that employs an up/down counter to "track" the incoming analog voltage.

Transistor-Transistor Logic (TTL). A type of bipolar logic that is very popular at the present time, distinguished by multiple-emitter transistors.

Transmission Gate. A CMOS circuit that can be used as a digital or analog voltage switch. Another type of transmission gate is used in a bubble memory to transfer data between the major and minor loops.

TRIGGER. An input to a digital circuit that starts an event.

Tri-state® Logic. A logic circuit whose output can be a 0, a 1, or an open circuit. (Also called three-state.)

Truth Table. A table that defines all possible outputs for all possible inputs of a digital circuit.

TTL. *See* Transistor-Transistor Logic.

UART (Universal Asynchronous Receiver/Transmitter). A device employing shift register memory that is used in digital communications systems.

ULA (Uncommitted Logic Array). *See* PAL.

Up Counter. A counter that counts in ascending order.

VF (Vacuum Fluorescent). A type of display that excites electrons in an evacuated tube to produce light.

Volatile. Memory that cannot retain data unless power is constantly applied.

Winchester Drive. A magnetic hard disk drive that uses Winchester flying heads.

Winchester Flying Head. A magnetic READ/RECORD head designed to float on a cushion of air as it "flies" over the surface of a disk.

WRITE Protect Notch. When covered with an opaque material, this notch prevents writing to a floppy disk.

APPENDIX D: BIBLIOGRAPHY

Barden, William, Jr., *How to Program Microcomputers.* Indianapolis, IN: Howard W. Sams and Co., 1977.

Bell, David A., *Solid State Pulse Circuits* (2nd ed.). Reston, VA: Reston Publishing Co., 1981.

Brey, Barry B., *Microprocessor/Hardware Interfacing and Applications.* Columbus, OH: Charles E. Merrill Publishing Co., 1984.

Cave, Frank E., and Cave, David L., *Digital Technology with Microprocessors.* Reston, VA: Reston Publishing Co., 1981.

COS/MOS Integrated Circuits Manual. Sommerville, NJ: RCA Solid State Division, 1979.

Deem, Bill, Muchow, Kenneth, and Zeppa, Anthony, *Digital Computer Circuits and Concepts* (2nd ed.). Reston, VA: Reston Publishing Co., 1978.

"Fifty Years of Achievement: A History," *Electronics,* April 17, 1980.

Floyd, Thomas L., *Digital Fundamentals* (2nd ed.). Columbus, OH: Charles E. Merrill Publishing Co., 1982.

Greenfield, Joseph D., *Practical Digital Design Using IC's.* New York: John Wiley and Sons, 1977.

Intel MCS-80/85 Family User's Manual. Santa Clara, CA: Intel Corp., 1983.

Intel Microprocessor and Peripheral Handbook. Santa Clara, CA: Intel Corp., 1983.

Lancaster, Don, *TTL Cookbook.* Indianapolis, IN: Howard W. Sams and Co., 1974.

Levine, Morris E., *Digital Theory and Practice Using Integrated Circuits*. Englewood Cliffs, NJ: Prentice-Hall, 1978.

Malvino, Albert Paul, *Digital Computer Electronics: An Introduction to Microcomputers* (2nd ed.). New York: McGraw-Hill, 1981.

Millman, Jacob, and Halkias, Christos C., *Integrated Electronics: Analog and Digital Circuits and Systems*. New York: McGraw-Hill, 1972.

Motorola MC6800 Programming Reference Manual. Phoeniz, AZ: Motorola, 1976.

Nashelsky, Louis, *Digital Computer Theory* (2nd ed.). New York: John Wiley and Sons, 1977.

O'Connor, Patrick, *Digital and Microprocessor Technology*. Englewood Cliffs, NJ: Prentice-Hall, 1983.

Porat, Dan L., and Barna, Arpad, *Introduction to Digital Techniques*. New York: John Wiley and Sons, 1979.

Schilling, Donald, and Taub, Herbert, *Digital Integrated Electronics*. New York: McGraw-Hill, 1977.

Shacklette, L. W., and Ashworth, H. A., *Using Digital and Analog Integrated Circuits*. Englewood Cliffs, NJ: Prentice-Hall, 1977.

Tocci, Ronald J., *Digital Systems: Principles and Applications*. Englewood Cliffs, NJ: Prentice-Hall, 1977.

Tocci, Ronald J., *Fundamentals of Pulse and Digital Circuits* (3rd ed.). Columbus, OH: Charles E. Merrill Publishing Co., 1983.

APPENDIX E: ANSWERS TO ODD-NUMBERED QUESTIONS

CHAPTER 1

1 High and low, 1 and 0, true and false, on and off.

3 10^6 = 1 million

5 a. 25
   b. 42
   c. 18
   d. 24
   e. 15
   f. 12
   g. 8
   h. 2

7 a. 10000 = 16
   b. 10000 = 16
   c. 100 = 4
   d. 10111 = 23

9 a. 1000
   b. 01
   c. 0110
   d. 1001

11 a. 1010 (-2)
   b. 1011 (-3)
   c. 10001 (-1)

13 a. 2
   b. 8
   c. none
   d. 6
   e. 86

15 a. 1011   f. 110000

b. 10000   g. 10100110

c. 10010   h. 11111

d. 10101010   i. 10010001

e. 1000001   j. 101000111010

17 a. 63   d. 4077

b. 2047   e. 2457

c. 256

19 a. 53   c. 13

b. 6   d. 333

CHAPTER 2

1 High and low, on and off, true and false.

3

a.    c.

b.   d.

A	B	C	NOR	AND	OR	NAND
0	0	0	1	0	0	1
0	0	1	0	0	1	1
0	1	0	0	0	1	1
0	1	1	0	0	1	1
1	0	0	0	0	1	1
1	0	1	0	0	1	1
1	1	0	0	0	1	1
1	1	1	0	1	1	0

5

A	B	C	X	Y
0	0	0	0	0
0	0	1	0	0
0	1	0	0	0
0	1	1	1	1
1	0	0	0	1
1	0	1	1	1
1	1	0	0	1
1	1	1	1	1

$$X = (A + B)\, C$$
$$Y = A + (BC)$$

7

$= A \cdot B$

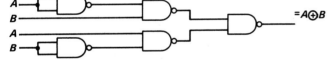

$= A \cdot B$

9

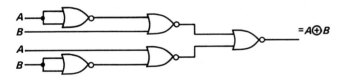

$= A \oplus B$

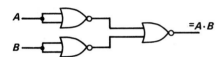

$= A \oplus B$

11

Y

13 a. 0   d. A

        b. 1   e. 1

        c. 0   f. A

15

$ABCDE$	Output
00000	1
00001	0
00010	0
00011	0
00100	0
00101	0
00110	0
00111	0
01000	0
01001	0
01010	0
01011	0
01100	0
01101	0
01110	0
01111	0
10000	0
10001	0
10010	0
10011	0
10100	0
10101	0
10110	0
10111	0
11000	0
11001	0
11010	0
11011	0
11100	0
11101	0
11110	0
11111	0

17

a.

b.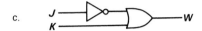

c.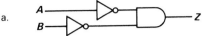

d.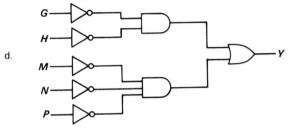

19

ABC	Z	Y
000	0	0
001	1	1
010	1	1
011	1	1
100	1	1
101	1	1
110	1	1
111	1	1

$(A + B) + C = Z$
$A + (B + C) = Y$
(Associative property)

CHAPTER 3

1   a. AC c. $\overline{A}C + AB$
    b. A d. 0

3   No. Only powers of two (i.e., 0, 1, 2, 4, 8, 16, etc.)

5 Yes, if properly used.

7

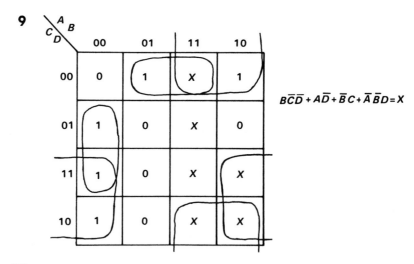

$$AB\bar{C} + ACD + \bar{A}C\bar{D} = X$$

9

	00	01	11	10
00	0	1	X	1
01	1	0	X	0
11	1	0	X	X
10	1	0	X	X

$$B\bar{C}\bar{D} + A\bar{D} + \bar{B}C + \bar{A}\bar{B}D = X$$

11 To minimize parts count, cut costs, reduce size of the printed circuit board, keep the circuit secret, change logic by replacing one part, etc.

13

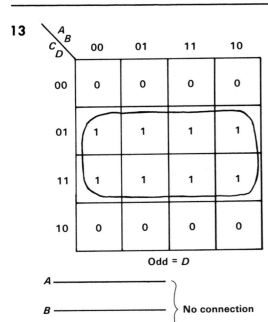

Odd = D

A ——————————⟩

B —————————— No connection

C —————————

D ———————————————— Odd output

$ABCD$	Odd
0000	0
0001	1
0010	0
0011	1
0100	0
0101	1
0110	0
0111	1
1000	0
1001	1
1010	0
1011	1
1100	0
1101	1
1110	0
1111	1

15

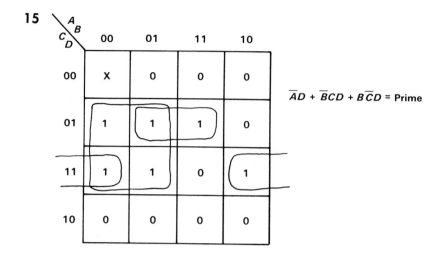

$\overline{AD} + \overline{B}CD + B\overline{C}D$ = Prime

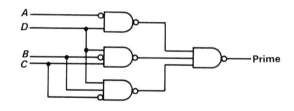

$ABCD$	Prime
0000	X
0001	1
0010	0
0011	1
0100	0
0101	1
0110	0
0111	1
1000	0
1001	0
1010	0
1011	1
1100	0
1101	1
1110	0
1111	0

CHAPTER 4

1 On, which is usually used to indicate a logic 0, and Off, which is usually used to indicate a logic 1.

3 Because a silicon diode will drop approximately 0.7 V when forward biased, the maximum output voltage is 4.3 V.

5

A	B	C	V_{OUT}
0	0	0	0.7 V
0	0	1	0.7 V
0	1	0	0.7 V
0	1	1	0.7 V
1	0	0	0.7 V
1	0	1	0.7 V
1	1	0	0.7 V
1	1	1	5.0 V

7 On occasion, the digital designer may find that one more AND or OR gate is needed. In this situation it may be more cost effective to use a diode logic gate than to add another integrated circuit.

9

11 The switching speed of a transistor switch is limited by the emitter-base junction capacitance.

13 If an additional input is required on a TTL circuit, the manufacturer adds additional emitters to the input transistor.

15 0.4 V.

17 The connection of 11 inputs to one TTL output would cause the output voltages to fall below the logic 1 level and above the logic 0 level.

19 2.

21 An open-collector TTL logic gate is a device that contains no internal pullup resistor or transistor. The circuit designer must include this pullup to ensure that the output will rise to a logic 1 level when required.

23 74133.

25 The 7432, Quadruple 2-input OR gate.

27 The Schmitt trigger input increases the noise immunity of a TTL system to about 1.25 V.

29 Current Mode Logic (CML).

31 ECL logic has a switching speed of about 2 ns, which is very fast.

33 The inputs to an IIL circuit are used to control transistor switches which act as steering networks for current from an internal constant current source.

35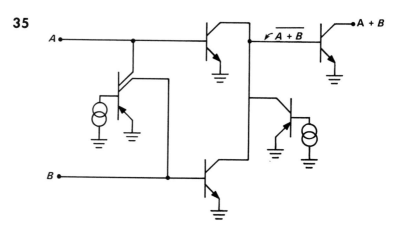

CHAPTER 5

1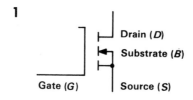

3 A positive V_{gs} turns an n-channel MOSFET on.

5

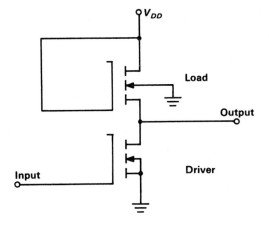

7

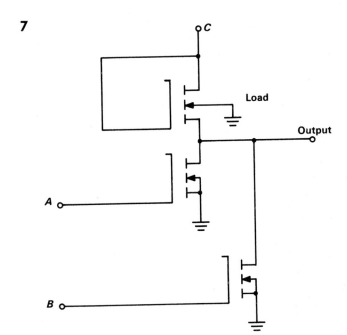

9

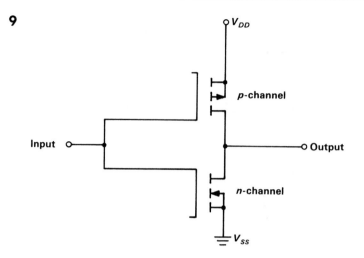

11 No, the power-supply range for CMOS is 3-15 V (minimum). This means that some drift on the supply voltage will not cause the supply to leave this range.

13

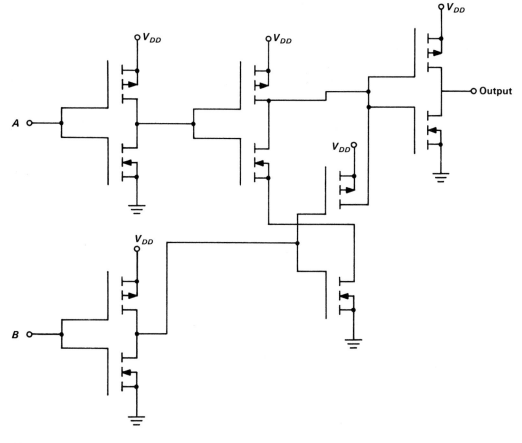

15 49.

17 If the input exceeds V_{DD}, the diodes conduct and the resistor dissipates the power, protecting the CMOS circuitry against destructive input spikes.

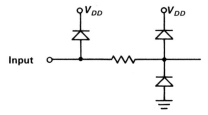

19 Voltage, because this term is squared in the P_{dyn} equation.

21 Indefinitely, the drain-to-source resistance of the output will limit the current.

23 Twice as much—2 μA as compared to 1 μA.

25 7486 or 5486.

27

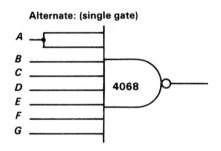

Alternate: (single gate)

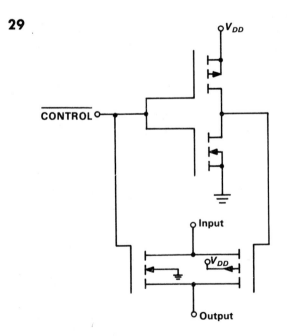

CHAPTER 6

1 The basic difference between a combinational logic circuit and a sequential logic circuit is that the combinational circuit works with present data only and the sequential circuit uses both present and past data.

3 A digital combination lock is an example because it must use past information—the combination—to generate an UNLOCK signal.

5

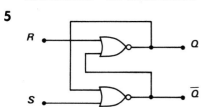

7 $R = 0$ and $S = 1$

9 Contact bounces look like many separate contacts when the electrical signal is viewed. These multiple signals could produce errors in a digital system.

11

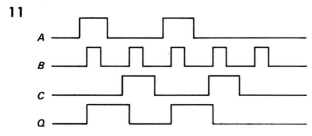

13

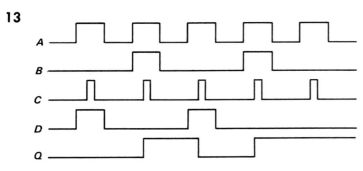

15 A race condition is an uncontrollable oscillation that can occur if the output of a gated or asynchronous latch is coupled back to the inputs.

17 0.

19

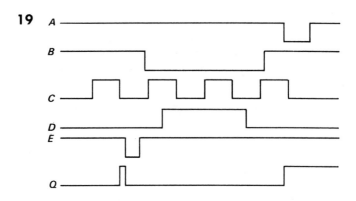

21 0.

23 No. Some master-slave flip-flops are positive edge–triggered.

25 If the master is receptive to data, noise can cause an unwanted change that will be passed on to the slave.

27 A monostable multivibrator is a device that has one stable state (usually 0). When it is triggered, the output will change states for a predefined period of time and then return to its stable state.

29 1.43 μF.

31

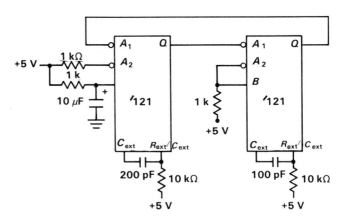

33 $C = 0.1\ \mu$F, $R_B = 6.8$ kΩ, $R_A = 500\Omega$.

CHAPTER 7

1 2-bit = 4 states, 3-bit = 8 states, 7-bit = 128 states, and 9-bit = 512 states.

3 The counter will repeat its counting sequence. For example, a 2-bit counter would count in the following order: 00, 01, 10, 11, 00, 01, etc.

5 Because a 6-bit counter has 64 states, it will divide the input frequency by 64.

7

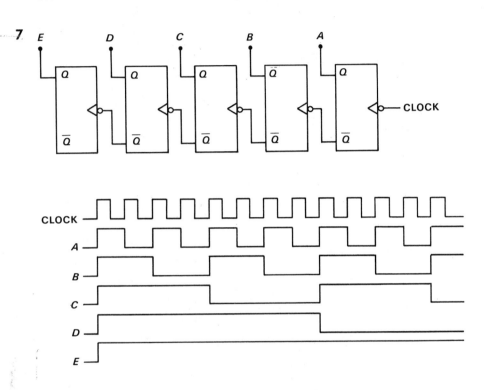

9 Five.

11 They are used to prevent any erroneous CLOCK inputs when the direction of the counter is changed.

13 The only difference is that in the synchronous counter the outputs all change together, and in the asynchronous counter the outputs do not change together.

15 The maximum operating frequency of a counter is determined by the settling times of the flip-flops used to construct it.

17 First, convert the modulus number to binary. This tells you how many flip-flops are required to build the counter. Next, construct a binary asynchronous up counter and tie all the CLEAR inputs together. Finally, NAND all of the 1 outputs in the binary modulus number together and apply the output of the NAND gate to the CLEAR inputs.

19

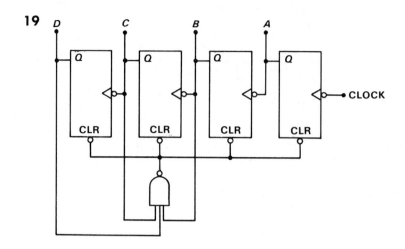

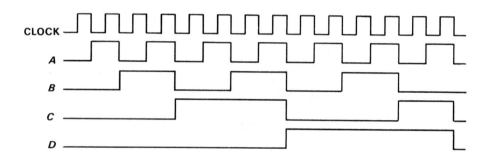

21 50.

23 A biquinary counter is constructed from a modulo-2 and modulo-5 counter. The modulo-5 counter is connected to the input; the modulo-2 counter's input is connected to the output of the modulo-5 counter; and the output is taken from the modulo-2 counter.

25 Parallel-to-parallel, parallel-to-serial, serial-to-parallel, and serial-to-serial.

27 A shift register shifts data because each Q output is connected to the next flip-flop's D input. When the CLOCK pulse occurs, the data from each bit position are moved to the next bit position.

29 Ten.

31

Present state			Next state	
A	X	Y	X	Y
0	0	0	0	1
0	0	1	1	0
0	1	0	1	1
0	1	1	0	0
1	0	0	0	0
1	0	1	0	1
1	1	0	1	0
1	1	1	1	1

33

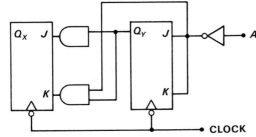

CHAPTER 8

1

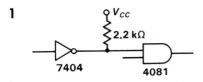

3

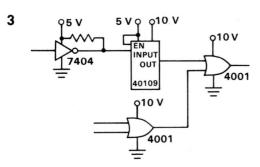

5

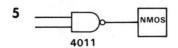

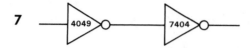

7

(No interface is necessary for driving a single TTL load.)

9   Not safely, for while under typical conditions, a 4049 can drive three TTL inputs, its "worst case" value is *one*. Therefore, for reliable results, connecting it to two inputs should be avoided.

11

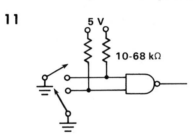

13

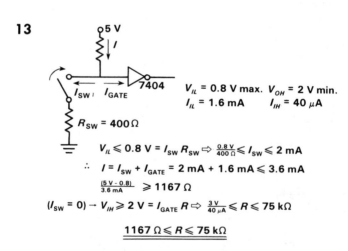

V_{IL} = 0.8 V max. V_{OH} = 2 V min.
I_{IL} = 1.6 mA I_{IH} = 40 μA

$V_{IL} \leqslant 0.8$ V = $I_{SW} R_{SW} \Rightarrow \frac{0.8\,V}{400\,\Omega} \leqslant I_{SW} \leqslant 2$ mA

$\therefore$ $I = I_{SW} + I_{GATE}$ = 2 mA + 1.6 mA $\leqslant$ 3.6 mA

$\frac{(5\,V - 0.8)}{3.6\,mA} \geqslant 1167\,\Omega$

$(I_{SW} = 0) \rightarrow V_{IH} \geqslant 2$ V = $I_{GATE} R \Rightarrow \frac{3\,V}{40\,\mu A} \leqslant R \leqslant 75$ kΩ

$\underline{1167\,\Omega \leqslant R \leqslant 75\,k\Omega}$

15

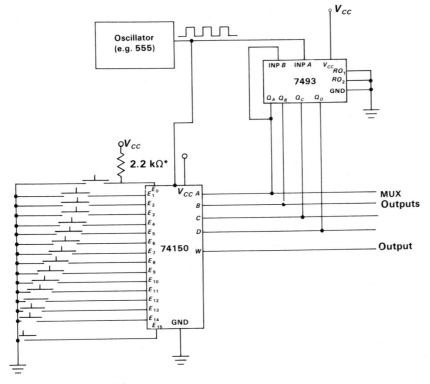

*pullup for each switch

17

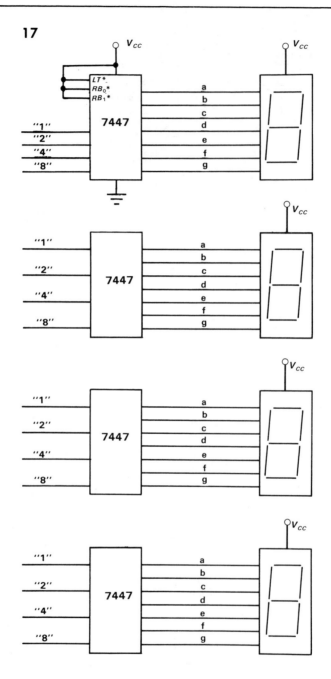

*Same connection on each 7447.

19 A CRT.

21 TTL. 15 V CMOS has much greater noise immunity.

CHAPTER 9

1

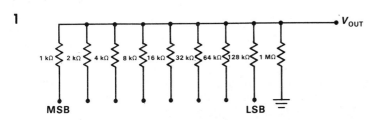

3 0.127 V.

5

	2^5	2^4	2^3	2^2	2^1	2^0	V_{OUT}
	0	0	0	0	0	0	−3.0 V
	0	0	0	0	0	1	−2.873 V
	0	0	0	0	1	0	−2.746 V
	0	0	0	0	1	1	−2.619 V
	⋮	⋮	⋮	⋮	⋮	⋮	⋮
	1	1	1	0	0	0	+4.111 V
	1	1	1	0	0	1	+4.238 V
	1	1	1	0	1	0	+4.365 V
	1	1	1	0	1	1	+4.492 V
	1	1	1	1	0	0	+4.619 V
	1	1	1	1	0	1	+4.746 V
	1	1	1	1	1	0	+4.873 V
	1	1	1	1	1	1	+5.0 V

7

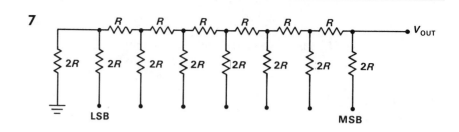

9 0.531 V.

11

2^4	2^3	2^2	2^1	2^0	V_{OUT}
0	0	0	0	0	−5 V
0	0	0	0	1	−4.469 V
0	0	0	1	0	−3.938 V
0	0	0	1	1	−3.407 V
0	0	1	0	0	−3.076 V
0	0	1	0	1	−2.545 V
⋮	⋮	⋮	⋮	⋮	⋮
1	1	1	0	1	+10.407 V
1	1	1	1	0	+10.938 V
1	1	1	1	1	+11.469 V

13 R_{fb} is used to set the gain of the external op-amp.

15 0.167 V.

17 $O_1 = 1$, $O_2 = 0$, $O_3 = 0$, $O_4 = 0$, and $O_5 = 0$.

19 The charge pump is a constant current source. If it is connected to a capacitor, the voltage on the capacitor will build linearly, developing a ramp voltage across the capacitor.

21 The greater the input voltage, the greater the length of the ramp.

23 The inherent error in an ADC is due to the comparator's inability to resolve the difference between certain voltage levels.

25 The main difference is that the tracking ADC uses an up/down counter to generate its ramp, and the ramp/slope ADC uses either an up or a down counter.

27 The tracking ADC hunts because of the inability of the comparator to resolve the difference between the two voltages applied to it.

29 800 ns.

31 It indicates that the ADC has completed a conversion on the negative edge.

33 Every time that the output of the 555 timer goes to a logic 0, the converter converts another voltage into binary.

CHAPTER 10

1 RAM is generally used for high-speed read/write memory.

3 Data are the information stored in a memory.

5 A byte is a group of eight bits.

7 No. In order to use an NMOS EPROM, "wait" states (delays) would have to be inserted to slow the memory access time to the speed of the EPROM used.

9

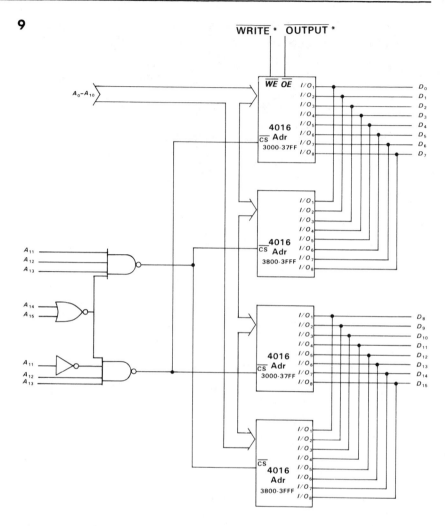

*Same connection for each 4016

11 The memory is stored in a capacitor. Since the charge can leak off the capacitor, it must be refreshed to hold the data.

13 A ROM is programmed with the mask from which the semiconductor is made.

15 PROMs are programmed with a programmer by the user.

17 An EPROM is erased by ultraviolet light, which penetrates to the floating gate through a quartz-glass window on the chip.

19 An EEPROM is erased by placing a voltage on the control gate.

21 The time you would most likely write to the EEPROM of a NOVRAM®
is as a power failure is occurring.

23 The standby current of CMOS is the lowest of any type of RAM, thus
assuring maximum battery life.

25

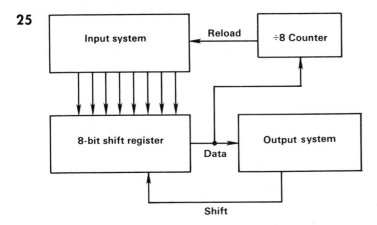

CHAPTER 11

1 If current is passed through the core from left to right the magnetic field
generated will be counterclockwise.

3 X_2 and Y_1.

5

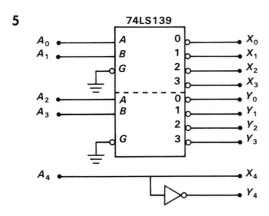

7 The sense amplifier cancels noise voltage because the voltage appears at both inputs in phase and at equal amplitude. The sense voltage, on the other hand, appears out of phase, so it is amplified.

9 The first WRITE changes all the cores in the selected memory location to 0 so that its contents can be read from the sense amplifiers. Because the prior contents of the memory are thus destroyed, the data from the outputs of the sense amplifiers must be rewritten or the contents of the location will be lost.

11 A core bit plane is a bit position in a core memory word.

13 A permanent magnet.

15 The bubble generator is used to create bubbles, and the replicator/annihilator is used to destroy them.

17 The major loops are used to transfer data from minor loops to the user. The minor loops are used to store information.

19 Bubbles are transferred between the loops by a transmission gate.

21 Whenever the applied bit of data is a logic 1, the input circuit passes a 2225 Hz tone through to the low-pass filter. Whenever the applied bit of data is a logic 0, the input circuit passes a 2025 Hz tone through to the low-pass filter. The low-pass filter is used to remove the harmonic content of the squarewave.

23 When the WRITE ENABLE input is active, a logic 1 on the data input causes Q_1 to conduct, generating a clockwise magnetic field. If a logic 0 is applied, Q_2 conducts and causes a counterclockwise magnetic field.

25 The complementary outputs of the amplifier are used to set and clear the latch.

27 A floppy disk is 8 inches, a mini-floppy disk is 5¼ inches and a micro-floppy disk is 3½ inches.

29 The index hole indicates the start of a track.

31 0001 111010 100001.

33 It indicates that the head has been moved to the outermost track, track 0.

35 If a logic 1 is to be written, it is written without a CLOCK pulse. If two or more logic 0s are written in a row, then a CLOCK pulse is written.

37 The head on a floppy disk drive is only accessible if the drive is taken apart, so it is much easier to use a cleaning disk.

39 A Winchester flying head.

41 The head number.

CHAPTER 12

1 A microprocessor is a miniaturized CPU that can be programmed to perform the function of almost any sequential logic circuit.

3 A program is an ordered listing of commands or instructions designed to perform a useful task.

5 The CPU is used to transfer data, perform arithmetic and logic operations, and make decisions. The acronym CPU indicates *central processing unit*.

7 Magnetic disk and magnetic bubble memory.

9 ALU, register array, instruction register, and control logic.

11 Fig. 12-2.

13 +33 = 0010 0001, -59 = 1100 0101, +99 = 0110 0011, -127 = 1000 0001, -30 = 1110 0010, and +77 = 0100 1101.

15 The program counter is used by the microprocessor to locate the next instruction in a program. It is called a "counter" because it accesses the next instruction in the next sequential location of the memory.

17 The data, control, and address buses.

19 The bit-slice microprocessor can be modified to have as many bits as needed per word, whereas the fixed-word-width microprocessor has a predefined word width.

21 1973. This is when Intel introduced the 8080, the first NMOS microprocessor.

23 Fig. 12-4.

25 The keyboard, used to enter programs and data; the LED displays, used to display answers and instructions; and the memory, used to store programs and data.

27 THINK!

29

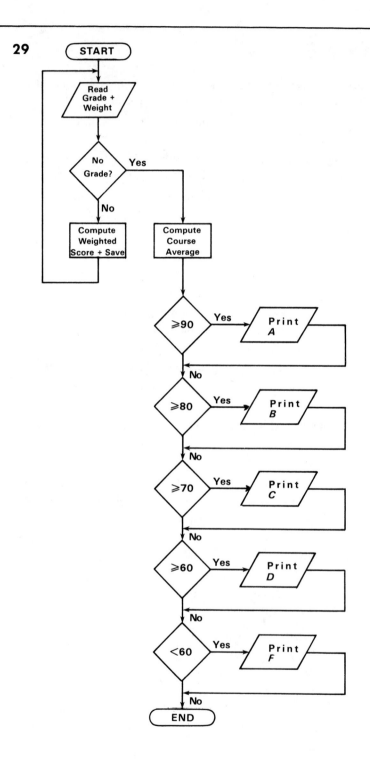

INDEX